Reading Natural Philosophy

HOWARD STEIN

Howard Stein was born on January 21, 1929 in New York City. He received a B.A. from Columbia College in 1947, a Ph.D. in Philosophy from the University of Chicago in 1958, and an M.S. in Mathematics from the University of Michigan in 1959. After teaching in the Natural Sciences Collegiate Division at the University of Chicago (1949–1958) and the Mathematics Department at Brandeis University (1959–1962), he spent five years in private industry, in the Systems Analysis and Computer Products divisions of Honeywell. A Professorship of Philosophy at Case Western Reserve University (1967–1973) brought him back to the academic world, after which he moved to Columbia University (1973–1980), and then returned to the University of Chicago, this time in the Department of Philosophy and the Committee on the Conceptual Foundations of Science. He retired in 2000. During his long and distinguished career, Howard Stein received fellowships from the National Science Foundation and the Guggenheim Foundation, and was elected to the American Academy of Arts and Sciences. A complete list of his many influential papers in the history and philosophy of science and mathematics is included in this volume.

Reading Natural Philosophy

Essays in the History and Philosophy of Science and Mathematics

EDITED BY
David B. Malament

OPEN COURT
Chicago and La Salle, Illinois

Cover illustration: *Newton* by William Blake.

To order books from Open Court, call toll-free 1-800-815-2280.

Open Court Publishing Company is a division of Carus Publishing Company.

© 2002 by Carus Publishing Company

First printing 2002

Printed and bound in the United States of America.

Library of Congress Cataloging-in-Publication Data

Reading natural philosophy : essays in the history and philosophy of science and mathematics / edited by David B. Malament
      p. cm.
    Includes bibliographical references and index.
    ISBN 0-8126-9506-2 (alk. paper) — ISBN 0-8126-9507-0 (pbk. : alk. paper)
    1. Science—History—Congresses. 2. Science—Philosophy—Congresses. 3. Mathematics—Philosophy—Congresses. 4. Stein, Howard, 1929—Congresses. I. Malament, David B.
    Q124.6 .R43 2002
    509—dc21   2002066281

To Howard Stein
on the Occasion of His 70th Birthday

Contents

Preface

In May of 1999, many friends, colleagues, admirers, and former students of Howard Stein gathered at the University of Chicago for a conference in honor of his seventieth birthday. All but four of the papers in this volume—the introductory essay by Abner Shimony, and the contributions of Nancy Nersessian, Robert Palter, and myself—were first presented there. (Three of the speakers—Jon Jarrett, Martin Klein, and Charles Parsons—did not submit papers for publication.)

The *Steinfest* was the occasion for many tributes to Howard's extraordinary work on Newton, on the philosophy of space and time, on the technical foundations of relativity theory and quantum mechanics, on developing conceptions of field theory in the nineteenth century, on the history and foundations of mathematics, and much, much more, but also for more personal expressions of gratitude for his generous contributions as a teacher, discussion partner, and careful reader of friends', colleagues', and students' manuscripts. Indeed, more than one speaker confessed that he never considered a paper finished until he had received and dealt with Howard's comments (and for this reason felt some discomfort at the prospect of presenting in his honor a paper that Howard had not yet seen!).

In addition to scholarly talks, the *Steinfest* included an afternoon recital, with mezzo-soprano Nancy Nersessian and pianist Geoffrey Hellman. Nancy added to her program a piece on the "ABCs of Relativity," sung to the tune of the Habanera from *Carmen*, and Geoffrey added one by a mystery composer who turned out to be Christiaan Huygens! There was also a dinner in Howard's honor, hosted by Ted Cohen, at which many good stories were told. Ted recalled Howard's exchange in the *New York Review of Books* with the great musicologist and pianist Charles Rosen. Though he had never studied Italian, Howard was apparently bothered by Rosen's translation of a passage from the *Marriage of Figaro*, and wrote this letter to the editors:

> The erudite and instructive Charles Rosen has made a surprising mistake in his review of the *Oeuvres* of Beaumarchais . . . He quotes a passage from the last act finale of Mozart's *Le Nozze di Figaro*, in English translation, as follows: *All is tranquil and placid/The beautiful Venus has gone in/She can take with wanton Mars/The new Vulcan of the age/In her net.* At first glance, this seems

impossible. The Homeric story, of course, is that Hephaistos ("Vulcan") took in *his* net the adulterous pair Aphrodite ("Venus") and Ares ("Mars"), *in flagrante delicto*—thus provoking the famous "Homeric" laughter of the gods when he summoned them to view the spectacle. On second thought, it seems barely possible that Figaro intends an ironic reversal: that he is "caught" by the infidelity of his affianced. But this reading seems tortured and unconvincing; and the Italian text shows that first thoughts are here best. The third through fifth lines of the passage read in fact—*Col vago Marte prendere/Nuovo Vulcan del secolo,/In rete la potrò*—that is: "I—the new Vulcan of the age—will be able to take her in a net, together with charming Mars."

(Rosen replied, "I am grateful for this correction of my inexcusable error.") The remarkable thing, as Ted explained, is that Howard reads *everything* with this sort of care. Many other toasts that evening returned to this theme.

I myself have known Howard for almost thirty years, first as a student, then as a colleague. In all this time, I have never once had the sense with him that I was about to touch bottom. He continues to startle me by the things he knows, by the things he understands, by his ability to ask just the right question after a colloquium talk, no matter what the topic. He still shows me things, even in my own relatively narrow area of supposed expertise, that I have never seen before, and would probably never come to see otherwise.

Life is complicated, time is short, and for most of us it is often all we can do just to muddle through somehow. We cut corners, make accommodations, and hope that in the process we never cross important lines. Howard does less corner cutting, less muddling through, than anyone I know. He has a way of taking even small tasks, such as writing comments on a student paper (or reading an article in the *New York Review of Books*), and investing them with so much care and concentration and intelligence that they become significant accomplishments in their own right. And when important matters of moral or intellectual principle are involved, Howard never looks away. He takes a stand and, if necessary, fights. He has been, at times, fierce in his defense of students and younger faculty.

It has been one of the great good fortunes of my life that I have been able to count Howard as my teacher, my colleague, and my friend.

Acknowledgments

I wish to thank the Department of Philosophy, the Committee on the Conceptual Foundations of Science, the Division of Humanities (all at the University of Chicago), and an anonymous donor for financial support that made the *Steinfest* possible. André Carus and Abner Shimony also deserve special thanks for their help with this volume. André, the President of Open Court Publishing Company as well as Howard's doctoral student, was enthusiastic from the beginning about the prospect of publishing a *Festschrift* for him, and smoothed the way for me at every turn. Abner, Howard's close friend and collaborator, agreed without hesitation when I asked him to write a (second) essay for this volume to serve as its introduction. His appraisal of the special character of Howard's work is surely one of the highlights of the volume.

[1]

Introduction

The Character of Howard Stein's Work in Philosophy and History of Physics

ABNER SHIMONY

The publication of this Festschrift to Howard Stein is an occasion for celebration in several professions: philosophy of science, history of science, general philosophy (especially epistemology, methodology, and metaphysics), and physics. His masterful interweaving of considerations commonly parceled out to these disciplines is a major reason for the unique value of his papers. Unfortunately, it has also been responsible for the fact that he is more widely admired than read, because his analysis and writing make demands that readers cannot meet without some knowledge of all of these disciplines, or at least a willingness to make an effort to follow argumentation drawn from them all. This Festschrift should draw in potential readers by acknowledging the inspiration of Howard's papers, by citing and amplifying some of his ideas, and in some cases by showing concretely how to use his suggestions for further research. It should be said, however, that Howard is the best commentator on his own work—as he somewhere said of Newton. Consequently, an even more useful guide than this Festschrift will be the collection, projected for the near future, of Howard's papers on Newton and other physicists and philosophers of that epoch. The unity of themes that bind the individual papers together ensure that one paper supports another, and passages that are condensed in one paper (usually because of limitations of time and space in conference proceedings) are made more accessible by leisurely expositions elsewhere in the collection. It is, of course, to be hoped that Howard's work on later science, especially relativity theory, quantum mechanics, and foundations of mathematics, will also be collected in the not-too-distant future.

Anyone who reads Howard's papers with even moderate attention cannot help but be impressed by the closeness of Howard's analysis of classical texts of physics and classical writings on natural philosophy. Howard is, to my knowledge, the *only* analyst of classical physical texts from a

1

philosophical point of view whose standards are comparable to those of the critics who have raised the close reading of literary texts to a high art.

There are specific procedures which deserve to be mentioned in discussing Howard's practice of analysis, and I shall come to them soon. But the excellence of his analysis depends primarily on two features that are not procedures, but rather analogues to the love of poetry in a fine poetic critic. One is that Howard is profoundly respectful of the intellectual achievements of, for example, Galileo, Huygens, Lagrange, Riemann, Maxwell, Lorentz, Einstein, and above all Newton, and he strongly feels that the reflections and interpretations which these giants offer concerning their own work deserve careful study (contrary to well-known depreciations of the philosophical acumen of certain great scientists). The other feature is his passion to understand—where the object of understanding may be a linguistic expression, a thesis, a mathematical demonstration, a scientific discipline, or a commentary. An indication of this passion is the amount of time I have seen him bestow on reworking proofs of generally accepted mathematical theorems, in order to achieve versions that seemed transparent to him—with apparently no worry that the effort thus expended might distract from research that could lead to publishable novelties. In a highly competitive world, his choice of understanding over the prestige of discovery is extremely rare. But I am happy to say that his virtue was rewarded, for his passion to understand passages that have been dismissed as idiosyncratic or naive in classical writings (for example, Newton's "Absolute, true, and mathematical time, of itself, and from its own nature, flows equably without relation to anything external") has often led him to fine innovations of interpretation and in some cases to genuine conceptual discoveries.

As to procedures of analysis, each of the following has been essential to Howard's work, and their combination—rare and probably unique—has been extraordinarily illuminating.

(1) He freely uses relevant parts of modern mathematics and physics in order to explicate texts and to comment on disputes of previous centuries. For example, "Newtonian Space-Time" begins with two characterizations of the structure of space-time implicit in Newton's dynamics: one uses concepts of affine geometry articulated in twentieth-century geometry (especially by Cartan), and the other uses the concept of Galilean invariance in a way that was not articulated fully until the end of the nineteenth century (by Lange). These characterizations are not window-dressing, but are used to clarify efficiently the points at issue between Newton and Leibniz and between Newton and Huygens, and in the case of the latter dispute to show how both disputants were partially correct and partially incorrect in their claims regarding absolute and relative motion. The efficient deployment of modern geometry and physics obviously requires an understand-

ing of these disciplines, and it also presupposes rejection of the widespread opinion among historians of science that it is anachronistic and "Whiggish" to bring contemporary scientific knowledge to bear upon disputations of an earlier epoch. Tangentially, in the first lecture of one of his courses at the University of Chicago, Howard reminded the class of the maxim supposedly inscribed over the gate of Plato's Academy, "Let no one unacquainted with geometry enter here"; he said that he did not make that requirement, but did demand that no one *emerge* from the course unacquainted with geometry.

(2) Howard is nevertheless as obsessive as a professional historian in seeking out the context of intellectual debates of a previous epoch. An amusing case of this obsession, of no importance for the conceptual issues under discussion, is his identification of a farce by Feydau to which Poincaré referred quite frivolously. By contrast, Howard's emphasis (in "Newtonian Space-Time" and elsewhere) upon Newton's sustained critique of the theories of space and motion of Descartes—whom he does not mention by name—is conceptually very important, throwing great light on passages in the scholium at the beginning of book 1 of the *Principia*, which are puzzling if they are read without attention to historical context. (Incidentally, Howard is a strong critic of Descartes's natural philosophy. He somewhere cites Harvey's famous remark that Francis Bacon writes philosophy like a Lord Chancellor, commenting that Bacon at least had the excuse that he was a Lord Chancellor, whereas Descartes had no such excuse!)

(3) Howard is extraordinarily attentive to language. When the classical texts are in English he habitually consulted the Oxford English Dictionary in search of meanings which are now obsolete but may have been intended in the seventeenth century. When English or German or French translations are available for Latin texts, he used them warily, always alert to the possibility that a puzzling phrase could be the result of mistranslation, which he then investigated with the aid of lexicons and grammar books, in spite of never having been instructed in Latin. A splendid example is his thoroughly persuasive recovery of Newton's ontological conception of space, by correcting (in "On Metaphysics and Method in Newton") the translation by the Halls of a passage in "De Gravitatione et Equipondio Fluidorum." I sometimes was enlisted as an assistant in these philological exercises, because of my course of high school Latin. Typically, the decipherment of a difficult sentence would proceed in the following way: as I tried laboriously to parse a Latin passage, Howard would study it and then offer the conjecture, "Couldn't it mean . . . ?". His suggestion was always based on the thrust of the argumentation—what Newton, or Leibniz, or whoever, would have needed to say, in the light of ideas previously expressed. And almost invariably Howard's suggestion was not only com-

patible with the rules of Latin grammar, but also more natural than other grammatically possible readings. Incidentally, he performed a similar feat of translation of a passage of Parmenides, without instruction in Greek. As philological *tours de force* these feats were awe inspiring. But these dazzling translations are secondary compared to another aspect of Howard's attention to language: his refusal to let antecedent biases obscure the intention of an author. The paper "On Metaphysics and Method in Newton" examines Newton's statement "I never intended to show wherein consists the nature and difference of colours, but onely to show that de facto they are originall and immutable quallities of the rays which exhibit them, & to leave it to others to explicate by Mechanical Hypotheses the nature & difference of those qualities." Howard makes a strong case that Hooke and Huygens, and incidentally some modern historians of optics, failed for doctrinal reasons to appreciate Newton's compact but clear discriminations. There is an obverse side to Howard's attention to distinctions of a careful writer: namely, his attention to unclarities in an insufficiently careful writer. For example, he makes the apparently novel observation (in "Some Philosophical Prehistory of General Relativity") that Mach's critique of Newton's inference of absolute acceleration from the water bucket experiment actually contains three different arguments—which Mach does not explicitly discriminate—with different premises and different implications. Since Mach is generally considered to have raised the level of critical thinking on the foundations of physics, and to some extent deserves this reputation, the ambiguities in a central portion of Mach's work seem to me a symptom of pervasive carelessness in the professions of history and philosophy of physics, which can be remedied only by Howard's kind of careful analysis and passion to understand.

(4) Howard is a careful student of the works of the great philosophers and applies his knowledge to physical texts and interpretations of physics. Obvious difficulties attend such applications—there have been radical changes of world view since the Greek philosophers and even since those of the seventeenth and eighteenth centuries; there are radical differences in the textures of analysis and exposition in classical philosophy and in later physics; and there is an admixture of largely discredited physical speculation in the metaphysics and epistemology of early philosophers like Plato and Aristotle that is disconcerting when one tries to extract whatever is perennially valuable in their writings. It takes strong critical judgment, combined with sympathy and imagination, to make good use of classical philosophy when reading post-Renaissance physics. Howard has these qualities and uses them to achieve a remarkable perspective. For example, to me it was surprising and illuminating, and yet after it was said quite obvious, to read his comment on the fate of Aristotle's causes in the light of Newtonian and post-Newtonian physics (in "How Does Physics Bear

upon Metaphysics, and Why Did Plato Hold that Philosophy Cannot Be Written Down?"). Prevailing philosophical opinion since Hume has made the efficient cause the most respectable intellectually of the four causes. But when one considers that for Newton impressed forces do not precede changes of motion but are simultaneous with them, and that in both Newtonian and relativistic physics explanations are fundamentally applications of laws of interaction, the efficient cause dwindles to a characterization of certain types of initial conditions. Hence, "the Newtonian forces of nature—*and* their successors—are in effect most analogous to Aristotelian *formal* causes."

In the last few decades, with the almost total eclipse of logical positivism and the partial eclipse of other varieties of analytic philosophy, there has emerged a widespread advocacy of the traditional, but somehow suppressed, idea that discoveries of the natural sciences throw light upon fundamental philosophical problems. Howard was an independent pioneer of this renaissance, beginning with his doctoral dissertation in 1958 for the Philosophy Department of the University of Chicago, "An Examination of Some Aspects of Natural Science." More important than pioneering, however, is the solidity, judiciousness, and subtlety of his applications of scientific discoveries to problems that are conventionally classified as philosophical. A few quotations will support this claim.

"On the Notion of Field in Newton, Maxwell, and Beyond" has a discussion of Carnap's distinction between "internal" and "external" questions (which he specializes to questions of existence, but Howard treats more generally). Internal questions are posed relative to a linguistic framework, whereas external questions are posed with the intention of transcending reference to this or that framework—and Carnap condemns them as ill-posed and vague precisely because of this intention. Howard grants that often precision is achieved by posing a philosophical question as an internal question and illustrates this assertion by considering the question of the empirical content of a theory, treated "internally" by Newton concerning notions of space and time and by Maxwell and Hertz concerning the concepts of electromagnetism. But he qualifies his approbation by saying

> Where Carnap's notions . . . seem to me deficient, is in the treatment of the large-scale evolution of theories. . . . If . . . it is agreed that the program for a definitive 'language of science has at least *not yet* achieved its aim, and that new theories may require new frameworks, then there is a danger that the internal/external distinction may lead to the neglect of important large questions that span the development of theories—on the grounds that these are questions external to the frameworks, and that only within framework are clear criteria of meaning and truth available. . . . The general (although unsystematic) point of view that I would urge as the correct one here I have already tried to

suggest—in distinguishing philosophically specious positivist criticism from analyses of constructive value like those of Newton and Hertz. No attempt to delimit, systematically and globally, the procedures and notions that are empirically legitimate—from 'Hypotheses are not to be regarded in experimental Philosophy' [a criticism of Howard's hero Newton] to the verifiability theory of meaning and beyond—has really succeeded. To say this is not to depreciate the efforts . . . which have contributed much of value though short of success; but it is to deprecate the appeal to programmatic notions as if the program had been realized: this leads to specious criticism. On the other hand, '*hypotheses non fingo*' and the verifiability theory meaning both had a valid core; this I earnestly hope we do not forget. It has been possible for scientists, in creating, criticizing, modifying, and revolutionizing their theories, to apply what is valid in these principles, despite the lack of an adequate precise general formulation. There is no obvious reason why philosophers of science cannot do the same.

Two problems that are much debated in contemporary philosophy of science—the ontological commitments of scientific theories and the possibility that historically successive theories are "incommensurable"—are treated tacitly throughout Howard's paper "After the Baltimore Lectures: Some Philosophical Reflections on the Subsequent Development of Physics," with explicit judgments expressed at its conclusion.

In this transformation of the ether problem, the presuppositions of Kelvin have undeniably been left behind: it would be absurd to raise the question of whether the ether or anything like an elastic medium, "really exists." But it is equally true, although often ignored, that the old notion of "space," that empty and quiescent container within which bodies exist and forces are propagated, has also been left behind. First, with special relativity, we were led to space-time as the frame whose structure constrained the form of all interactions; then, with general relativity, we were led to the view that space-time is not a quiescent container but is itself interactive; finally, with quantum electrodynamics, we have been led to the view that even "empty" regions of space-time are seething with—I almost said "physical activity," but I suppose it would be more correct to say physical possibilities. . . .

As to the other pole of Kelvin's pair, the atoms or molecules . . . some of what had seemed their most fundamental properties have fallen away, but their recognizable conceptual descendants have continued to play a basic role in our theories.

Finally, a word as to the character of this "recognizable conceptual descent": What is in fact "recognizable" is a distinct relationship, from older to newer theory, of *mathematical forms*—not a resemblance of "entities." . . . I do not suggest a philosophical "explanation" of this fact; I cite it, on merely historical evidence, just as a fact. But I think that, in its turn, this fact helps to "explain" why such a "conservative" as Lorentz, who was willing to borrow the mathematical structures suggested by older theories and to explore their application in contexts where the presumed "substrates" of those structures were lacking (Should one call this "realism," or should one call it a purely "instrumentalist" use of theory?) was able so greatly to advance our understanding of the world. (See also Howard's "On Locke, 'the Great Huygenius,' and the Incomparable Mr. Newton'," 57–58)

The hint in this passage that 'realism' or 'instrumentalism' are equally good characterizations is made explicit in "Yes, but . . . —Some Skeptical Remarks on Realism and Anti-realism," where he says "what I really believe is that between a cogent and enlightened 'realism' and a sophisticated 'instrumentalism' there is no significant difference—no difference that makes a difference." One should recall at this point the remark in a previous quotation about the valid core of the verifiability theory of meaning.

Howard Stein's world view is one (of a rather large family) that regards successful natural sciences as revealing approximately the real structure of the world, recognizes the role of common sense in the initiation of scientific investigation, and acknowledges the great discrepancies between common sense and current scientific theory. It is a philosophical problem of considerable importance for such a world view to account scientifically for the obduracy of common sense (despite partial educability). Toward the end of "On Relativity Theory and Openness of the Future" Howard offers a convincing solution to one aspect of this problem—why the concept of a "present" throughout all space is so "intuitive." He explains as follows, meaning by the 'contemporaneity' of an event e and a set of events S that mutual signals or influences can occur between them:

> the set of events contemporaneous with a specious present will always be a spatially extended one. And it is, I think, of very great relevance to the misconception I am trying to dispel, that this spatial extent—although finite—is in fact *and in principle, as a matter of physics,* always, in a certain sense immensely large. . . . The Minkowski metric can be taken to assign a ratio of lengths not only to a pair of space-like intervals or a pair of time-like intervals, but also to a space-like and a time-like one. The ratio, obtained in this way, of the spatial extent of our bodies to the temporal length of a specious present is exceedingly small: we are temporally long and spatially thin. And the same is true of all the ordinary objects with which we deal—including the earth. Why should this be so?
>
> The question . . . has a simple answer. . . . For although we know little about the physiological conditions required for consciousness to occur, one thing is pretty certain: these conditions involve the coordinated functioning of some part of the central nervous system. And it is clearer still, so far as perceptions of our surroundings are concerned, that the things we perceive must possess a degree of stability (and must interact with us in stable patterns). . . . But according to relativity theory, interactions are not instantaneous: they are propagated with a time delay—with a speed at most equal to that of light. Now, for stable configurations of particles to be established, and for processes with sta-

ble patterns to occur . . . it will in general be necessary for very many interactions back and forth to take place throughout the system in question. . . . And from this it immediately follows that the "graining" of time with respect to which a percipient organism can experience conscious interaction with its environment must be such that the "moments" of time (the specious presents) are long enough to allow such signals . . . to travel very many times the maximum spatial dimensions of the organism *together* with its (relevant) environment. . . .

But then it is entirely clear why we should have developed "intuitions" of something like "cosmic simultaneity," or a "cosmic present": in all our ordinary experience, the time that we experience as a "moment," a specious present—is in the exact sense already explained contemporaneous with events as far distant, spatially, as we ever normally have to do with at all. . . . But these intuitions are quite . . . illusory.

I consider this analysis to be a gem of naturalistic epistemology. More of Howard's gems could be exhibited, but they would be appreciated better in their settings.

I wish now to say something about my relation to Howard. We were classmates during my brief stay at the University of Chicago from 1948 to 1949, where we were both admiring members of Carnap's circle, though by no means disciples. Our friendship ripened further during the years from 1961 to 1967, when both of us were in the Boston area. We played enthusiastic bad tennis together, studied quantum mechanics together, collaborated on a paper concerning the quantum mechanical measurement problem, discussed his work on Newton during the year of his NSF Fellowship and my work on inductive logic. He was the godfather of my sons, who were born during that period. When he left the Boston area for Case Western Reserve, then Columbia University, then the University of Chicago, we stayed in contact by calls, letters, and visits. He read almost all of my papers and gave wonderfully incisive and constructive criticism, to the extent that I have come to regard him as a second intellectual conscience. We shared literary and musical tastes, and I relied upon his immense collection of recordings and fine judgment for recommendations of performances. We shared political anxieties. In sum, he was as nearly a brother as I have ever had. Having said all this—with gratitude—I must add that my high assessment of Howard's contributions to the philosophy and history of science, expressed in the body of this essay, is objective and independent of my strong personal affection for him.

PART I

Ancient and Seventeenth-Century Science

[2]

Noēsis: Plato on Exact Science

W. W. TAIT

1. Introduction

There are two places in Plato's *Dialogues* in which he discusses his conception of scientific explanation: the passages on the 'second best method' in the *Phaedo* and the passages on *noēsis* in the Divided Line simile in book 6 of the *Republic*. I have written about the first of these in (1986) and I want to discuss the second of them here. The conception in question is of what we would call *exact science*. Some exact sciences, the so-called *mathēmata*, were already in existence in the fourth century B.C. in Greece and Plato was concerned to argue for a proper foundation for them. The reason why is part of my story of the Divided Line. The Line itself, I will argue, is a rhetorical argument for foundations.[1] Plato was also concerned with extending the scope of exact science to other domains, including political

Earlier versions of this paper were read in the spring of 1997 in the Philosophy Colloquium at the University of Chicago; at the Pacific Division meeting of the APA in March 1998; and in the Philosophy Colloquium at the University of California-Riverside in November 1998. All of these were based on a manuscript composed in 1986. An even earlier version received valuable criticisms from Henry Mendell, as did the version read at the APA. When I recently returned to the study of Plato, the hard copy of the 1986 manuscript that I found was one returned to me with comments by the late Joan Kung. Those who knew her will guess the state of the manuscript: barely a margin remained without her useful, sometimes quite critical, but always generous remarks. My paper profited very much from them. I would like to acknowledge also the valuable comments sent to me by David Glidden on the penultimate draft. Above all, I thank Howard Stein for our many years of philosophical discussion and, in particular, discussion of Plato. I can only regret his occasional lapses, brought on by an excessive and unseemly admiration for Aristotle.

science; but that is not part of my story, although it should be a substantial part of any accurate account of the *Republic* as a whole.

My reading of Plato, compared to most contemporary commentaries, is deflationary: I understand him to be saying things that we understand quite well, at least in the case of his conception of science, and can agree with, although they were novel in his time. But also, on my reading, and again in contrast with many contemporary commentaries, Plato was a brilliant man of his times. Whether or not he was first to see the need for foundations, he certainly understood it very well and was the one spokesman for it whose writings have come down to us.[2] Often when I read present-day discussions of Plato's conception of science, I am reminded of Marc Antony's funeral oration. Plato was indeed a very great man, a genius: they all affirm this; but then they go on to attribute to him views that would have been as foolish or unintelligible in his time as they are in ours.

In any case, my story begins with book 4 of the *Republic*. At 435d, in the course of discussing the nature of justice and concluding that it consists of the right proportions of courage, moderation, and wisdom, Socrates remarks that the method of analysis that they had been employing up to that point is not entirely adequate for precise understanding, but that the correct way to proceed is more arduous. At that point he and Glaucon agree to continue the discussion at the level that they had been. But in book 6, at 504b, while discussing the education of the future guardians, Socrates refers back to the more arduous way as the one appropriate for them. At this point in the dialogue they are interested not just in the nature of justice in the state and in the soul, but in the various subjects that the guardians should understand; and the longer way refers to a particular conception of knowledge or, better, science, which yields a deeper and more precise understanding of these subjects. Socrates is implying, both in the earlier passage in book 4 and in the present one in book 6, that only by the canons of science in this sense will we really understand the nature of justice or anything else. The argument here reflects one in the *Phaedo* (95a6–97b), where he points out that even inexact science presupposes notions such as that of magnitude or quantity from exact science, and then goes on, in the passages on the 'second best method', to discuss exact science.

Socrates's argument that the guardian must take the arduous way is not just that he must obtain a precise understanding of justice and the like: at 504c9 he states that it is also necessary or else "he will never come to the end of the greatest study and that which most properly belongs to him." He is referring here to the study of the Good. The Sun Analogy then follows, interposed between his complaint about inexact science and his discussion of exact science—just as, in the *Phaedo*, the idea of the best order of things is interposed between his expression of dissatisfaction with inexact science and the 'second best method'. Plato's fullest account of the Good in the *Dialogues* is in the Sun Analogy, though perhaps the *Philebus* is the best place to look for a hint of how he proposed to give a rational account of it. But fascinating as this subject is, I will have to limit myself to a brief description of the role that the idea of the Good plays in Plato's conception of true knowledge.

Exact science presupposes a rational order or structure which the phenomena at least roughly exemplify. Why do the phenomena exemplify this rational structure and how is it that *we* should be able to discern just *this* structure in terms of which to understand the phenomena? In the *Phaedo*, the first question was answered in terms of the 'best order of things' and the second in terms of the doctrine of recollection. In the *Republic*, the doctrine of recollection is abandoned (it was never a very good idea)[3] and the idea of the best order of things is incorporated into the idea of the Good, which is intended to answer both questions. Before the theory of evolution by natural selection, these questions seemed to admit of no naturalistic answer and they rightly taxed philosophers up to the time of Leibniz (whose solution resembles Plato's in many respects) and Kant (whose solution is quite different).

But although Plato thought that the efficacy of exact science pressupposed the Good, he did not think that it presupposed *knowledge* of the Good. This is clear from Socrates's disclaimer in the *Phaedo* to knowledge of the best order of things (which would yield the 'best' method of explaining the phenomena). For turning to the second best method would be of no avail if that method required knowledge of the Good. And in the *Republic*, for example at 533a, he is at least ambiguous on the question of whether he knows the nature of the Good. "But that something like this is what we have to see, I must affirm."

Having accounted for the possibility and efficacy of exact science in the Sun Analogy, Plato goes on to discuss it in more detail.

2. Opinion and Knowledge

At 509d–511d we consider a divided line segment

AC represents the sensible domain and CB the intelligible domain of Forms. Correlatively, AC represents the domain of the opinionable (*to doxaston*) and CB the domain of the knowable (*to gnoston*). Plato has already argued in book 5 (477–78) that there is such a correlation between kinds

of cognition and their objects: he speaks of the faculty or power of opin-
ing or knowing, and argues that such a faculty can be distinguished only
by "that to which it is related and what it effects." Science or true knowledge is of that which is and opinion is intermediate between knowledge
and ignorance: it is about that which both is and is not. He then argues
that sensible things both are and are not whereas Forms are absolutely.
Vlastos (1965) argues, correctly I think, that "are" here should not be
interpreted intransitively as "exist," since Plato's argument in this connection is that a sensible thing S is both f and not-f, whereas the Form Φ corresponding to f is always simply f. *Sophist* 259a–b makes clear that Plato
regards "S is" as incomplete, just as "Simmias is small" (*Phaedo* 102b3–d2)
is incomplete: the latter requires completion to "Simmias is smaller than
Phaedo," the former to "S is f." Plato is thus saying that true propositions
about sensibles are never entirely true but true propositions about Forms
are absolutely true.[4]

The tendency to read "is" or "are" as "exist" is closely connected with
another tendency, namely to read Plato as holding that knowledge of the
Forms is not propositional (knowledge about) but a kind of knowledge by
acquaintance (knowledge of). Thus, belief and knowledge are of the existence of objects rather than of facts. Belief is of objects which change and
which come into existence and pass out, and so belief is of that which both
exists and does not and is both true and false. Knowledge is of objects
which are changeless and eternal and so is of that which exists absolutely
and hence is true absolutely. I have argued against this view (1986) and
won't repeat the argument here.

But it should be noted that the matter is capable of some confusion
because of the ease with which the notion of a fact can be absorbed under
that of an object—or perhaps it is a matter of the verb 'to be' in Greek having a wider scope than translators have respected. Thus, at 476e8–477a1,
Glaucon asserts that to know is to know something, and then is asked
whether 'it' is something that is or something that is not.[5] Another example is at *Theaetetus* 159b, where Socrates distinguishes between the objects
Socrates-ill and Socrates-well. Thus, "Socrates is well" is true just in case,
or perhaps better, to the extent that the object Socrates-well exists. In
other words, in Plato's writings, *facts* or *states of affairs* seem to be easily
included under the title "object." This observation will be important in
section 6, where we attempt to identify the objects corresponding to each
of the four segments of the divided line.

Plato does not explicitly say in his initial description of the Line that the
intelligible domain consists of the Forms—he merely refers to it as the
intelligible domain. This is supposed to give some credence to the view
that Plato's ontology contained the so-called *intermediates* or *mathematicals* that Aristotle attributed to him. But the argument at 477–78 in book

5 seems fairly clear on this point. "But in the case of a faculty, I look to one thing only—that to which it is related and what it effects, and it is in this way that I come to call each one of them a faculty, and that which is related to the same thing and accomplishes the same thing I call the same faculty, and that to another I call other." The faculties explicitly mentioned there are opinion and knowledge. Since the Forms are clearly objects of knowledge, I don't see that there is room for intermediates.

I do not want to discuss here precisely what Plato meant by the Forms. If one goes passage-hunting through the dialogues, as Ross did in his *Plato's Theory of Ideas* (1951), one will find references to Forms or, probably better, uses of the same terms that Plato used to refer to Forms throughout his writings. Plato himself showed signs in the later dialogues, for example, the *Sophist* and the *Parmenides*, that he felt that the 'friends of the Forms' had gotten out of hand. But there is a central role that the Forms play in the *Phaedo* and *Republic* which does concern me here. True knowledge or exact science cannot have as its object sensible things. Plato argues for this in the *Phaedo* (e.g., "Two logs are never exactly equal"), but the conclusive basis for the argument was the discovery of incommensurable line segments in the late part of the preceding century. Reasoning in geometry cannot be founded on what we can see and measure, since measurements cannot distinguish between those lines commensurable with a given one and those which are not. More generally, as Whitehead was later on to put it, nature has ragged edges.[6] The terms in which we describe it in exact science don't literally apply. Then what *is* exact science about? What are the grounds for calling the theorems of geometry true, for example? Neugebauer (1969), in his discussion of this situation suggests with an almost charming innocence that the Greeks simply introduced axiom systems in which the phenomena were idealized and then based truth on provability from the axioms. A wonderful idea! But, unfortunately, not one available to the Greeks in fourth century B.C.: it was to be more than twenty-three centuries before the idea of a formal axiomatic theory would be invented. For example, Frege did not even understand it: for him, as for the Greeks, axioms have to be *true*. But what are they true *of*? What are, to use Plato's terms, the corresponding objects? We might take from Neugebauer the suggestion that they are true of an 'idealization' of the phenomena. But I think that if we try to spell out what this means, we are led to the view, which I think was essentially Plato's, that they are true of a certain structure which the phenomena in question roughly exemplify, but which, once grasped, we are capable of reasoning about independently of the phenomena which, in the causal sense, gave rise to it. The theorems of geometry are not literally true of sensible things: indeed, they do not even literally apply to them. No sensible figure can be a point or a line segment or a surface or solid in the sense of geometry. Yet the assumptions made in geometric proofs are also not arbitrary; something provides traction for them. We have the *idea* of a point, a line segment, a surface, whatever, which we can, by a process of analysis or, as Plato called it, *dialectic*, come to understand purely rationally, stripped free of its empirical source.

I believe that it is this which provided motivation for Plato's reference to Forms and against which attempts to understand his so-called 'doctrine of Forms' should be measured. I believe also that this conception of autonomous reason in the aid of natural science was Plato's great contribution. (Certainly, before him, Parmenides had emphasized the autonomy of reason; but the evidence suggests that he did not conceive it in aid of natural science.)

We have here the conception which Aristotle refers to as the *separate Forms*. A contrasting view is that the structure studied in exact science is obtained by *abstraction* from the phenomena. This is a very different idea: if I abstract the color of my shirt from my shirt, then what I say of the color is true just in case it is a true statement about my shirt, though it be restricted to the language of color. This seems to be Aristotle's view of geometry: when we speak of geometric figures, we are really speaking about sensible substances, except that what we say is restricted to the language of extension. This is a very different idea from Plato's; and his argument in the *Phaedo* already refutes it. If comparison in magnitude of sensible objects never yields exact results, then abstraction can't make the results more precise.[7]

In any case, the faculty of opinion is our power to ascertain the truth of propositions about sensibles in the rough sense in which these are true. The faculty of knowledge is our power to ascertain the truth of propositions about Forms in the absolute sense.[8]

On Plato's conception, all scientific explanation of phenomena begins with the recognition that they exemplify a certain structure or, as he would say, 'participate in' a certain Form. The sensible thing S is f in virtue of participating in the corresponding Form Φ: that is the 'ignorant' or 'naive' explanation of why S is f (*Phaedo* 100d). As Aristotle put it: "The Forms are the causes of all other things" (*Metaphysics* 987b19). But for Aristotle—as it was to be later on for other philosophers, such as Leibniz and Whitehead— this was a criticism of Plato's conception, whereas for us, it is precisely the way exact science proceeds.

Having recognized that S is f, one may be further warranted in asserting that S has some other property, for example, that S is g. That is the 'sophisticated' kind of explanation (*Phaedo*). The sophisticated explanation rests on the naive explanation: the sophisticated explanation of why S is g is that S is f, in other words, that S participates in the Form Φ, together with the fact that Φ is g. For example, consider the propositions

(a) S is right triangular

S is f

(b) The squares on the sides of S equal the square on the hypotenuse

S is g

(c) The squares on the sides of a right triangle are equal to the square on the hypotenuse

Φ is g

As we noted, the sensible figure S is not really a right triangle: indeed, the terms "point," "line," and "angle" by which the notion of a right triangle is to be defined never perfectly apply to sensible things. Nor does the notion of equality (*Phaedo* 74–75). This does not mean, however, that (b) and (c) have no empirical content. The surveyor does indeed apply the Pythagorean theorem and gets good results. But the results, expressed in terms of empirical measurements and constructions, are only 'rough'. And one should not take 'rough' here to mean 'approximate'. For example, the circle can be approximated to any degree of accuracy by an inscribed regular polygon. But here the difference between the two figures is itself a precise magnitude, an area. But the sense in which the sensible figure S is roughly right triangular or in which the result of the empirical construction roughly corresponds to (c) is different from this. It is not a case of one geometric object (in our sense) differing from another by some precise amount: one of the terms in the correspondence is such that the geometric ideas do not perfectly apply to it. Thus it is in the nature of things that (c) applies only roughly to sensible figures and is never absolutely true of them. Plato's notion of *participation*, in spite of the logical positivists' attempts to analyze it in terms of coordinating definitions and the like, remains an essential ingredient in the story of how exact science works.

Proposition (b) is a consequence of (a) and (c) (the sophisticated explanation of (b)). (a) and (b) are about sensibles. What about (c)? It reads to us as a general proposition, about all right triangles. But what, for Plato, is a right triangle? As we noted, Aristotle and, following him, many later commentators, attribute to Plato the view that, besides sensibles and Forms, there are perfect instances of the Forms, the so-called 'intermediates'; and, in particular, there are perfect instances of the geometric Forms, such as the Form Right Triangle. On this view, (c) could be read as a general statement about all perfect instances of Φ. Indeed, the doctrine of intermediates is frequently made an integral part of the interpretation of the Divided Line: intermediates are taken to be the objects corresponding to CE. But, as I have already indicated, I think that there is little merit in the view that Plato held that there are such things. (We shall shortly encounter another objection to the idea of the intermediates inhabiting CE.) So, what are the right triangles, the instances of Φ? The only instances there can be for Plato are the imperfect ones, the sensible figures. But now the notion of (c) as a universal proposition in our sense breaks down. For its scope, the right triangles, would have to consist of the imperfect exemplifications of Φ; but when does a sensible figure count as an exemplification of Φ? How 'straight' do the sides have to be and how thin, for example? For this reason, incidentally, it seems to me seriously misleading to speak of Forms as universals: for they determine no precise extensions. But, moreover, a universal proposition is true because each of its instances is true. On the other hand, the instances of (c), namely (b), are *never* perfectly true: they are in the domain of opinion. And, however one interprets the Line and, in particular, CE, mathematics is assigned to CB, the domain of knowledge. It is therefore impossible to understand (c) as a universal proposition in our sense: its truth does not derive from the truth of its instances. Rather, (c) is true of the Form Φ and the imperfect truth of the instance (b) is *explained* by the fact that S exemplifies Φ. (This is a sophisticated explanation.) Let me remark that it is on this point precisely that we have the starkest contrast with Aristotle's philosophy.

Thus true knowledge for Plato is of the Forms, and the faculty of knowledge is our power to discern truth about the Forms. This faculty is reason and does not involve sensible things. It is important to note here, though, that Plato does not deny a *causal* role to sense experience in our coming to know the Forms: the doctrine of recollection in the *Phaedo* and the Sun Analogy explicitly affirm this role. His point is only that, once given the Forms, they have their own internal logic which is the source of truth about them.

Notice that (c), which I assert to be about the Form *Triangle,* is an ordinary geometric proposition. Thus, as I am reading Plato, the propositions of exact science *are* the propositions about Forms. In this respect I am in disagreement with a tradition according to which the doctrine of Forms exists as a separate theory, distinct from and superior to the exact sciences. For example, a common reading of the 'second best method' passage in the *Phaedo* has it that the doctrine of Forms is an *example* of the method, rather than, as I have indicated, its underpinning. But I would challenge anyone to make any real sense of that reading.

It follows then from my reading that, when we say that geometric propositions are *about* the Forms, "about" cannot be understood in terms of the usual correspondence theory of truth. The theorem that every line segment contains two distinct points, if true of the Forms in the sense of correspondence, would demand the existence of two different Forms 'Two'. Indeed, we would clearly be led to the existence of infinitely many such Forms by Euclid's postulates. Aristotle's *Metaphysics*, books *M* and *N*, bears witness that there may have been some attempt to understand 'aboutness' in the sense of the correspondence theory of truth in the Academy, either by postulating that there are many Forms Point, Line, etc., or by postulating the existence of infinitely many perfect 'mathematicals' or 'intermediates' to serve as the reference of the terms occurring in the theorem of geometry. But there is simply no indication in his writings that Plato himself held such a view.

3. The First Proportion

The segments AC and CB are to be unequal. I have chosen to take AC<CB, expressing the higher status accorded knowledge over opinion. (Plato, realizing that this is arbitrary, does not specify which alternative we should choose.) Socrates says that AD represents images and DC the corresponding sensibles of which they are images. At 510a–b he asks if Glaucon would be willing—and Glaucon agrees—to express the ratio between AC and CB "in respect of truth or falsity" by the proportion

(1) AD:DC = AC:CB.

Earlier, at 509d–e, Socrates describes (1) as an expression of relative clarity or obscurity. At 511e, he explains that the comparison with respect to clarity concerns the kinds of cognition corresponding to the four subsegments of the Line and the comparison with respect to truth has to do with their objects. But what does this mean? One difficulty is that, if AC and CB each correspond to two kinds of cognition and to two kinds of objects, then which of the four possibilities is the right-hand side, AC:CB, of (1) supposed to represent either with respect to clarity or with respect to truth? This, indeed, is a difficulty for *any* view which would have the subsegments of either AC or CB consist of distinct objects. (This is another difficulty for the view that CE consists of intermediates.)[9]

Some commentators would explain all the ratios involved in the Line, including (1), simply in terms of the image metaphor: just as the things in AD are images of things in DC, so sensibles (AC) are images of Forms (CB). Thus, each segment and subsegment must be understood to represent a kind of object and the ratios express a comparison of image to model. We have already mentioned one difficulty with this view: with which of the image and sensible object is the Form being compared in AC:CB? But another difficulty is that it requires that the relation between CE and EB be understood as one between image and model. But the only image/model relation that Plato suggests with the Forms as models has sensible objects as the images; and sensible objects do not inhabit CE.

So let's think about (1) in a different way. Given that we have agreed to correlate length of line segment with degree of truth, the rough truth of "S is g" compared with the absolute truth of "Φ is g" yields AC<CB.

On the other side of the equation (1), AD:DC could reasonably be thought to concern the relation between a *particular* image I and *its* sensible model S, e.g., the reflection of a sensible figure on a surface and the figure itself. But, in what respect are they being compared? In analogy with the principle division, we might suppose that we are comparing "I is g" with "S is g"; but there are objections to that. If "is g" is predicable only of solid objects, then "I is g" is either absolutely false or meaningless; and in either case, it is hard to see what sense can be given to (1) (since in AC:CB, AC is represented by "S is g," which is neither absolutely false nor meaningless). If "is g" is also predicable of images, as in the case of "is right triangular," then there is no reason to think that in general "I is g" is less true that "S is g." The most serious objection, however, is this: if AD is represented by I is g and DC by "S is g," then with which of these two is "Φ is g" being compared in AC:CB?

In view of these considerations, it seems to me that, in both AD and DC, we must be considering the sensible object S and, although we shall only discuss this later on, in both of CE and EB we must be considering the Form Φ. The difference in both cases between the two subsegments must concern *how* we judge the objects in question. Namely, in AC we are concerned with sentences (b) about the sensible figure S. But the grounds for (b) may be of two kinds: in DC we observe S directly, making the necessary constructions and measurements, and in AD we make the observations on an image I of S, say a reflection of S on some surface. Thus, I am suggesting that the difference between AD and DC does not reflect a difference in the sentence in question but rather a difference in the *evidence* for the same sentence. In other words, Socrates is distinguishing two ways of judging in AC, judging about a sensible directly and judging about it on the basis of an image. So if we agree to also correlate length of line segment with degree of *evidence*, then we have the inequality AD < DC.

So we have AC<CB and AD<DC. But why do Socrates and Glaucon go on to agree to (1), to the assertion that the two pairs are in the same ratio? It is clear that no literal sense can be given to this equation: the left hand term is the 'ratio' of evidence for (b) as judged from the reflection of S and the evidence for (b) as judged from S itself. On the right hand side we have the 'ratio' of the truth of (b) and the truth of (c). On neither side do we in any sense have ratios between like magnitudes. Socrates secures Glaucon's explicit agreement to (1) at 510*a* on the basis of the image/model metaphor: I is to S as S is to Φ. But according to the best

sense we can give to this metaphor, in AD:DC we are comparing the evidence I with the evidence S for the same proposition (b); whereas in AC:CB we are comparing different propositions (b) with (c) with respect to truth. Thus, the relation of image to model plays quite different roles in the two sides of (1).[10]

It seems more reasonable to look for a reason for asserting (1) which is independent of the fact that, literally, it makes no sense. It is true that the Line is just a simile; but it is a very elaborate one and, if it is not just a pretentious bit of mathematical nonsense, then it—and in particular (1)—must have a point. I believe that it indeed does, namely a *rhetorical* point: from (1) we shall derive

(2) AC:CB = CE:EB

so that AC:CB is a measure of the superiority of *noēsis*, EB, over *dianoia*, CE. Thus, I propose to read the Line simile not as an illustration of certain relationships between kinds of cognition and truth, which are meaningful in their own right and need only to be pointed out, but as a dramatic argument: *To the extent that the superiority of DC over AD is a measure of the superiority of CB over AC, it is also a measure of the superiority of EB over CE.* The superiority of exact science, that is of knowledge of the Forms, is not the issue here: that is generally agreed upon. The issue is *how* to reason about the Forms; and Plato is arguing rhetorically for the superiority of *noēsis* over *dianoia*.

Of course, in Socrates's original description of the Line at 509d, he specifies that both (1) and (2) are to hold. But at that point he is simply describing the geometric structure of the Line: he has not yet told us how to interpret it. When he explains the interpretation of the segments AC, CB, AD and DC, he immediately gains Glaucon's acceptance of (1) on the basis of the imprecise image/model metaphor. Then, on the basis of (1) and the interpretation of the other segments, he gains his acceptance of (2). When Glaucon agrees to the appropriateness of the simile at 511e, he is agreeing to an argument that began with and depends on the *stipulation* of (1).

4. *Dianoia* and the 'Hidden Equality'

But, to substantiate this reading, we have to go on to consider CE, the domain of *dianoia*. Concerning CE Socrates says

> there is one section of it [CE] which the soul is compelled to investigate by treating as images the things imitated in the former division [DC], and by

means of assumptions from which it proceeds not up to a first principle but down to a conclusion.

Glaucon does not fully understand and Socrates tries again:

> students of geometry and reckoning [i.e., algebra] and such subjects first postulate the odd and the even and the various figures and three kinds of angles and other things akin to these in each branch of science, regard them as known, and, treating them as absolute assumptions, do not deign to render any further account of them to themselves or others, taking it for granted that they are obvious to everybody. They take their start from these, and pursuing the inquiry from this point on consistently, conclude with that for the investigation of which they set out. . . . they further make use of the visible forms and talk about them, though they are not thinking of them but of those things of which they are a likeness, pursuing their inquiry for the sake of the square as such and the diagonal as such, and not for the sake of the image of it which they draw. (510b3–e1).

Thus, CE is concerned with the Forms, with propositions 'Φ is g' and not with 'S is g'. This is in conformity with Socrates's general description of the Line, according to which all of CB is concerned with the Forms. What distinguishes CE from EB is that, in the former and not in the latter, we make use of sensibles and we reason from hypotheses without giving an account of them. As I read the above passage, the reasoning from the hypotheses is entirely rigorous; and so we may assume that the appeal to sensibles arises only in the choice of hypotheses. It is reasonable to suppose that Plato has in mind here the sort of reasoning illustrated by the slave's proof of a special case of (c) in the *Meno*. Starting with a drawn figure, further constructions are made from which the equality of certain areas becomes evident, perhaps by appeal to certain symmetries, for instance. But these are sensible constructions which, being special, cannot prove a general proposition and, being sensible, in any case do not perfectly exemplify the kind of structure in question. For example, the argument involves the construction of the square on a line segment (in a given half-plane): how do we know that this square exists? (See Euclid's *Elements*, book 1, proposition 46.) The assumption that it does is the kind of 'absolute assumption' to which Socrates is referring here. This assumption is extracted from the drawn figure, which is inadequate on the two grounds that we have noted. The construction of the square on a given side is an example of what Socrates means by the postulation of a figure. Note also that it is precisely such constructions that the postulates in Euclid's *Elements* provide for.

In DC, (b) is about sensible figures and we establish it by appeal to such figures. Of course, (b) is not about sensible objects *simpliciter*: it is about them as right triangles, with 'sides' and 'vertices' distinguished. But, as we have already noted, no sensible object is really a right triangle. So, with respect to the comparison of (b) and (c), we have that DC is less than CE. Indeed, in this respect, the ratio is identical with the ratio between AC and CB: for this comparison has to do with the kind of object, sensible or Form, that the proposition is about. However, in CE, although (c) is about Φ, we again establish it by appeal to sensible figures in framing our hypotheses. We may reason ever so carefully, but the starting point of our reasoning is tainted by reference to the phenomena. So, although the two sections, DC and CE, are concerned with the different propositions 'S is g' and 'Φ is g', we invoke the same evidence in both of them. Thus, in exactly the same sense in which we have AD<DC with respect to the evidence we invoke, we have the equation

(3) DC = CE.

In other words, given (1) and the interpretation of the segments AC, CB, AD, DC, and CE, the point E is determined. This 'hidden equation', so called because it follows from (1) and (2) but is not explicitly stated, does not seem to me to be adequately accounted for in the literature on the Line. It is either assumed that Plato did not notice it or that he noticed it but it plays no role in his simile, convicting him of a considerable inelegance. But in book 7 at 533e–534a Socrates explicitly states the equations CB:AC = EB:DC = CE:AD as though he were simply restating (1) and (2). And to see this, he would have had to know (3): indeed, the first of these equations and (2) immediately imply (3); and the conjunction of (1) and (2) follow from these equations only under the assumption of (3). It therefore makes no sense at all to suppose that Plato was unaware of (3). And to propose a reading of the Line which makes no sense of (3) is to attribute to Plato a simile that, as he must have been perfectly aware, *immediately* breaks down. It would seem preferable to infer from the fact that, on a given reading, no sense can be made of (3) that one does not have the right reading.[11]

But, in any case, it seems to me that (3) has a very important meaning for Plato. It is clear from the passage just quoted from the Line and from book 7 that he was assuming agreement, at least among his immediate audience, that the source of truth in arithmetic, algebra and geometry is not in the objects of sense but in the Forms. In this respect these subjects are contrasted in book 7 (530–31) with astronomy and music theory, against the practitioners of which Plato complains that they mistake the objects of their study to be the paths of the stars and, say, the string on the monochord, rather than the curves and line segments in which these participate.[12] But what (3) tells us is that it doesn't matter what we take science to be about, the phenomena or the forms of structure which they (imperfectly) exemplify: if we draw on the sensible figure as evidence for (c), then it is no more reliably established than is (b). The point is not that (3) is a consequence of (1) and (2). Rather it is that, given the interpretations of the subsegments of the Line, (3) stands on its own feet and (2) follows from (1) and (3). This, I believe, is the point of an otherwise contrived-appearing simile: once Socrates and Glaucon agree on (1), all else follows. If we agree that CB:AC measures the superiority of proving (b) about S by considering S itself over proving it by considering an image of S, then we are forced to accept CB:AC as the measure of the superiority of *noēsis* over *dianoia*. And it is at this that the Divided Line simile aims. It is, as I suggested in the beginning of my paper, an argument for foundations of exact sciences. To understand the sense of 'foundations' I have in mind, we should consider what Plato has to say about *noēsis*.

5. *Noēsis*

What is EB, the domain of *noēsis*? At 510b, after the first description of *dianoia*, Socrates continues

> there is another section in which [the soul] advances from its assumptions to a beginning or principle that transcends assumptions, and in which it makes no use of the images employed by the other section [i.e., CE], relying on ideas only and progressing systematically through ideas.

This is the point at which Glaucon confesses his failure to fully understand. Indeed, it is not quite clear what the above passage means. Is Socrates saying that, in the advance to a principle, no use of images is made? Grube translates the passage as explicitly saying that. But it doesn't make sense read in that way. From what point would the advance begin, if not with sensible images? The matter is clarified when Socrates reformulates his description of EB at 511b:

> by the other section [EB] of the intelligible I mean that which the reason itself lays hold of by the power of dialectic, treating its assumptions not as absolute beginnings but literally as hypotheses, underpinnings, footings, and springboards so to speak, to enable it to rise to that which requires no assumption and is the starting point of all, and after attaining to that again taking hold of the first dependencies from it, so to proceed downward to a conclusion, making no use whatever of any object of sense but only of pure ideas moving on through ideas to ideas and ending with ideas.

It is clear from this formulation that Socrates has in mind two stages:

UP STAGE. Up, by means of dialectic, from hypotheses to a principle which transcends hypotheses.

DOWN STAGE. Down from the principle to a conclusion, by means of rigorous reasoning and without appeal to sensibles.

Thus, the answer to the question raised by Socrates's first description of EB seems to be that it is in the Down Stage that there is no appeal to sensibles. There is no reason to suppose that Plato conceived the process of dialectic as not at least beginning with hypotheses that are suggested by sensory experience or 'recollected' from it. Indeed, if I am right that Plato is expanding here on his discussion of the second best method in the *Phaedo*, then it is clear from the latter discussion that he does understand the process of dialectic to begin with what is suggested by empirical experience. The difference between the hypotheses in EB and in CE has to do with the fact that, in the latter case, they are treated as absolute; whereas, in the former case, we analyze them, pushing them back until we come to propositions that transcend hypotheses (or, in the *Phaedo*, until we come to something 'adequate'). Notice that, on this analysis of the passage, it is a mistake to identify, as van der Waerden (1963) does, dialectic with deductive reasoning. The former, although it may involve deduction, is the method of the Up Stage, the latter is the sole method of the Down Stage.

So, to sum up my reading of the Divided Line: It is the embodiment of an argument for the deductive method in exact science, for finding the first principles, that is, the definitions and axioms, which define the structure in question, and then proceeding purely deductively to investigate it. Plato is arguing that the practice of beginning deductions with premises drawn from the consideration of empirical examples is inadequate, because the empirical examples do not adequately represent the structure.

Probably the most serious objection to this interpretation arises from the question of motivation: why should Plato at that time have been concerned with foundations? I have mentioned one motive, the discovery of incommensurable line segments. The earliest evidence of the discovery of these is about 435 B.C. One might suppose that the reaction to the discovery would have come somewhat earlier than Plato's middle dialogues; but it should be pointed out that the most important reaction to the discovery, a geometric theory of proportion, had likely still not been discovered by the time of the *Republic*.

Behind the question of motivation lies a fairly common assumption that the deductive method was already in place at the time that Plato wrote. But the evidence for this assumption is exceedingly thin. It is certainly true that proof, in the sense of deriving something less obvious from something more obvious, had been around for a very long time and was not the invention of the Greeks. Cut-and-paste proofs of the kind found in books 1 and 2 of Euclid existed long before fourth century B.C. and in other cultures besides Greek. One thing characteristic of Greek mathematics is the definition of terms and the ordering of theorems according to dependence, so that earlier ones may serve as lemmas in the proofs of later ones. But when Proclus, who is the main source of information about the development of geometry in classical Greece, wrote of others before Euclid who wrote 'elements', there is no reason to think that he was referring to more than this. In particular, there is no reason to think that the idea of geometry as a deductive science based on *primary truths*, as represented by the Common Notions and Postulates in Euclid's *Elements*, preceded the *Republic*.[13]

Indeed, there is evidence in the *Republic* itself to the contrary. First, there is Plato's criticism of the geometers at the end of book 6 and in book 7 (533b–d). This is often taken to indicate that Plato believed geometry, along with the other exact sciences, to be intrinsically inferior to 'real knowledge', namely knowledge of the Forms, and so confined to the realm of *dianoia*. But it is far more plausibly taken to be an indication that geometry had not yet been sufficiently founded on primary truths. Further evidence for this may be found in book 7 (527a), where the geometers' use of terms such as "squaring," "adding," etc., which suggest physical construction, is criticized, although Socrates states that "they cannot help it." A plausible interpretation of this is that Plato is calling for a precise foundation for the notion of geometric construction. The plausibility is moreover reenforced by the fact that the Common Notions and Postulates in Euclid's *Elements*, written some fifty years later, *do precisely that*.

6. The Four Kinds of Objects

There are two places where Plato refers to the objects corresponding to the subsegments of the Line. One is at the very end of book 6 (511e) where he is speaking of the segments as representing affections of the soul, and then says that they should "participate in clearness and precision in the same degree as their objects partake of truth and reality." For, indeed, all of the ratios are equal to AC:CB. But the passage at 534a in book 7 seems unambiguously to associate with each of the four subsegments its own object. He speaks of the objective correlates of AC and CB and then of the "division into two of each of these." In other words, he seems to be saying that both the domain of the sensibles and the domain of Forms are to be subdivided corresponding to the subsegments of the Line. But he also says "let us dismiss [this], lest it involve us in discussion many times as long as the preceding." What precisely Plato had in mind here is perhaps unimportant in relation to the main point of simile, since he does, after all, leave it aside. But also, since he leaves it aside, it is reasonable to suspect that there is a twist in the answer. In view of all this, the most straightforward possibility seems to me to be that the objects in AD are the same objects as in DC, but qualified. Thus, in DC, we judge the sensible object S on the basis of S itself; but in AD, we judge on the basis of S-as-imaged-by-I. That is, in AD the basis, the 'objective correlate', of the judgment about S is not S as it is in itself, nor is the basis of judgment simply the image I. (For example, I may be elliptical; but we nevertheless judge that S is circular.) In the same way, the objects in CE are the same as those in EB, the Forms; but in EB we are judging about Φ as it is in itself, in other words, our judgments are based on deductions from first principles. But in CE we are judging, not on the basis of Φ as it is in itself. Nor are we judging about the sensible S which images Φ. For example, if Φ is the Form, Triangle, the ratio of the sides of S are irrelevant to its representation of Φ. (For example, the sides of S may be uneven, though what we are judging must apply to isosceles triangles, too.) The objective correlate is Φ-as-imaged-by-S.[14]

NOTES

1. My account of the Line, as an argument for foundations of the exact sciences, is in substantial agreement with the excellent discussion in Nicholas White 1976, 95–99. The main difference, also substantial, between our accounts lies in the fact that, whereas White understands the new foundations to be a new and separate science of *dialectics*, with its own axioms and theorems, on my account the foundations are to consist in adequate first principles for, say, geometry, itself, to be found by a *process* of dialectic. The difference in our views reflects a difference in our judgment of the state of geometry itself at the time that Plato was writing.

2. I am excluding Aristotle's theory of demonstrative science in the *Posterior Analytics* for two reasons: first, tied as it was to his conception of logic as syllogistic, it lost its relevance to exact science. This is manifest in the historical distinction between the geometric mode and the syllogistic mode of reasoning, surviving perhaps in Kant's distinction between demonstration and discursive reasoning. But my second reason is at least equally important: For Aristotle, the primary truths, the first principles, are general empirical truths. Aside from the difficulty that we shall raise in section 2 with this in the case of exact science, it means that the original *motive* for foundations of exact science, for finding first principles, is completely lost. For Plato, as we shall see, the goal of foundations is to make explicit the rational structure we are studying and so to define what is true of that structure—namely, what can be deduced from those principles. For Aristotle, the goal of foundations can only be organizational: to organize empirical truths in a deductive

system (where here I am ignoring the fact that deductive systems based on syllo-gistic are in any case puny things).

3. See Leibniz, *New Essays*, 78–79.

4. It has been pointed out, for example in Owen 1957, 109, that Plato's case for propositions about sensibles never being perfectly true is favored by taking examples involving geometric concepts and looks less plausible when we consider propositions such as "Socrates is a man." But it was the context of exact science that was Plato's concern—although the issue is muddied by the fact that he clearly felt that all science, including political science, could be modeled on the exact sci-ences that he knew.

5. See Paul Shorey's note *c* (3) to this passage in the Loeb edition of the *Republic*, volume 1.

6. In *Metaphysics* I vi 2–3, Aristotle traces the motivation for Plato's doctrine to the influence of Heraclitus's view that "the whole sensible world is always in a state of flux."

7. The most sophisticated attempt to found exact science and in particular geometry on abstraction is Whitehead's, in his method of 'extensive abstraction'. But, whatever version of this one takes, the fundamental relation of extensive con-nection between the objects of perception—events or regions, as he variously iden-tified them—must be taken to be well defined, in the sense that it is determined whether or not two such objects are extensively connected. Otherwise, contrary to Whitehead's claim, the ordinary Euclidean geometry of empirical space cannot be derived. But this determinacy of extensive connection is hardly compatible with his view that nature has ragged edges.

8. It is clear that Plato's use of the term *doxa* does not entirely correspond to our use of the terms 'opinion' or 'belief', since, as we use these terms, that con-cerning which we can have opinions or beliefs we can also have knowledge and, conversely, we can have opinions (short of knowledge) about what is knowable: one and the same proposition may be an expression of opinion and an expression of knowledge. However, I shall continue to translate 'doxa' as 'opinion'.

9. There is a tendency to regard AB as a 'continuous scale'—an ordered set of points like a thermometer, whose points correspond to degrees of reality or truth or knowledge. This picture is sometimes embellished by placing the sun at C and the Good at B. (See Grube 1974, 164, fn. 16). But I agree with Fogelin 1971 that no sense can be made of this picture. A segment is not a set of points representing a range of degrees; rather, it is a geometric object representing, in ratio at least, one degree. Otherwise, no sense can be made of the ratios. AD:DC is a ratio of *mag-nitudes*, not quantities. If the segments are not sets of points, then it follows that the sun and the Good do not occupy points on it.

10. There is another difficulty with the image/model metaphor: in the sense that I is an image of S, they share a form of structure. For example, if I is the reflec-tion of S in a pool, then they share the form of structure that is preserved by the projection of S on the pool from the sun. If S is a sensible triangle, then it has ver-tices and sides, and I contains images of these. Φ, on the other hand, *is* a form of structure and, as I have argued in 1986 quite independently of the arguments in the *Parmenides*, it is unlikely that Plato held that Φ was also an exemplification of

itself. The Form, Triangle, is triangular in a different sense than S is triangular. S participates in Φ (i.e., has the Form Φ), but Φ does not participate in itself. In particular, it does not have vertices and sides. Thus, I is an image of S in a quite different sense than S is an image of Φ.

One might want to argue that Plato simply failed to distinguish two different senses of *paradigme*, namely our sense of 'paradigm' and the sense of a form of structure which a paradigm perfectly exemplifies. This would be analogous to attributing confusion to him because he asserts both that Φ is f and that S is f, which I have argued in 1986b would be unjust. There might be some grounds for attributing confusion to him if the equation (1) otherwise made literal sense. But we have already noted that it does not; and so it would be entirely gratuitous to assume that Plato was confused on this point.

11. It is sometimes argued that Plato cannot have intended the hidden equation because, right before the above cited passage, at 533d, Socrates indicates that *dianoia* involves more clearness than opinion and more obscurity than *noēsis*. Socrates actually uses the term *episteme* here; but he immediately makes it clear that he means *noēsis*. But note that the comparison of *dianoia*, of CE, is not with DC but with AC; and this makes sense only as a comparison with respect to truth and falsity of (c) with (b); for, with respect to the kinds of evidence, AC includes two cases, AD and DC. Thus, in the only reasonable sense of comparison, we have that AC is less than CE. On the other hand, CE and EB are compared with respect to the kind of evidence invoked; and in that sense, we have CE<EB. So the passage at 533d makes perfectly good sense and is compatible with the hidden identity which compares DC with CE with respect to the kind of evidence invoked.

12. This passage is widely misinterpreted as a call to give up empirical science in favor of 'mathematics'. But, as I noted in 1986b and above, this is an anachronism: Plato had no conception of mathematics in our sense. For example, geometry *was* for him the study of sensible figures and measurement of them. What he understood, put in terms natural for us, is that this subject advanced by idealizing the phenomena that it studied, and he was advocating that astronomy and music theory proceed in the same way—by studying the structure imperfectly exemplified by the phenomena in question.

In fact, this is a slight oversimplification, since Plato was also advocating the study of a more general form of structure than that exemplified by the phenomena: steriometry in general and numerical proportions. Perhaps in part this may be explained by his recognition of the added insight gained from studying the general case: think of Newton studying the dynamics of central forces in general in developing his theory of gravitation. But there is a strong hint (531c–d) that Plato also believed that this study would lead to the discovery of connections among the forms of structure exemplified by the various phenomena—e.g., between the proportions exemplified by harmonies and those exemplified by the periods of the superimposed circular motions of the heavenly bodies—and so to the discovery of the proportions which express the best order of things: the Good.

13. Indeed, he refers to 'elements' written by the fifth-century geometer Hippocrates of Chio, who would have been too early to have been motivated by the discovery of incommensurable lines.

14. This approach to the problem of 'objective correlates' is similar in some respects to that of N. D. Smith in his excellent paper (1981). What I am calling 'S-in-itself', 'S-as-imaged-by-I', etc., Smith refers to as *intensional* objects. An essential difference is that he takes CE to be inhabited, not by 'Φ-as-imaged-by-S', but by 'S-as-image-of-Φ'. He considers but rejects my alternative on grounds of the image metaphor: He writes (p. 134), "Unless it can be shown that the poorly conceived Forms of the geometer image the properly viewed Forms of the dialectician, we have not, by making this move [viz. adopting my alternative], generated a contrast analogous to the one between *eikasia* and *pistis*." But that is so because Smith fails to carry the analysis through to the subsegments of AC. For him, the objective correlate of AD is not S-as-imaged-by-I, but I. Thus, he not only admits an asymmetry in his interpretation of the Line, but in my opinion, fails to solve the problem of what the ratio AC:CB is to mean: Which of S and I in AC is being compared with which of Φ and S-as-image-of-Φ in CB?

REFERENCES

Fogelin, R. (1971). "Three Platonic Analogies." *Philosophical Review* 80, no. 3 (July 1971): 371–82.

Grube, G. (1974). *Plato's Republic.* Indianapolis: Hackett Publishing Company. Revised edition by C. D. C. Reeve in 1992.

Neugebauer, O. (1969). *The Exact Sciences in Antiquity.* New York: Dover. A slightly corrected republication of the second edition published in 1957 by Brown University Press.

Owen, G. (1957). "A Proof in the *Peri Ideon.*" *Journal of Hellenic Studies* 77: 103–11. Reprinted in Owen 1986.

———. (1986). *Logic, Science, and Dialectic: Collected Papers in Greek Philosophy.* Edited by M. Nussbaum. Cornell: Cornell University Press.

Ross, D. (1951). *Plato's Theory of Ideas.* Oxford: Clarendon Press.

Smith, N. (1981). "The Objects of Dianoia in Plato's Divided Line." *Apeiron* 15: 129–37.

Tait, W. (1986). "Plato's Second Best Method." *Review of Metaphysics* 39: 455–82.

van der Waerden, B. (1963). *Science Awakening.* New York: John Wiley and Sons.

Vlastos, G. (1965). "Degrees of Reality in Plato." In *New Essays on Plato and Aristotle,* ed. R. Bambrough. London: Routledge and Kegan Paul. Reprinted in Vlastos 1981.

———. (1981). *Platonic Studies.* 2nd edition. Princeton: Princeton University Press.

White, N. (1976). *Plato on Knowledge and Reality.* Indianapolis: Hackett Publishing Company.

[3]

From the Phenomenon of the Ellipse to an Inverse-Square Force: Why Not?

GEORGE E. SMITH

Even though I focus less on the conceptual side of science than does Howard Stein, and more on the evidential side—especially on why certain sciences have been so much more successful in turning data into evidence than others—no philosopher has influenced my approach to philosophy of science more than he has. This is particularly true of three papers he published within the last decade, one on Locke, Huygens, and Newton (Stein 1990a), one on Newton's deduction of universal gravity from phenomena (Stein 1990b), and most of all "Some Reflections on the Structure of Our Knowledge in Physics" (Stein 1994). One theme that runs through these papers is how extraordinarily different Newtonian science was from anything that went before, so different that the failure of Huygens, for example, to appreciate it is not reason to respect him less, but instead to marvel at Newton. As the following excerpt from a letter of Huygens to Leibniz in 1690 makes clear, although Huygens did understand the *Principia* itself, he definitely did not understand the approach to science that Newton was pursuing:

> Concerning the Cause of the tides given by M. Newton, I am by no means satisfied, nor by all the other Theories that he builds upon his Principle of Attraction, which seems to me absurd, as I have already mentioned in the addi-

I wish to thank Kenneth G. Wilson and, as always, Curtis Wilson for reading and commenting on an earlier draft of this paper; I have ceased being able to distinguish what I have learned over the years about science from these two and what I have managed to sort out for myself. I must also acknowledge helpful suggestions beyond those cited below from Babak Ashrafi, Jody Azzouni, David Bloor, I. Bernard Cohen, Markus Fierz, Bill Harper, Michael Nauenberg, Eric Schliesser, and India Smith, as well as the assistance of Benjamin Weiss of the Burndy Library.

tion to the *Discourse on Gravity*. And I have often wondered how he could have given himself all the trouble of making such a number of investigations and difficult calculations that have no other foundation than this very principle. (Huygens 1901, 538)

The why-not question of my title brings out the extraordinary character of Newtonian science and its revolutionary departure from the science preceding the *Principia* in a way that I hope to show complements and supplements what Stein has said in his writings on Huygens and Newton.[1]

Why Not?

The standard textbook approach is to deduce the inverse-square variation of celestial gravity from Kepler's ellipse. Some textbooks, and some philosophers as well, even say that this is what Newton did in the *Principia*. Stein has been careful to point out that this is what Newton *could* have done, but did not do. The textbook tradition probably derives from Laplace, for he obtains the inverse-square from the ellipse in the chapter entitled "The Law of Gravity Deduced from Observation" in volume 1 of his *Celestial Mechanics*, published in 1798 (Laplace 1966, 239ff.). Why he chose to do this rather than follow Newton is an interesting question; he does not attribute it to Newton. The tradition of saying that this is what Newton did appears to date from sixty years earlier, starting with Pemberton's popular account of the *Principia*, followed by those of Voltaire and Maclaurin.

Very early in the *Principia*, and in the *De Motu* tracts preceding it, Newton does prove two propositions about bodies moving under centripetal forces that describe an ellipse: (1) if the centripetal force is directed toward the center of the ellipse, then the centripetal acceleration varies linearly with the distance from the center (as if the body were connected to the center by a spring); and (2) if instead the force is directed toward a focus of the ellipse, the centripetal acceleration varies inversely with the square of the distance from the focus. Newton never uses this second proposition to infer the inverse-square variation from the Keplerian ellipse. In the *De Motu* tracts, he instead uses the law of force for uniform circular motion, which he and Huygens had independently established, to infer the inverse-square for both the planets and the satellites of Jupiter from Kepler's 3/2 power rule. He offers this same line of reasoning in book 3, proposition 3 of the *Principia*, but then adds that the inverse-square for the planets can be proved with "the greatest exactness from the fact that the aphelia are at rest." Newton is here invoking his "precession theorem" (book 1, proposition 45) which establishes a systematic relationship between the exponent of *r* in the centripetal acceleration rule and the rate of precession of the line of apsides of the orbit in the case of orbits that are nearly circular. He undoubtedly came upon this extraordinary theorem while worrying about the distinctly non-Keplerian orbit of the Moon, and not while searching for a better way of establishing the inverse-square for planets.[2] But he took advantage of the precession rule once he had it.

What Newton instead does with his result for the Keplerian ellipse in *De Motu* is to infer the ellipse from the inverse-square:

> *Scholium.* The major planets orbit, therefore, in ellipses having a focus at the center of the Sun, and with their *radii* drawn to the Sun describe areas proportional to the times, exactly as Kepler supposed. (Newton 1974a, 49)

He draws this same inference in book 3, proposition 13 of the *Principia* after having established the more general result in book 1 that motion in a conic-section governed by centripetal force directed toward a focus entails an inverse-square variation of the force. Huygens states Newton's conclusion exactly in the notes he made while reading the *Principia*: "the excentrics necessarily become elliptical" (Huygens 1944a, 143). Newton's reasoning here has been the subject of a long controversy, with some even suggesting that he confused the conditional with its converse. When Newton heard that Johann Bernoulli was claiming that he had no proof that the ellipse is the sole bounded trajectory under the inverse-square, he added a brief proof in the second edition.

I am not going to dignify the silliness that Newton became confused between the conditional and its converse. Instead I am going to ask why he proceeded as he did rather than just infer the inverse-square from the Keplerian ellipse. He knew perfectly well that the planets do not move uniformly in concentric circular orbits. So, why not infer the inverse-square from the ellipse, which he knew at least approximates the true orbits better than the simple circle?

In fact, Newton knew more than this. He knew that four distinct approaches to planetary orbits were being taken at the time, all achieving the same level of accuracy, and that Horrocks had proposed a fifth. These are summarized in the table on the next page:[3]

Only Kepler and Horrocks employed the area rule. Boulliau and Streete following him instead used a geometric construction, which Mercator had shown closely approximates the area rule; and Wing had initially used an equant oscillating about the empty focus, later replacing it with another geometric construction. Horrocks and Streete following him were the only ones to use Kepler's 3/2 power rule to infer the length of the major axis from the more accurately known period. In fact, Kepler himself treated the 3/2 power rule as what we now call an accidental, not a

	ORBITAL TRAJECTORY	LOCATION VS. TIME	MEAN DISTANCE FROM SUN
KEPLER	ellipse	area rule	from observations
BOULLIAU	ellipse	geometric construction	from observations
HORROCKS	ellipse	area rule	via 3/2 power rule
STREETE	ellipse	Boulliau's construction	via 3/2 power rule
WING	ellipse	oscillating equant	from observations

nomological, generalization, reflecting a particular sequence of planet densities that God had chosen (Kepler 1995, 65ff.).

So, the two phenomena from which Newton originally inferred inverse-square centripetal accelerations were not even generally accepted among the leading orbital astronomers of the time. The ellipse, by contrast, was the one thing on which they all agreed. Hence the natural move, if he were trying to persuade people to accept the inverse-square, would have been to proceed from the accepted ellipse. This is precisely how Leibniz proceeded in his *Tentamen de motuum coelestium causis*, published in 1689, adding in the later revised version that "what follows is not based on hypotheses, but is deduced from phenomena by the laws of motion" (Aiton 1972, 132). That Newton did not proceed this way suggests that he did not see himself as simply trying to persuade others to accept the inverse-square. What then did he see himself as doing? This is the other question I am going to try to answer.

Let me start with some plausible, but uninteresting answers to why Newton did not infer the inverse-square from the ellipse. The problem Hooke originally put to him in 1679, and that Halley presumably put again in the summer of 1684, was to determine the curve a body describes under inverse-square centripetal forces. Hooke even intimated that the curve is not the ellipse of the astronomers: "This curve truly calculated will show the error of those many lame shifts made use of by astronomers to approach the true motions of the planets with their tables" (Newton 1960, 309). So, one possible answer is that Newton derived the ellipse from the inverse-square, and not the other way around, because this gives the solution to the problem put to him. This answer may well be the correct one in the case of the *De Motu* tract of November 1684. It seems a bit petty, however, in the context of the *Principia*.

This is not to deny that the question whether the ellipse is exact or only approximate was very much open at the time and Newton saw himself as resolving it. That this question remained open, however, did not by itself stop Newton from inferring the inverse-square from the ellipse. Questions about whether the area rule and the 3/2 power rule are exact or only approximate were no less open, and Newton saw himself as resolving them too, yet he drew inferences from these two features of Keplerian motion, but not the ellipse.

A second uninteresting answer is that the orbit of our Moon, which everyone at the time knew definitely not to be a Keplerian ellipse, stood in the way of inferring the inverse-square from the ellipse. The Moon's orbit surely did pose a threat to Newton's overall line of reasoning in the *Principia*, and I am sure he was pleased when he came upon a common way of inferring the inverse-square for it and for the planets from the precession theorem. Nonetheless, the Moon did not pose a counterexample to inferring the inverse-square from the ellipse of the planets, and regardless of which way Newton ran this inference, the Moon was still going to demand special treatment.

I want to propose an answer that seems to me much better than these, and also more interesting. Newton was more careful in drawing inferences from phenomena than we sometimes notice. Consider first the inference from the area rule to the conclusion that the acceleration and hence the force is centripetal. The second of the two theorems of Book 1 underlying this inference includes the following two corollaries:

> *Prop. 3, Corol. 2.* And if the areas are very nearly proportional to the times, the remaining force will tend toward body E very nearly.
> *Prop. 3, Corol. 3.* And conversely, if the remaining force tends very nearly toward body E, the areas will be very nearly proportional to the times.

These provide the basis for concluding that a small deviation from the area rule implies nothing more than a small additional acceleration superposed orthogonally on the predominate centripetal acceleration. In other words, Newton took the trouble to show that even if the area rule holds only to high approximation—or, to use his phrase, *quam proxime* (very nearly)— the direction of the acceleration is still centripetal to a high approximation.

He goes even further in the case of the precession theorem for nearly circular orbits by deriving a strict relationship between the apsidal angle θ—the angle at the force-center between the apocenter and the pericenter—and the square root of the index n, namely $n = (180/\theta)^2$, where the centripetal acceleration varies as $r^{(n-3)}$. This relationship confirms not only that the exponent of r is exactly -2 when the apsidal angle is 180 degrees, but further that it is very nearly -2 when the apsidal angle is very nearly 180. As Ram Valluri, Curtis Wilson, and Bill Harper have shown, this inference (though not the formula for n) continues to hold even when the eccentricity is large and the trajectory is therefore not nearly circular (Valluri et al. 1997); further, I have shown that Newton's derivation put him in a perfect position to know this (Smith forthcoming).

The same is true of the inference from the 3/2 power rule for concentric circular orbits to the inverse-square. Newton provides corollaries to his theorem on uniform circular motion to make clear that the exponent of r is very nearly -2 in the rule giving the variation of the centripetal accelerations from orbit to orbit so long as the square of the periods varies very nearly as the cube of the radii—in particular:[4]

> *Prop. 4, Corol. 7.* And universally [in the case of concentric circular orbits], if the periodic time is as any power R^n of the radius R, and therefore the velocity is inversely as the power of R^{n-1}, the centripetal force will be inversely as the power R^{2n-1} of the radius; and conversely.

And, though he does not bother to say this, small deviations from uniform motion in concentric circular orbits can be attributed to small accelerations superposed on the predominant inverse-square accelerations. In short, wherever Newton did draw inferences from phenomena, the conclusion still holds *quam proxime* even when the premise holds only *quam proxime*; and Newton took the trouble to include mathematical results supporting this.

Now consider the ellipse. Newton showed that the centripetal acceleration varies as r to the minus 2 power when sweeping out equal areas with respect to a focus, and it varies as r to the plus 1 power when sweeping out equal areas with respect to the center. Both of these inferences continue to hold to high approximation so long as the body describes an ellipse to high approximation, for, as with the circle, any small deviation from the ellipse can be attributed to a small motion superposed on the predominant motion. But what if the center of force is not exactly at either the center or a focus of the ellipse? In particular, consider the case of an ellipse of small eccentricity in which the center of force lies not at a focus, but slightly off it along the line between it and the center. The center of force can then be both very nearly at the focus *and* very nearly at the center; the implied exponent in the centripetal acceleration rule, however, cannot be both very nearly -2 and very nearly $+1$. This gives reason to suspect that the inverse-square is not guaranteed to hold *quam proxime* whenever the orbit closely approximates Kepler's ellipse.

Within four years after the first edition was published in 1687, Newton undertook what Tom Whiteside has called a "radical restructuring" of the *Principia* (Newton 1974b). Why Newton did so is a matter of speculation. The intervening events I am inclined to single out are the publication of Leibniz's *Tentamen*; Huygens's visit to England during the summer of 1689, when he and Newton had two long conversations; and the publication of Huygens's *Discourse on the Cause of Gravity* in February 1690, which includes a detailed evaluation of the *Principia*. In the original *Principia* uniform circular motion is treated as a special case of conic section motion. One of the main features of the restructuring undertaken in the early 1690s was to use curvature to treat conic section motion as a generalization of uniform circular motion, with the ellipse resulting from the body sliding from one osculating circle to the next in somewhat the way Leibniz's *Tentamen* had it sliding from one circular vortex stream to the next. This curvature approach yielded results comparing the rules of centripetal force for a single trajectory with two different centers of force. The proposition displayed in figure 1 treats the case of an arbitrary orbit. By the time the second edition finally appeared in 1713, Newton had abandoned the idea of a radical restructuring. Some results from twenty years earlier nevertheless do show up in it, including both a slightly modified form of this proposition and its corollary, which compares the rules of central force for two different centers of force in the case of an ellipse with period fixed.[5]

In the second and third editions Newton uses this comparison result to prove the theorem on the Keplerian ellipse and the inverse-square in a second way. In a scholium at the end of section 3 on conic-section trajectories, however, he restates it in the form of a general rule: "If a body P, under the action of a centripetal force tending toward any given point R, moves in the perimeter of any given conic whatever, whose center is C, and the law of centripetal force is required, let CG be drawn parallel to the radius RP and meeting the tangent PG of the orbit at G; then the force will be as CG^3/RP^2." As figure 2 makes clear, Newton is here allowing for centers of force other than the focus and the center. His CG^3/RP^2 rule thus gives a way of determining whether the inverse-square fails to hold *quam proxime* when the center of force is slightly off the focus.

In particular, let C be the center of an ellipse, S a focus, R the center of force, CS/CA the eccentricity e of the ellipse, and CR/CA the excentricity ϵ of the center of force (see figure 3).

Then, with a little work CG can be given in terms of RP, e, ϵ, and a, the length of the semi-major axis:

$$CG = \frac{RP}{(1 - \frac{\varepsilon^2}{e^2}) + \frac{\varepsilon}{e} \sqrt{\frac{RP^2}{a^2} - (1 - \frac{\varepsilon^2}{e^2})(1 - e^2)}}$$

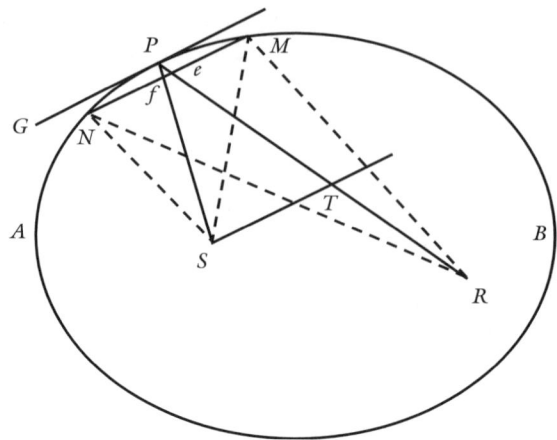

Proposition. The force whereby any body P can revolve in any orbit APB whatever round the center S of force is to the force whereby another body P can revolve in the same orbit and in the same periodic time round any other center R of force as the product of the height of the first body and the square of the height of the second body, SPxRP², to the cube of the straight line PT which the straight line ST parallel to the orbit's tangent cuts off from the height of the second body in the direction of that body.

FIGURE 1
A Proposition from the "Radical Restructuring" of the *Principia* in the Early 1690s

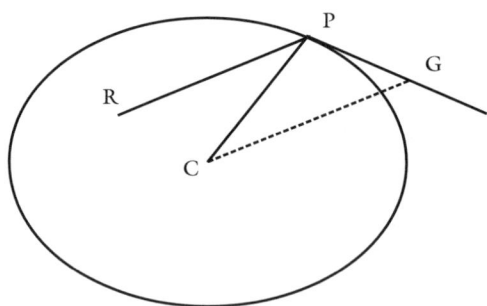

FIGURE 2
Diagram from the Scholium Added at the End of Book 1 Section 3 in the Second Edition

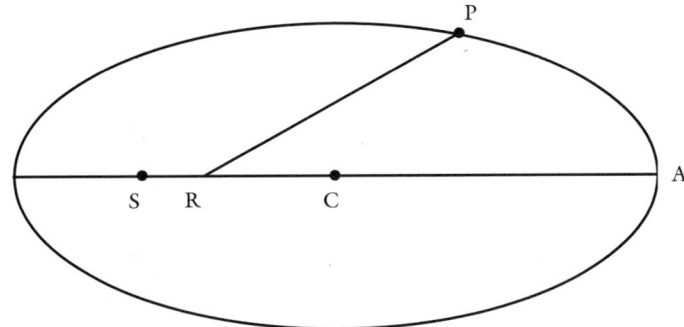

FIGURE 3
An "Off-Keplerian" Excentric Ellipse

Using Newton's CG^3/RP^2 for the rule of centripetal force and inverting, we find that the force and hence the acceleration varies inversely as the cube of a complicated quantity involving RP, all divided by RP:

$$\frac{\left\{ (1 - \frac{\varepsilon^2}{e^2}) - \frac{\varepsilon}{e} \sqrt{\frac{RP^2}{a^2} - (1 - \frac{\varepsilon^2}{e^2})(1 - e^2)} \right\}^3}{RP}$$

When the excentricity equals the eccentricity, this expression has the force varying inversely as RP squared, as it should; and similarly when the excentricity is 0, the force varies inversely as $1/RP$, or just RP. Moreover, when the eccentricity is large and R is near S, the force varies inversely as RP squared *very nearly*, for the two terms not involving RP in the expression are then very small. But when e is small and ε is an intermediate fraction of e, it is no longer true that the force varies very nearly as the inverse of RP^2. The rule of force then becomes complicated, involving four terms, none of which vary precisely as RP^2. This confirms that the exponent in the force rule ceases to be very nearly –2 for the ellipse when the eccentricity is small and the center of force is between the center and focus of the ellipse.

    Did Newton know this result? I do not know. I used his CG^3/RP^2 rule in deriving it, employing math entirely within his command (I sometimes wonder if I am capable of employing math beyond Newton's command). Still, nothing like my expression for CG occurs in his surviving papers. So, perhaps he never did carry through this analysis to the point of proving that the inverse-square need not hold *quam proxime* when the trajectory is

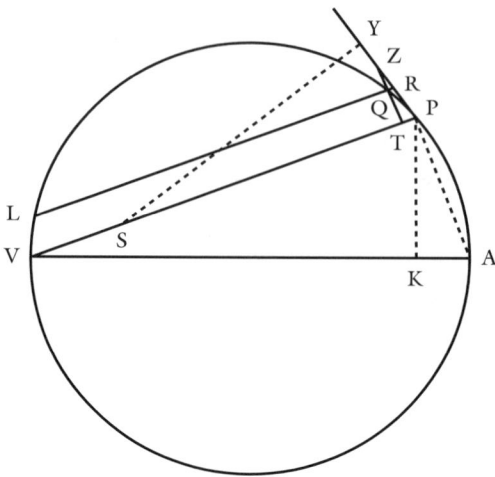

FIGURE 4
Excentric Circle Diagram from Book 1 Proposition 7, Added in the Second Edition

only *quam proxime* a Keplerian ellipse. No matter. For Newton had a much better way of showing this.

In the second edition of the *Principia* he inserted a new proposition (book 1, proposition 5), which appears in his papers from the early 1690s, that gives the rule of centripetal force for an excentric circle—that is, a circle in which a body sweeps out equal areas in equal times relative to a point S off the center. The rule is that the centripetal acceleration and force vary inversely as the product of SP^2 and PV^3. A corollary to this result then provides the basis for the CG^3/RP^2 rule I used above. Now, PV can readily be expressed in terms of SP in the case of a circle. I have instead chosen to apply the CG^3/RP^2 rule to the excentric circle. Either way, the upshot is that the centripetal force and hence acceleration vary inversely as a combination of four terms in which SP occurs respectively in powers of 5, 3, 1, and −1:

$$\left(\frac{SP}{a}\right)^5 + 3(1-\varepsilon^2)\left(\frac{SP}{a}\right)^3 + 3(1-\varepsilon^2)^2\left(\frac{SP}{a}\right) + (1-\varepsilon^2)^3\left(\frac{SP}{a}\right)^{-1}$$

Notice that SP to a power of 2 is nowhere to be found in this expression.[6]

The evidence at the time that the planetary orbits are ellipses was confined to Mercury and Mars; and even in the case of Mercury, the most elliptical of the orbits then known, the minor axis is only 2 percent shorter than the major axis. The orbits really are *nearly* circular—so much so that an excentric circular orbit together with Kepler's area rule gives results for Venus, Jupiter, and Saturn of the same level of accuracy as Kepler's ellipses gave. In other words, to the level of precision of the data at the time, the orbits of Venus, Jupiter, and Saturn were not observationally distinguishable from excentric circles in which the planets sweep out equal areas in equal times with respect to the Sun. But then, postulating that the orbits are Keplerian ellipses and inferring a force rule exponent of –2 from these ellipses was a risky move.[7] If some of these orbits are Keplerian ellipses only *quam proxime* and excentric circles exactly, then an exponent of –2 will not hold even remotely *quam proxime.*

Did Newton know this? I don't know if he took the trouble to derive the full formula, but his published result for the excentric circle is enough to make clear that the inverse-square need not hold *quam proxime.* The more interesting question is not whether he knew it, but when he knew it. The excentric circle proposition first appears in Newton's surviving papers from the early 1690s. But the corollary to it—the force varies inversely with the fifth power when the center of force is on the circumference— appears in the version of *De Motu* registered by the Royal Society in December 1684. Newton scholars have long found this proposition strange, for why would Newton or anyone else have asked what the rule of force is when the center of force is on the circumference? The proof of the full excentric circle proposition is easy, and the diagram for it is virtually the same as the one for the special case that appears in *De Motu.* Perhaps Newton had the full result early on, before writing *De Motu,* and then included only the simple limiting case in the tract. If so, he had already explored the excentric circle before the first edition of the *Principia* and would have seen that the inverse-square need not hold *quam proxime* when the orbit approximating the ellipse too closely approximates an excentric circle as well.

This then is what I regard as the best answer to my "why not" question: even though the Keplerian ellipse entails the inverse-square, one cannot always infer that the inverse-square holds *quam proxime* when the Keplerian ellipse holds only *quam proxime.* In particular, observations at the time were unable to distinguish clearly between the Keplerian ellipse and the excentric circle for Venus, Jupiter, and Saturn. The centripetal acceleration rule for a body in a circular orbit sweeping out equal areas in equal times about any point off center is far removed from inverse-square. What Newton did instead was to infer the inverse-square for the planets from their nearly circular, nonprecessing excentric orbits. This licensed the further inference that the orbits must be exact ellipses in the absence of any further components of acceleration. To quote more of the entry in Huygens's notebook from which I quoted earlier:

The famous M. Newton has brushed aside all the difficulties together with the Cartesian vortices; he has shown that the planets are retained in their orbits by their gravitation toward the Sun. And that the excentrics necessarily become elliptical. (Huygens 1944a, 143)

Huygens thus saw Newton as answering an important question—a question that would have been begged if the inverse-square had been inferred from the ellipse.

My answer has virtues beyond its putting the "why not" question to rest. It explains why Newton had no qualms inferring the inverse-square for the planets from the 3/2 power rule for concentric circular orbits even though he knew perfectly well that the orbits are not concentric circles. Newton was throughout employing approximative reasoning, from *De Motu* forward. This answer also responds to one of the objections put forward by Pierre Duhem (Duhem 1991, 190–95) and repeated by Karl Popper (Popper 1972) and Imre Lakatos (Lakatos 1978), namely, that Newton could not be deducing the law of gravity from phenomena, for this law, inductively generalized, entails that the premises from which the supposed deduction proceeds are false. Fair enough. But the noncentripetal planetary interaction forces implied by the law of gravity do not entail that the area rule ceases to hold *quam proxime*. Appreciating that Newton was engaged in a form of approximative reasoning takes the sting out of this objection.

So What?

It does not, however, take the sting out of the main Duhem-Popper-Lakatos complaint against Newton's deduction of the law of gravity. Suppose we grant that Newton, faced with the limited precision of measurement, was requiring only that the area rule, for instance, holds *quam proxime*. But any number of other rules locating the planets in their orbits hold *quam proxime* as well. Therefore, the inference from the area rule to central forces is not an hypothesis-free deduction, for the choice of the area rule amounts to a hypothesis that it is somehow more correct than these others. Moreover, some of the alternatives to the area rule may entail their own theoretical conclusions differing from the central force conclusion. But then doesn't the evidence for the central force conclusion consist only in a deduction from it of a rule that conforms *quam proxime* with the observed motions? Isn't Newton's deduction of central forces from the area rule nothing more than an expositional device masking the true logical force of the evidence, just as Duhem, Popper, and Lakatos claim? More generally, once we grant that Newton is engaged throughout in approximative reasoning, questions arise about what, if anything, his rigorous deductions of the law of gravity can be adding beyond a purely hypothetico-deductive construal of the evidence.

Part of the Newtonian answer to such questions is straightforward. First, granting the laws of motion, Newton wants to make sure that the inferred force rule holds at least *quam proxime* for the motions of the relevant bodies, at least over a particular period of time. Even though the force rule may not be the optimal way of describing the situation for purposes of inductive generalization, it nevertheless does describe it. Second, again granting the laws of motion, Newton wants to thwart hypothetical elements beyond the statement of the phenomena from slipping into the force rule. The idea is to confine the conjectural aspect of the conclusion, and hence its risk, so far as possible to its inductive generalization. This is much of what I take Newton to be saying in the scholium at the end of book 1, section 11:

> Mathematics requires an investigation of those quantities of forces and their proportions that follow from any conditions that may be supposed. Then, coming down to physics, these proportions must be compared with the phenomena, so that it may be found out which conditions of forces apply to each kind of attracting bodies. And then, finally, it will be possible to argue more securely concerning the physical species, physical causes, and physical proportions of these forces.[8] (Newton 1999, 588f.)

Now, Stein (1990b) has pointed out—as has Dana Densmore, independently (1995, 353), not to mention Cotes in 1713 (Newton 1975, 391ff.)—that the deduction of the law of gravity in the *Principia* violates these requirements by tacitly presupposing that the third law of motion holds between the Sun and the individual planets. This is tantamount to assuming that the total momentum of the Sun and Jupiter, for example, remains constant as they interact or, in other words, that no momentum is transferred to any ethereal fluid mediating their interaction. This presupposition is key to the inference that gravitational force is proportional to the mass of the "attracting" body. Again as Stein has pointed out, this step is precisely where Huygens drew the line between what Newton had established and further hypothetical conjecture (Huygens 1944b, 471). A conclusion Stein draws from this is that Newton's argument for universal gravity does not end with its initial deduction, but extends across the rest of book 3. I agree with this conclusion about the rest of book 3, but I nonetheless want to offer a small point here on Newton's behalf:

Newton in fact could have come close to deducing the law of gravity from the nearly circular, nonprecessing orbits without this tacit presupposition. He would still have needed an assumption, namely that all *external* forces acting on the Sun and the individual planets have negligibly differing effects on their respective motions.[9] Under this somewhat lesser assumption, the small "two-body" correction to the 3/2 power rule holds if and only if the third law of motion holds between the orbiting and central bodies—that is, if and only if the orbiting bodies are interacting with the central body. This correction does not presuppose the law of gravity, for it can be derived from and stated in terms of the strength of the centripetal accelerations directed toward the two bodies, for example, in terms of the ratio of the invariant quantities $[a^3/P^2]_J$ for Jupiter and $[a^3/P^2]_H$ for the Sun:

$$P_J^2 \propto r_{JH}^3 \; \frac{1}{1 + \dfrac{[a^3/P^2]_J}{[a^3/P^2]_H}}$$

Now, as Newton found out by querying Flamsteed at the end of 1684, the data for Jupiter and Saturn were not precise enough to show whether this correction holds for them. Hence he could not at the time have invoked this correction as a phenomenon and inferred the applicability of the third law of motion from it. Even so, Newton had a response to the lacuna Stein has called attention to, for he could have said that he was not so much conjecturing a two-body interaction between Jupiter and the Sun as he was taking out a promissory note on specific future data. Of course, this is not what Newton said in the *Principia*, so Stein is perfectly correct about *it*, and also in the claim that *some* further presupposition is needed.

Let me get back to the question of what Newton's deduction from phenomena provides beyond a hypothetico-deductive construal of the evidence. The answers given so far—namely, granting the laws of motion, the deduction assures that the force rule is true at least *quam proxime* of the motions before inductive generalization and it confines the conjectural risk primarily to that generalization—are satisfactory as far as they go, but they do not go far enough. In proposition 8, near the end of the deduction of universal gravity in book 3, Newton says:

> *Proposition 8.* . . . After I had found that the gravity toward a whole planet arises from and is compounded of the gravities toward the parts and that toward each of the individual parts it is inversely proportional to the squares of the distances from the parts, I was still not certain whether that inverse-square proportion obtained exactly in a total force compounded of a number of forces, or only nearly so. For it could happen that a proportion which holds exactly enough at very great distances might be markedly in error near the surface of the planet, because there the distances of the particles may be unequal and their situations dissimilar. But at length, by means of bk. 1, prop. 75 and 76 and their corollaries, I discerned the truth of the proposition dealt with here.

Clearly, then, Newton wants the conclusion of this deduction to be taken as holding exactly, not just *quam proxime*. The requirement that the conclusion still hold *quam proxime* when the premises hold *quam proxime* is a *constraint* he imposes on an exact deduction, not a better description of the deduction itself. The question remains, what, if anything, can the exact deduction be adding in the way of evidence? Indeed, since Newton was certain that the phenomena do not hold exactly, how can he conceivably be taking the deduction to establish its conclusion exactly?

What Is Newton Up To?

I find this question, which I take to be the heart of the Duhem-Popper-Lakatos complaint, both difficult and important—important because, ever since the *Principia*, "the structure of our knowledge in physics," to borrow Stein's phrase, has sprung from an *exact-approximate duality* in the way physicists construe physics that is not altogether unlike wave-particle duality. Years ago a physics teacher responded to my confusion by saying, "Electrons are waves when it is convenient to us for them to be waves, and they are particles when it is convenient for them to be particles. You might say that on Mondays, Wednesdays, and Fridays they are waves, on Tuesdays, Thursdays, and Saturdays they are particles, and on Sundays we pray for deeper insight." Somewhat analogously, physicists readily grant that their evidential practices can do no more than to show that theoretical claims hold to a certain degree of approximation; yet they seem generally to proceed—not just in response to challenges from outside the discipline, but from within the discipline as well—as if current "accepted" theory is unqualifiedly true. (Are Sundays the occasion for dreams of a final theory?) I want to take a stab here at answering the question of what Newton was up to, partly in the hope of shedding light on this exact-approximate duality. Equally, however, I do so because I think the answer to this question brings out the most important respect in which the science coming out of Newton's *Principia* was radically different from the mathematical physics of Galileo and Huygens.

The tradition of rational mechanics of Galileo and Huygens, as well as of the Dutch and French figures before 1650, was committed to mathematically exact descriptions of motions, but of ideal motions, not actual ones. Thus we find Galileo saying in *Two New Sciences*,

> As to the perturbation arising from the resistance of the medium this is more considerable and does not, on account of its manifold forms, submit to fixed laws and exact description. . . . Of these accidents of weight, of velocity, and also of shape, infinite in number, it is not possible to give any exact description;

hence, in order to handle this matter in a scientific way, it is necessary to cut loose from these difficulties; and having discovered and demonstrated the theorems, in the case of no resistance, to use them and apply them with such limitations as experience will teach. (Galileo 1991, 252f.)

After discovering that the cycloidal pendulum with a point-bob is isochronous, and then determining the centers of oscillation of solid bobs, Huygens similarly despaired when he found isochronism for pendulums with solid bobs to be beyond mathematical description (Ariotti 1972, 404ff.). Most likely Newton never read Galileo's *Two New Sciences*, nor may he have been aware of the letter to Mersenne in which Descartes similarly remarks, "As for the cause of the air resistance which you asked me about, in my view it is impossible to answer this question since it does not come under the heading of knowledge" (Descartes 1991, 9f). Still, Newton had read Descartes's *Principia* in which he suggests that the planetary trajectories are not mathematically exact, remarking that as "in all other natural things, they are only approximately so, and also they are continuously changed by the passing of the ages" (Descartes 1983, 98). Perhaps it is not entirely a coincidence that the two kinds of motions Newton treats in his *Principia*, planetary orbits and motion in resisting media, are ones for which Descartes had denied the possibility of exact science.

Still, my favorite statement of the limits of exact science, given the complexity of actual motions, is by Newton himself in the augmented version of *De Motu*, written shortly before he began drafting the text of what became book 1 of the *Principia*:

By reason of the deviation of the Sun from the center of gravity, the centripetal force does not always tend to that immobile center, and hence the planets neither move exactly in ellipses nor revolve twice in the same orbit. There are as many orbits of a planet as it has revolutions, as in the motion of the Moon.... But to consider simultaneously all these causes of motion and to define these motions by exact laws admitting of easy calculation exceeds, if I am not mistaken, the force of any human mind. (Newton 1974, 78)

Sometimes we do not emphasize as much as we should how acutely aware Newton was of the complexity of the actual motions of the planets. On the view I am about to put forward, the *Principia* is at least as much a response to this complexity as it is an effort to establish universal gravity in the face of the prevailing "mechanical philosophy."

In contrast to the rational mechanics of Galileo and Huygens, the science coming out of the *Principia* tries to come to grips with actual motions in all their complexity—not through a single exact solution, however, but through a sequence of successive approximations. The schematic shown in figure 5 provides a rough picture of how this works. The idea is

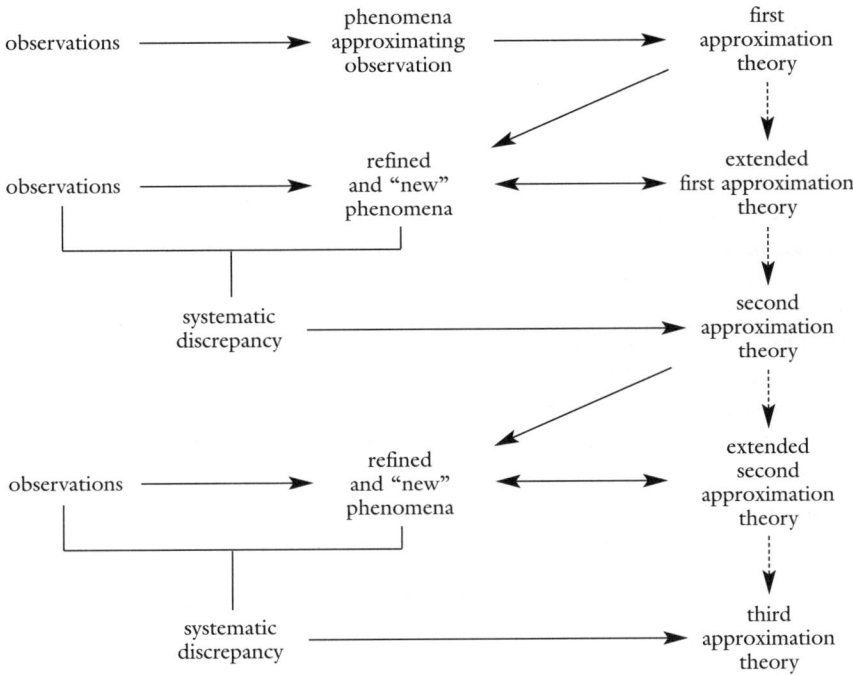

FIGURE 5
A Schematic of Exact Science Through Successive Approximations

to start with mathematically precise statements of phenomena that represent observed motions to a first approximation. These phenomena are required to provide crucial evidence for theoretical conclusions about forces, conclusions that, when inductively generalized, have implications beyond the original phenomena. Further phenomena such as the ellipticity of the orbits—and, more importantly, refinements of the original phenomena such as the small correction to the 3/2 power rule—are then deduced from the theory, constituting second and higher approximations to the actual motions. The crucial demand is that systematic differences between the idealized representations of the motions at any stage and observation themselves become evidence either for the original theory or for a refinement of it. Such systematic differences amount to "second-order" phenomena, for they do not arise from observation alone, but presuppose the theory and the idealized representations of motion deduced from it.

In other words, theory at any stage of approximation is being used to isolate ever more minute residual systematic deviations that can then be turned into crucial evidence for the next stage. If the complexities of the actual motions can be addressed in such a sequence of successive approximations, then they are scarcely reasons for despair, for they can in principle be exploited to yield increasingly strong evidence—evidence of a quality beyond anything the prior tradition of rational mechanics thought possible.

I do not mean to suggest that Newton was the first to adopt successive approximations as a strategy in evidential reasoning. Kepler's use of systematic deviations from an eccentric circular orbit for Mars (adapted from Ptolemy's solar theory) in arriving at his ellipse is a famous earlier example (Wilson 1968). Based on Noel Swerdlow's compelling reconstruction of how Ptolemy arrived at his bisection of eccentricity for the equant of Mars, a strong case can be made that even Ptolemy proceeded in successive approximations (Swerdlow manuscript). Newton nevertheless does seem to have been the first to turn to successive approximations in marshalling evidence for a theory of an underlying physical mechanism. More specifically, he was the first to adopt successive approximations in pursuing evidence for the underlying physics from observable motions *in the face of a complexity of these motions that might otherwise have stood in the way of ever obtaining high quality evidence from them.*[10] This is what I am saying was so revolutionary.

Not just any old first approximation will permit such a sequence of successive refinements, as it might if this were merely tantamount to curve-fitting. The theoretical claim for which Newton requires the first-approximation phenomena to provide crucial evidence is a generic force law—the law of gravity in the case of orbital motions and his law for the resistance force arising from the inertia of the fluid in book 2. Moreover, when the force in question is a net force acting on a macroscopic body, he requires a compositional account of it in terms of forces acting on the individual parts of the body—in terms of microstructural gravitational forces or, in his resistance case, in terms of the forces of impact of fluid particles on parts of the body. Finally, once inductively generalized, the force law ought to have as its consequences a host of idealized phenomena reaching beyond those providing the original evidence for the law. These consequences are not confined to such orbital phenomena as the exact ellipse of planets and the conic-section trajectory of comets in the absence of forces beyond the inverse-square force directed toward the Sun. They also include Newton's version of the variational orbit of the Moon, the complex motion of the lunar nodes, and even the precise nonsphericity of the Earth and the variation of surface gravity with latitude for the ideal case of a uniformly dense Earth. These idealized consequences are expected to agree with observation to increasingly high approximation, and to the extent they do, they of course provide further evidence for the law of force.

This approach thus tries to construe the complicated, irregular actual motions of which Galileo and Descartes spoke—and Newton, too, just before the *Principia*—as "compounded" (to use Hooke's term) of constituent motions each of which is itself lawlike and hence not irregular. As the more than fifteen hundred terms of the Hill-Brown theory of the orbit of our Moon shows, the number of such constituents can be large. The constituents, one hopes, will nevertheless fall into a small number of distinct kinds, with each kind characterized, and hence demarcated, by a generic force law. Newton expressly singles out three such basic kinds of motions in the *Principia*: uniform motions in a straight line, gravitational motions, and motions arising from the impact of matter with matter. In a famous passage in the preface to the first edition, he points the way toward further elements of this taxonomy:

> If only we could derive the other phenomena of nature from mechanical principles by the same kind of reasoning! For many things lead me to have a suspicion that all phenomena may depend on certain forces by which the particles of bodies, by causes not yet known, either are impelled toward one another and cohere in regular figures, or are repelled from one another and recede. Since these forces are unknown, philosophers have hitherto made trial of nature in vain. But I hope that the principles set down here will shed some light on either this mode of philosophizing or some truer one.[11]

Now, any actual motion may be compounded of constituent motions of several different kinds. Newton's theory does not, for example, require celestial motions to be purely gravitational; he even hints at one point that the Earth's magnetic field may be contributing to the precession of the line of apsides of the Moon's orbit.[12] The best hope for getting the sequence of successive approximations off the ground, however, is to find some actual motions, through experiment if necessary, in which a single constituent motion dominates to such an extent that it can provide evidence sufficient to establish a law of force characterizing an entire kind. A fundamental claim of book 3 of the *Principia* is that the orbital motions of the planets and their satellites meet this requirement for one such kind, and similarly in book 2 of the second and third editions that vertical fall of spheres from sufficient heights in water and air do so for a second kind.

I realize that this picture of Newtonian science is sketchy and abstract, but hopefully it will suffice to allow me to contrast different categories of systematic discrepancies between theory and observation that can emerge—categories of "second-order" phenomena, as I called them before. One category consists of second-order phenomena that turn out to arise from systematic observational error; an example is Bradley's discovery of the aberration of light, which provided evidence for a much more precise value for the speed of light than Römer had managed. A second category consists of discrepancies arising because the deduction from the theory had not been carried through fully; the most prominent examples are Clairaut's account of the mean precession of the Moon's apogee and Laplace's discovery of the "Great Inequality" in the motions of Jupiter and Saturn, both of which provided strong further evidence for Newtonian gravitation. A third category consists of discrepancies arising from not taking into account theretofore unnoticed gravitational forces; the discovery of Neptune is an example, as is Bradley's discovery of the nutation of the Earth, which in the hands of d'Alembert became important additional evidence for Newton's compositional account of gravitational forces, as well as for the gravitational interaction of the Moon and Earth.[13] A fourth category consists of discrepancies arising from non-gravitational forces, such as the secular acceleration of the Moon and the slowing rotation of the Earth stemming from viscous forces of our oceans.[14] A fifth category is illustrated by the forty-three arc-seconds per century of the precession of the perihelion of Mercury. No one can observe these forty-three arc-seconds per century. This number is what remains of the observed 574 arc-seconds per century of precession after the Newtonian gravitational effects of Venus, Jupiter, and so on are subtracted.[15] Of course, much the same can said for the other "second-order" phenomena as well. I am singling out the forty-three arc-seconds because it is the one "second-order" phenomenon in our planetary system so far to provide decisive evidence that Newton's law of gravity is not exact.

There is still another possible category—one that has arguably yet to be exemplified in the case of celestial mechanics, though it surely has been exemplified in the case of resistance forces—namely residual discrepancies that recalcitrantly refuse to admit of any systematic characterization. I will discuss this possibility in a moment.

First, let me return to the question of what is gained in taking the law of gravity to hold exactly when its deduction from phenomena seemingly showed only that it holds *quam proxime*. As indicated earlier, not just any first approximation is well suited for initiating a sequence of successive approximations in science. The preferred starting point is a phenomenological regularity that *would* hold exactly in certain identifiable circumstances, for then observed deviations from it would indeed reflect specific physical factors, and not just imprecision in a description. Let me limit the term 'idealization' to approximations that would hold exactly under identifiable circumstances. A natural hope in taking any phenomenon to be evidence for a theoretical claim intended to initiate science by successive approximations is for it to be an idealization in this sense. Now suppose one can deduce a theoretical claim from this phenomenon, taken to hold exactly, and one can then deduce from this theoretical claim conditions under which the original phenomenon would hold exactly. Then, at least from the point of view of the theory, the phenomenon would be well suited to initiate science by successive approximations, and one would have justification for viewing every deviation from it as physically significant.

This is precisely how Newton reasons when he finally gets around to conclusions about the planetary orbits in book 3. Consider, for example, what he says when he finally asserts that the trajectories are Keplerian in proposition 13:

> We have already discussed these motions from the phenomena. Now that the principles of motions have been found, we deduce the celestial motions from these principles a priori. Since the weights of the planets toward the Sun are inversely as the squares of the distances from the center of the Sun, it follows that if the Sun were at rest and the remaining planets did not act upon one another, their orbits would be elliptical, having the Sun in their common focus, and they would describe areas proportional to the times. The mutual actions of the planets upon one another, however, are so very small that they can be ignored and they perturb the motions of the planets in ellipses about the mobile Sun less than if those motions were being performed about the Sun at rest.
>
> Yet the action of Jupiter upon Saturn is not to be ignored entirely.

He ends up formulating the conclusion carefully in the subjunctive, and then turns immediately to the most prominent deviation from it, going on to make what proved to be an erroneous stab at the precise character of the departure of Jupiter and Saturn from Keplerian motion. Similarly, in the very next proposition he returns to the phenomenon invoked in "deducing" the inverse-square, concluding:

> The aphelia are at rest, by bk. 1 prop. 11, as are also the planes of the orbits, by prop. 1 of the same book; and if these planes are at rest, the nodes are also at rest. But yet from the actions of the revolving planets and comets upon one another some inequalities will arise, which, however, are so small that they can be ignored here.

In a scholium to this proposition added in the second and third editions, Newton offers a set of what appear to be hypothetical values to illustrate the extent to which perturbing forces from Jupiter and Saturn may be causing the aphelia of the inner planets to advance.[16] In both of these examples Newton is as attentive to the departure from the ideal as he is to the ideal.

On my view of what Newton is up to, then, taking the law of gravity to hold exactly for the planets by virtue of its deduction from phenomena amounts to an evidential strategy. The deductions from the law of gravity constitute an idealized world—or, more correctly, a sequence of increasingly complex idealized worlds. Any such idealized world does more than just offer a description, in the manner of Galileo and Huygens, of the way the observed world would be if it were more rational. Its main purpose is to bring to light ways in which the observed world systematically deviates from the ideal, and these deviations are then to provide evidence feeding the process of successive approximation. This is why it is so appropriate that the very next thing Newton does, after identifying circumstances in which the nonprecessing ellipse would be exact, is to turn to deviations from this ideal. This is one place where the adage, "the exception proves the rule," actually makes sense. Leibniz, by contrast, makes no mention at all of deviations from the ideal following his deduction of the inverse-square in his *Tentamen*.

How much risk was there in taking the law of gravity to hold exactly of the planets when its deduction showed only that it holds *quam proxime*? If I understand the residual forty-three arc-seconds per century of precession of the orbit of Mercury and how it provides evidence for the theory of gravity in general relativity, the answer is: not all that much risk. Even though general relativity entails that Newton's law of gravity is not exact, the inaccuracies in question do not undercut or nullify any of the second-order phenomena that have been deployed as evidence during the sequence of successive approximations leading up to general relativity, including the second-order phenomenon of the forty-three arc-seconds itself. The great risk in doing science in successive approximations is arriving at a point where a revision to theory nullifies all or much of prior evidential reasoning. This would amount to saying that the second-order phenomena brought to the surface by the original theory were merely artifacts stemming from the error of this theory, and the sequence of successive approximations based on them was all along nothing but a garden path. The sixth category that I mentioned earlier—excessive residual discrepancies that recalcitrantly refuse to admit of any systematic characterization—is symptomatic of such a garden path.

Insofar as Newton's law of gravity had been shown to hold at least *quam proxime* for the planets, the main risk of a garden path arising from it lay not so much in its turning out not to be exact, but in its turning out to be like the Titius-Bode "law" for the planets—not suitable for inductive generalization at all.[17] In short, the risk in the evidential strategy of taking the law of gravity to hold exactly for the planets was minor compared with the risk of inductively generalizing it into the law of *universal* gravity.

Before turning to the question of how one might protect against being led down a garden path, let me pause here to insert two brief comments on what I have just said. First, the forty-three arc-seconds example calls attention to a second type of idealization that has been central to physics, namely approximations that would hold exactly not in certain specifiable circumstances, but instead in a certain limit. Thus, with general relativity, Newton's gravitational mechanics comes to be taken as an approximation that would hold exactly in the limit as the strength of the gravitational field approaches zero, and the forty-three arc-seconds and other such systematic deviations from Newtonian gravitational mechanics, like the time-delay effect, are second-order phenomena serving to pick out one among a range of possible, more general theories of which it is a limiting case. Relativity theory and quantum mechanics have made this second type of idealization especially important in twentieth-century physics.

The logic of evidential reasoning predicated on these two types of idealizations is not the same. In particular, both relativity theory and quantum mechanics seemed much more like starting science all over again, in part because the initially available data so under-constrained the task of selecting a preferred generalization of classical mechanics. The point to emphasize, however, is that the science extending from the second type of idealization in the twentieth century has no more nullified prior evidential reasoning than did the science extending from the first type during the eighteenth and nineteenth centuries. The extraordinary conceptual shifts they precipitated notwithstanding, neither quantum mechanics nor relativity theory has implied that classical mechanics was proceeding down a garden path.

Although it has gone largely unnoticed, this second type of idealization is present in Newton's *Principia*. Newton does not construe the uniform gravity of Galileo and Huygens as a mere approximation holding only because the variation in inverse-square gravity near the surface of the Earth is too small to notice. Instead, he construes it as the limiting case of a gravitational centripetal force that varies linearly with r, specifically the limit as the curvature of the (equipotential) surface approaches zero. He first indicates this in a scholium to book 1, proposition 10, but the point emerges fully only with propositions 51 and 52, which establish the isochronism for the generalized cycloidal pendulum under centripetal forces that vary as r and the corresponding law relating the strength of the centripetal acceleration to the length and period of the pendulum; in a corollary to the latter, Newton announces:

Corollary 2. Hence also there follows what Wren and Huygens discovered about the common cycloid. For if the diameter of the globe is increased indefinitely, its spherical surface will be changed into a plane, and the centripetal force will act uniformly along lines perpendicular to this plane, and our cycloid will turn into a common cycloid. But in that case...the descent of heavy bodies during the time of one oscillation will be the descent which Huygens indicated.[18] (Newton 1999, 555)

As this corollary goes on to say, the law of universal gravity entails that gravity varies linearly with r up to the surface of a uniformly dense Earth.

Newton has thus taken the trouble to show that the Galilean-Huygensian theory of motion under uniform gravity holds as a limiting case of his theory at the surface of the Earth, namely the limit as the curvature of (an idealized) uniformly dense Earth approaches zero. In the process he has legitimated Huygens's theory-mediated measure of the strength of surface gravity from the point of view of his own theory. This measure subsequently is crucial in two places in book 3: first, in establishing (in proposition 4) that the centripetal force holding the Moon in its orbit is terrestrial gravity; and then (in propositions 19 and 20) in using the variation of surface gravity with latitude, in combination with the oblateness of the Earth, to provide the only then tractable empirical contrast between a theory of macroscopic inverse-square gravity among celestial bodies and the theory of universal gravity.[19] That Newton would take the trouble to legitimate Huygens's measure in the manner he did is one more sign of how deeply he thought through issues concerning evidential reasoning.

My second comment before turning to the question of how one might protect against a garden path is more polemical. The paucity of extended garden paths in post-Newtonian physics is not a minor point. Our attitude toward physics—where 'our' includes both physicists and those of us who view it from outside—would be very different if the advances of the last three hundred years had again and again entailed that the evidence for what had gone before was illusory. In his description of our inductive practice, the late Nelson Goodman put great emphasis on the way in which continuing success of research predicated on a theoretical framework has the effect of *entrenching* that framework. I think it a truism to say that the success of research in celestial mechanics and physical geodesy produced an increasingly deep entrenchment of Newtonian gravitational mechanics during the eighteenth and nineteenth centuries. In other words—generalizing—when doing science in successive approximations, the failure to expose recalcitrant discrepancies indicative of a garden path itself becomes a kind of evidence, ultimately eclipsing the more limited evidence entering into each stage along the way. Now, the controversial point I want to add: precisely because neither quantum mechanics nor relativity theory have nullified the evidential reasoning underlying classical mechanics, their advent has in fact further entrenched classical mechanics, for classical mechanics *anchors* quantum mechanics and relativity theory in the very same way that Keplerian motion and Galilean-Huygensian theory anchor Newtonian celestial mechanics.

The complex ways in which evidence accrues to a theory through success of research predicated on it deserve more extensive discussion than I have space for here. Suffice it to say that it behooves those of us who are preoccupied with the nature and structure of knowledge in physics—and I here refer as much to post-Kuhnians engaged in "science studies" in the spirit of the Edinburgh School as to those who investigate science in the spirit of Howard Stein—it behooves all of us to look closely at the long-term development of physics, asking how it has managed to avoid garden paths to the extent it has. I submit that it is this, more than anything else, that induces physicists to regard external challenges to their discipline as so ill-conceived.

Enough polemics. Let me get back to what Newton was up to and the question, how does one protect against being led down a garden path by data that misleadingly invite inductive generalization? Huygens and Leibniz would presumably have said that the only way is to demand a plausible underlying mechanism. To this Newton replies in the General Scholium added at the end of the second edition of the *Principia*,

> I have not as yet been able to deduce from phenomena the reason for these properties of gravity, and I do not feign hypotheses. For whatever is not deduced from the phenomena must be called a hypothesis; and hypotheses, whether metaphysical or physical, or based on occult qualities, or mechanical, have no place in experimental philosophy. In this experimental philosophy propositions are deduced from the phenomena and are made general by induction. (Newton 1999, 943)

Frankly, I would love to find a Reichenbachian argument that deductions from phenomena provide some sort of safeguard against inductive generalizations leading down garden paths. I would love to find such an argument, if for no other reason than because Newton himself ended up abandoning his demand for a deduction from phenomena in the second and third editions of book 2 of the *Principia* and instead relied on hypothetico-deductive evidence for his proposed law for the resistance force arising from the inertia of fluid media; and, sure enough, his approach to addressing fluid resistance in a sequence of successive approximations turned out to be unmistakably a garden path within forty years after he died.[20] As much as I would delight in such an argument, however, I do not see how one is possible. The only way I see to protect against being led down a garden path when inductively generalizing is to push the theory immediately for all it is worth, deducing conclusions from it as far removed as one can find from the phenomena that provided the original evidence, before inductively generalizing.

This is precisely what Newton proceeds to do in the rest of book 3, deducing in sequence conclusions about: (1) the nonspherical shape of the Earth and the variation of surface gravity with latitude; (2) the variational orbit of the Moon, the motion of its nodes, and its fluctuating inclination;

(3) the tides; (4) the precession of the equinoxes; and (5) the trajectories of comets, which even Hooke had said do not fall under the Sun's "virtue." Like Stein, then, I conclude that Newton's argument for universal gravity does not end with its deduction from phenomena at the beginning of book 3. It extends across all of book 3, providing evidence that the step from the law of gravity for the planets—which, recall, Huygens was ready to grant—to the law of universal gravity was not going to result in a garden path.

Although the deductions that occupy the rest of book 3 were all extraordinarily innovative, none of them were straightforward predictions of novel phenomena of the sort that Huygens regarded as the strongest form of evidence for empirical theories.[21] In every case some further, problematic assumption was needed beyond Newton's theory— if only the assumption that no other forces are at work besides gravity. Newton's deduction of the seventeen-mile oblateness of the Earth and the correlative variation of surface gravity with latitude, for example, requires the idealizing assumption that the density of the Earth is uniform. As Newton made clear in the first edition (in a passage at the end of proposition 20 that was dropped in subsequent editions), departures from his calculated values of oblateness and variation of gravity could provide an evidential basis for conclusions about how the Earth's density varies. Thus, neither this deduction, nor any of the others, provided a simple test of the theory of gravity, for discrepancies could be taken not as falsifying the theory, but as pointing the way to replace the gratuitous assumptions. The theory was nonetheless being put to test because revisions to the assumptions had to meet constraints. In the case cited, Huygens himself showed in his *Discourse on the Cause of Gravity*, which appeared some thirty months after the *Principia*, that the relationship between oblateness and variation of gravity differs for different rules of gravity (Huygens 1944b, 467–71, 476f.); hence any modification to the assumption of uniform density introduced in order to obtain an observed oblateness would not automatically satisfy an observed variation in gravity as well.

A good deal of effort went into measuring the oblateness of the Earth during the eighteenth century, yet inconsistencies among the results were not straightened out until well into the nineteenth century, yielding the modern value of a fraction more than thirteen miles. The discipline of physical geodesy given rise to by these efforts is still engaged in a sequence of successive approximations in which inferences about the internal structure of the Earth are being drawn from the shape of the Earth and variations in surface gravity.[22] In this respect the "test" of gravitation theory is ongoing. Of course, the continual improvement the discipline has achieved over the three centuries has it now preoccupied with residual discrepancies in much higher significant figures. Using such continual improvement as the gauge, the theory has been passing the test all along.[23]

At the time Newton died in 1727, however, one year after the third edition of the *Principia* and thirty-two years after Huygens, major loose ends still remained in book 3—and I do not just mean the small number of comets that had been examined and the lack of a quantitative treatment of the tides. The average precession of the orbit of the Moon is a factor of two greater than Newton was deducing, and judging from Machin's remarks, Newton saw no way of dealing with this.[24] The data on the variation of surface gravity with latitude that had accumulated over fifty years were wildly scattered, so wildly that in the third edition Newton decided to throw all of them out except Richer's very first data from the expedition to Cayenne in 1672. Worse, Newton's solution for the period of the precession of the equinoxes in the second and third editions bordered on being a hoax, as he could not have helped but recognize; like a student stuck on an exam, he had contrived a way to produce the already known correct answer, but not even his editor Cotes could follow the reasoning.[25] And finally, by the early 1720s it was clear to orbital astronomers that what Newton was saying about the inequalities in the orbits of Jupiter and Saturn could not begin to explain them.[26]

Clairaut, d'Alembert, and Euler devoted a great deal of effort from 1735 to 1765 to tying up these loose ends, as did Lagrange and Laplace over the next two decades.[27] In the process, they recast Newton's physical theory from the mathematical framework of an extended geometry that he had employed into that of the Leibnizian symbolic calculus. This reformulation in its own right was no small contribution to the science coming out of the *Principia*. For, because it allows perturbations of increasingly high order to be generated in a systematic fashion, it greatly facilitates science by successive approximations.

The successes of these eighteenth-century giants provided evidence for Newton's theory of gravity sufficient to remove all doubts even for Euler, who to his dying day remained committed to planetary vortices. The importance of this evidence may have been best expressed by Euler in a letter remarking on Clairaut's discovery in 1749 that the factor of two discrepancy in the motion of the lunar apse disappears once higher-order terms in eccentricity are taken into account:

> the more I consider this happy discovery, the more important it seems to me, and in my opinion it is the greatest discovery in the Theory of Astronomy, without which it would be absolutely impossible ever to succeed in knowing the perturbations that the planets cause in each other's motions. For it is certain that it is only since this discovery that one can regard the law of attraction reciprocally proportional to the squares of the distances as solidly established; and on this depends the entire theory of astronomy. (Bigourdan 1930, 34)

By the time Laplace began his *Celestial Mechanics* at the end of the century, this further evidence had so eclipsed Newton's original evidence that a deduction of universal gravity from phenomena really *had* become just an expositional device.

The deduction was not just an expositional device for Newton, however. In no small part because he did not deduce the inverse-square from the ellipse, Newton's deduction showed that the motions of the planets and their satellites were in fact arising, at least to high approximation, from inverse-square centripetal accelerations. And this in turn showed that the Keplerian ellipse is the preferred first approximation to these motions. Huygens found this much beyond dispute. Had data been available filling the third law lacuna that Stein has pointed out, Huygens might also have conceded that a Newtonian law of interactive celestial gravity holds at least *quam proxime* for the motions of the planets and their satellites about their central bodies. But Huygens would still not have granted license for the further steps of inductive generalization by which Newton obtained the law of universal gravity.

I do not think we should belittle Huygens for this. Even granting Newton's laws of motion (for which the evidence at the time was at best limited), the overall argument of book 3 does not show that the law of universal gravity is true. Newton tells us as much in his fourth rule of reasoning, added to the *Principia* only in the third edition:

> *In experimental philosophy, propositions gathered from phenomena by induction should be considered either exactly or very nearly true notwithstanding any contrary hypotheses, until yet other phenomena make such propositions either more exact or liable to exceptions.*
>
> This rule should be followed so that arguments based on induction may not be nullified by hypotheses. (Newton 1999, 796)

Rather, as this rule so carefully states, the overall argument gave compelling reasons for *considering* it to be true, at least very nearly, if not exactly. That is, the argument gave compelling reasons to *take* universal gravity to be true as a basis for ongoing research in which discrepancies between observation and conclusions deduced from it *might* be turned into evidence of a quality theretofore unknown. Newton was not clairvoyant. He had no way of knowing for sure that further research predicated on universal gravity was going to work out as well as it did—any more than he had a way of knowing that further research predicated on his law of inertial resistance for fluid media was going to turn out to be a dead end. What he could see was the extraordinary promise of a new kind of science.[28]

Concluding Remarks

My debt to Howard Stein is large. I first began looking carefully at the history of science after a prominent experimental psychologist, concerned with how to strengthen the quality of evidence in his field, asked me how physics and chemistry first came to have high quality evidence, and I realized that the only answers I could give him were superficial. I did not begin focusing my research and teaching in philosophy of science on specific historical episodes, however, until after one of Stein's former students told me about how he uses classics from the history of science in his teaching, and I then read his papers on Newtonian space-time (Stein 1970a), the philosophical prehistory of relativity theory (Stein 1977), and the notion of field in Newton and Maxwell (Stein 1970b). The second of these papers convinced me of the importance of trying to read the *Principia* as Huygens had read it, and to think of Newton as writing with Huygens in mind. How truly disappointed Newton must have been when he found that this giant, on whose shoulders he saw himself standing, had not found the line of reasoning for universal gravity laid out in the *Principia* compelling.

Huygens's private remark in response to Leibniz, quoted at the beginning of this paper and repeated here, was his most outspoken rejection of Newton's theory:

> Concerning the Cause of the tides given by M. Newton, I am by no means satisfied, nor by all the other Theories that he builds upon his Principle of Attraction, which seems to me absurd, as I have already mentioned in the addition to the *Discourse on Gravity*. And I have often wondered how he could have given himself all the trouble of making such a number of investigations and difficult calculations that have no other foundation than this very principle. (Huygens 1901, 538)

Huygens's public rejection of the theory in his *Discourse on the Cause of Gravity*, however, was no less adamant: "This I could not concede, because I believe I see clearly that the cause of such an attraction is not explicable either by any principle of mechanics, or by the laws of motion" (Huygens 1944b, 470). Eric Schliesser and I have shown that Huygens had not just conceptual but also what he regarded as compelling empirical reasons, based on the variation of surface gravity with latitude, for balking at universal gravity (Schliesser and Smith, forthcoming). The second sentence of his remark to Leibniz, however, convinces me that Huygens also never saw the potential in the new kind of science Newton was putting forward—a science, as it were, of delayed gratification. Had he, he might have been more receptive to Newton's theory, these other reasons notwithstanding.

Book 2 of the *Principia* shows us that Newton had no more the gift of being in tune with the natural world than had Huygens—who, we should keep in mind, was deservedly the most renowned scientist in the world at the time the *Principia* was published. What made Newton extraordinary was his capacity to see so far down novel, complex pathways of reasoning opening before him—in the case of the *Principia*, a pathway of evidential reasoning. I prefer to liken Newton to a great chess champion in a challenge match with the natural world. With the *Principia* he introduced a radical new opening that ended up changing the way the game is played forever. Lacking the benefit of hindsight that we have, Huygens's bewilderment was scarcely inappropriate.

NOTES

1. I should also acknowledge that the question in my title was put to me by Matt Frank, one of Stein's students, when I was visiting the Conceptual Foundations of Science program at the University of Chicago in 1997, and neither of us found the answer I offered then satisfying.

2. Professor Michael Nauenberg has argued that this theorem "must have emerged from his [Newton's] early geometrical-numerical study of orbits under the action of general central forces which he presented first in his December 13, 1679 letter to Robert Hooke" (private communication). See Nauenberg 1994 for details.

3. For the details lying behind this table, see Wilson 1989a.

4. This corollary giving the fully generalized form of the relationship was added to the *Principia* in the second edition.

5. Michael Nauenberg has argued that the curvature-based results of the early 1690s were "a return to his earlier concepts of curvature which he developed during the 1660s prior to his correspondence with Hooke" (private communication). See Nauenberg 1994. For a detailed analysis of Newton's curvature results and the relationship between those that appeared in the *Principia* and those that did not, see Brackenridge 1995.

6. As noted in the text, I have derived both this expression and the preceding one for the "off-Keplerian" ellipse by using Newton's CG^3/RP^2 rule, first deriving an expression defining CG in terms of RP. Because the circle is a special case of an ellipse, the derivations for the two expressions start out the same, but then depart to avoid a singularity when e, but not ϵ, is zero. I have nondimensionalized the expression for the excentric circle, but not the one for the off-Keplerian ellipse, only to make each easy to read; because both express proportionalities, this contrast is not one of substance.

Should the contrast between these two expressions and the inverse-square rule seem strange, consider the inference that is being drawn: the exact functional form (up to a multiplicative constant) of the variation in acceleration toward a central point over 360 degrees around this point. For one thing, the total variations from minimum to maximum are not large. For Venus, Jupiter, and Saturn, the ratios of maximum to minimum centripetal accelerations are 1.028, 1.212, and 1.251; even for Mercury it is only 2.303. Furthermore, the contrast drawn in the text concerns the *comparative* variations in centripetal accelerations toward two points near one another. A plot of two such variations would yield two curves differing only a little from one another. The exact functional forms of two such nearby curves can nevertheless be totally different from one another, especially when the curves are comparatively flat.

Newton, by the way, would have considered a centripetal acceleration that varies as the sum of different powers of r as resulting from different physical mechanisms. This is clear from his treatment of resistance forces in book 2 of the *Principia*, where he allows for different powers of velocity for viscous and inertial fluid effects. That the motions of the planets might be arising from a combination of more than one physical mechanism was not at all an unusual idea before Newton's *Principia*: both Kepler and Leibniz had the ellipse arising from a pair of distinct mechanisms.

7. In a private communication Michael Nauenberg has challenged this statement, arguing that Newton knew the inference from the ellipse to the inverse-square was not risky, the eccentric circle notwithstanding. Nauenberg contends that Newton knew as early as 1679 "that the $1/r^2$ force law is the *only* central force law decreasing with distance which has closed orbits for a range of *different initial conditions. This fact is crucial to Newton's theory of planetary orbits. While a special gravitational force law can account for the nearly circular eccentric motion of some planet moving around the sun, having a given mean radius, mean velocity, and eccentricity, the *same* solar force would in general fail to account for the motion of another planet with a *different* mean radius, mean velocity, and eccentricity." Be this as it may, for Newton to have invoked such reasoning to eliminate eccentric circular orbits from consideration at the beginning of book 3 would clearly have been question-begging. To begin with, so long as the possibility that the actual orbits result from multiple mechanisms, there was no compelling reason to insist that some single centripetal acceleration rule must cover all the planets. Consider, for example, Kepler's two-mechanism proposal in which noncircularity results from the changing alignment of the magnetic fibers in each planet as it orbits the Sun, with variations in the intensity of this effect producing different eccentricities. Why should a single centripetal acceleration rule be mandated when minor vagaries in such magnetic fibers and their alignments are all that is needed to undercut a single rule? More important, as discussed later, by the time Newton was drafting the *Principia*, he was certain that the actual planetary orbits are not closed and that treating them as closed was at best an idealized approximation motivated at first glance more by mathematics than by physics. This would have raised not only the issue of whether claims inferred from such an approximation themselves equally hold to reasonable approximation, but also the issue of which idealized, approximate closed orbit is to be preferred, the ellipse, the eccentric circle, or some other, yet to be specified alternative.

Still, Nauenberg's suggestion about the desirability of a unified account for all planets is a point well taken, for otherwise the prospect of marshalling evidence from the phenomena of planetary motion dims. This is why the inference of the inverse-square from the 3/2 power rule for concentric circular orbits was important, notwithstanding its limitations, noted in the text. What Newton needed was a second inferential path to the inverse-square for the individual orbits that is free of these limitations and that does not presume too much about the orbit or the physics underlying it. My claim in the text is that, for good reasons, Newton regarded the inferential path from the Keplerian ellipse to the inverse-square as unsatisfactory for this purpose.

8. The quoted passage—indeed, the entire scholium at the end of book 1, section 11—remained word-for-word the same in all three editions of the *Principia*.

9. I am indebted to Bill Harper for calling my attention to the need for this assumption, which I had missed in an earlier draft of this essay. Harper discusses the matter in detail in his essay in this volume. The assumption in question, concerning the differing effects of forces external to the planetary system, is more germane to Newton's thinking in book 3 than the published record shows. For I. Bernard Cohen has discovered a stray unpublished manuscript of Newton's in which he considered whether some combination of external forces acting on the Sun and individual planets could be producing their observed relative motions *in a Tychonic system*. In other words, Newton himself showed concern over whether combinations of external forces could nullify the reasoning in book 3 upholding the Copernican system.

10. I thank Peter Galison for asking the question to which this paragraph provides an answer. In the text I focus on successive approximations in the science coming out of the *Principia*. For more on how the "Newtonian style" within the *Principia* itself involves successive approximation, see Cohen 1980 and Smith 2000a.

11. The proposed investigation of fundamental forces—more accurately, as Stein has emphasized, fundamental interactions—is discussed in much more detail in Newton's too much neglected drafts of the preface and conclusion for the first edition of the *Principia* (Newton 1962, 302–8 and 320–47). One quote from the unpublished portion of the preface can serve to illustrate:

> I therefore propose the inquiry whether or not there be many forces of this kind, never yet perceived, by which the particles of bodies agitate one another and coalesce into various structures. For if Nature be simple and pretty conformable to herself, causes will operate in the same kind of way in all phenomena, so that the motions of smaller bodies depend upon certain smaller forces just as the motions of larger bodies are ruled by the greater force of gravity. It remains therefore that we inquire by means of fitting experiments whether there are forces of this kind in nature, then what are their properties, quantities, and effects. For if all natural motions of great or small bodies can be explained through such forces, nothing more will remain than to inquire the causes of gravity, magnetic attraction, and the other forces. (Ibid., 307)

That Newton chose to include neither the speculative portion of the draft preface nor the draft conclusion I take to be one more sign of his uncompromising commitment to not tainting established scientific results with hypothetical conjectures.

12. The primary passage in which Newton calls attention to the possible effect of the Earth's magnetism on the Moon is in a brief paragraph following the corollaries added in the second edition to book 3, proposition 37—specifically, the corollaries recalculating the force of the Earth's gravity on the Moon. Earlier, in book 3, proposition 6, corollary 5 Newton contrasts inverse-square gravitational forces with magnetic forces, mentioning among other points of difference, that magnetic force "decreases not as the square but almost as the cube of the distance, as far as I have been able to tell from certain rough observations" (Newton 1999, 810). As is clear from book 1, proposition 44, a centripetal force that varies as the inverse-cube of distance is precisely what is needed to produce a precession of the line of apsides. Robert Palter has pointed out that Newton undoubtedly did attempt to measure the decrease in magnetic force with distance, getting the effect he should have from a magnetic dipole; see Palter 1972.

13. D'Alembert remarks: "The nutation of the terrestrial axis, confirmed by both observations and the theory, furnishes, it seems to me, the most complete demonstration of the gravitation of the Earth toward the Moon, and consequently of the tendency of the principal planets toward their satellites. Previously this tendency had not appeared to be manifest except in the tides of the sea, a phenomenon perhaps too complicated and too little susceptible to rigorous calculation to be able to reduce to silence the adversaries of reciprocal gravitation" (d'Alembert 1749, xxviii).

14. As Curtis Wilson first pointed out to me, the slowing rotation of the Earth and the corresponding transfer of angular momentum to the Moon is a far more complex subject than this sentence suggests, with many distinct mechanisms involved and with a history stretching from Halley and Laplace through a sequence of episodes of controversy down to the present time. See Lambeck 1980 and Stephenson 1997 for general reviews of the current state of the topic. It has been an active area of research over the last thirty years, in large part because more precise measurements of the Moon as a consequence of lunar laser ranging have provided data that allow pursuit of heretofore underspecified discrepancies between theory and observation. Whether some of these discrepancies fall into the fifth or even the sixth of my categories is still an open question, at least among some investigators. For purposes of this paper, the key point is modest: deviations in motion from an idealized condition have provided compelling evidence that any theory of the angular momentum of the Moon and the rotation of the Earth must take into account nongravitational viscous forces in the Earth's fluids.

15. The 574 arc-seconds per century of precession of Mercury are equally not a simple matter of observation, for the apparent rate of precession observed from the Earth is 5601 arc-seconds per century. All but 574 and a fraction of these arc-seconds are owing to the precession of the equinoxes, the precise value for which is a matter to be determined at least in part from Newtonian theory. G. M. Clemence has remarked that, because the observed rate includes the effects of the precession of the equinoxes, "the determination of the precessional motion [of the perihelia] is one of the most difficult problems of observational astronomy, if not the most difficult" (Clemence 1947, 361).

16. The seemingly hypothetical numbers Newton lists for the precessions of the aphelia of the inner planets caused by Jupiter appear to have resulted from dou-

bling the rounded-off magnitudes implied by his precession formula. Presumably Newton, knowing that the formula falls short of giving the precession of the apogee of the Moon by roughly a factor of two, silently introduced such a correction in his calculation of the hypothetical precessions of the aphelia of the inner planets. For more details, see Smith forthcoming.

17. The Titius-Bode "law" generalizes an arithmetical pattern in the mean distances from the Sun of the planets from Mercury through Saturn. The discovery of Uranus initially contravened it, but the discovery of asteroids shortly thereafter and the conjecture that the asteroids originally formed a single planet then "saved" it. It ended up being falsified by Neptune, although both Leverrier and Adams used it in their analyses entailing the existence of Neptune—used it in a way that its erroneous prediction of the mean distance of Neptune turned out to make little difference. For details see Hoskin 1995 and Morando 1995, both in Taton and Wilson 1995.

18. The ellipticized portion of the quotation points out that Wren's result on the length of the cycloidal arc and Huygens's result on the evolute required to produce the cycloid both hold in the limit of Newton's generalized cycloid.

19. The role of Huygens's measurement of surface gravity and its precise variation with latitude in distinguishing between universal gravity and alternatives to it, especially Huygens's alternative, is discussed in detail in Schliesser and Smith forthcoming.

20. Newton's approach to resistance forces in the second edition tried to establish a law for the resistance force from the inertia of the fluid alone—that is, the force in a purely inviscid fluid—and then use discrepancies between theory based on this law and observation in vertical fall experiments to isolate the contribution viscosity and surface friction make to resistance. This approach fails because of d'Alembert's paradox, first announced in 1752: in a purely inviscid incompressible fluid, of the sort Newton had assumed, the resistance force on any moving body is exactly zero, regardless of shape; for, the force on the rear portion of the body must always exactly cancel the force arising on the front portion. For details of Newton's approach to fluid resistance forces and why it formed a garden path, see Smith 2000a and Smith 2000b.

21. In the preface to his *Treatise on Light*, published in 1690 together with his *Discourse on the Cause of Gravity* commenting on Newton's *Principia*, Huygens offers as clear a statement of the hypothetico-deductive approach to evidence as perhaps has ever been offered:

> One finds in this subject a kind of demonstration which does not carry with it so high a degree of certainty as that employed in geometry; and which differs distinctly from the method employed by geometers in that they prove their propositions by well-established and incontrovertible principles, while here principles are tested by the inferences which are derivable from them. The nature of the subject permits of no other treatment. It is possible, however, in this way to establish a probability which is little short of certainty. This is the case when the consequences of the assumed principles are in perfect accord with the observed phenomena, and especially when these verifications are numerous; but above all when one employs the hypothesis to predict new phenomena and finds his expectations realized. (Huygens 1937, 454)

22. Extended discussions of the historical sequence of successive approximations and current state of physical geodesy can be found in Lambeck 1988 and Torge 1991.

23. I am here borrowing the phrase 'continual improvement' from Kenneth G. Wilson, who has argued that the continual (and still continuing) improvement in the precision with which theory in physics agrees with observation is itself an important form of evidence supporting modern physics; see, for example, Wilson and Barsky 2001. Such continual improvement, however, should not be confused with what I am calling the general (though not universal) success of physics in pursuing a program of successive approximations. This success is a special subcategory of Wilson's continual improvement, namely, one in which the improvement that is achieved does not nullify prior evidential reasoning. Because prior evidential reasoning is conserved, all (or almost all) of the evidence for a prior theory carries over as evidence for a new theory, once qualifications are made for levels of precision and approximation; this is illustrated by the manner in which all the evidence for Newtonian mechanics became evidence for the special theory of relativity. Such carry-over of evidence in turn produces a continual growth of the total body of evidence bearing on any area of theory—a continuity of evidence even in the face of sometimes immense differences between the conceptual structures of old and new theories. This kind of success, I submit, provides stronger support for physics than other ways of achieving continuing improvement in the accord between theory and observation do by virtue of its entrenching (in Goodman's sense) the taxonomy underlying inductive generalizations from initial through later successive approximations.

24. John Machin's "The Laws of the Moon's Motion according to Gravity" was published as an appendix to Motte's translation of the *Principia* published in 1729. In it, Machin remarks, "Neither is there any method that I have ever met with upon the commonly received principles, which is perfectly sufficient to explain the motion of the Moon's apogee" (Machin 1729, 31). Machin was the astronomer in closest personal contact with Newton in the 1720s, and hence this statement broaches on having as much authority as if Newton had said it himself.

25. In a private communication Professor Markus Fierz has questioned whether I am being fair to Newton in my comment about his analysis of the precession of the equinoxes. In particular, he argues that Newton's analogy between the precession and the motion of the lunar nodes is well-founded, save that a ring of mass around the Earth must drag the Earth along with it; and even here, although Newton had no theory of the gyroscope, he saw that MR^2 for the Earth is the essential factor. Thus Newton had the driving forces, the sign, and the general magnitude of the precession correct, but was not in a position to derive the precise rate of precession. I do not dispute any of this, and hence I may well be unfair to Newton in what I say. My picture of what happened is that Newton honestly thought he had managed to derive the rate of the precession, with at worst negligible fudging, in the first edition of the *Principia*, only to discover that the value he had used for the ratio of the lunar to the solar gravitational force was seriously mistaken. Because he feared that dropping this seemingly exact calculation from the second edition would hand his opponents an ill-founded argument against his gravitation theory that might carry disproportionate weight,

he scrambled for a way to generate the known rate of precession from the new value of the ratio of the lunar to solar force. My sense that this new calculation was a contrivance stems in part from Newton's inability to explain to Cotes what he was doing, in part because the ratio of lunar to solar force Newton used was still seriously mistaken, and in part from d'Alembert's outspoken critique of the calculation (d'Alembert 1749). Newton undoubtedly deserves more credit for the precession of the equinoxes than my remark suggests; still, my remark is not out of line with the view of Newton's calculation in the middle decades of the eighteenth century. For more details on this topic, see Westfall 1973 and Wilson 1987.

26. In particular, it was clear to Flamsteed that the period of the vagary in the motion of Saturn did not coincide with the nearly twenty-year period of Jupiter's conjunction with it, as Newton's proposal required. See Wilson 1985, 42–53.

27. The chapters in sections 5 and 6 of Taton and Wilson 1995 describe these efforts.

28. In conversation Kenneth Wilson has suggested that Newton was luckier in the success of his gravitation theory than I have made clear—in particular, lucky in that his theory treats gravity as a nonlocal phenomenon, and one would not expect a simple power law to characterize it even *quam proxime* in the absence of locality. I do not want to understate the extent to which the success of Newton's gravitation theory depended on factors beyond what he could see for sure, whether the factor Wilson points to or such factors as the extreme dominance of gravitational forces in the motions of the planets, their satellites, and comets, and the extremely long characteristic time of "chaotic" divergence in our planetary system. Luck played little role, however, in Newton's seeing the *possibility* of his gravitation theory leading to evidence of a quality beyond anything theretofore thought within the reach of empirical inquiry.

REFERENCES

Aiton, E. J. (1972). *The Vortex Theory of Planetary Motions.* London: Macdonald.

Ariotti, Piero E. (1972). "Aspects of the Conception and Development of the Pendulum in the 17th Century." *Archive for History of Exact Sciences* 8: 329–410.

Bertoloni Meli, Domenico. (1993). *Equivalence and Priority: Newton versus Leibniz.* Oxford: Oxford University Press.

Bigourdan, G. (1930). "Lettres inédites d'Euler à Clairaut." In *Comptes rendus du Congrès des sociétés savantes de Paris et des départments tenu à Lille en 1928, Section des sciences.* Paris: Imprimerie Nationale. Translation by Curtis Wilson in Wilson 1980, 143.

Brackenridge, J. Bruce. (1995) *The Key to Newton's Dynamics.* Berkeley, CA: University of California Press.

Clemence, G. M. (1947). "The Relativity Effect in Planetary Motions." *Reviews of Modern Physics* 19: 361–64.

Cohen, I. Bernard (1980). *The Newtonian Revolution*. Cambridge: Cambridge University Press.

D'Alembert, Jean. (1749). *Recherches sur la Précession des Equinoxes et sur la Nutation de l'Axe de la Terre dans le Systêm Newtonian*. Paris: David. Translation of quoted passage from Wilson 1987, 239.

———. (1752). *Essai d'une Nouvelle Théorie de la Résistance des Fluides*. Paris: David.

Densmore, Dana. (1995). *Newton's Principia: The Central Argument*. Santa Fe, N.M.: Green Lion Press.

Descartes, René. (1983). *Principles of Philosophy*. Translated by Valentine Rodger Miller and Reese P. Miller. Dordrecht: D. Reidel Publishing.

———. (1991). *The Philosophical Writings of Descartes*. Vol. 3 (correspondence). Translated by John Cottingham, Robert Stoothoff, Dugald Murdoch, and Anthony Kenny. Cambridge: Cambridge University Press.

Duhem, Pierre. (1991). *The Aim and Structure of Physical Theory*. Translated by Philip P. Wiener. Princeton: Princeton University Press.

Galilei, Galileo. (1991). *Dialogues Concerning Two New Sciences*. Translated by Henry Crew and Alfonso de Salvio. Buffalo: Prometheus Books.

Goodman, Nelson. (1973). *Fact, Fiction, and Forecast*. 3rd edition. Indianapolis: Bobbs-Merrill.

Hoskin, Michael. (1995). "The Discovery of Uranus, the Titius-Bode Law, and the Asteroids." In Taton and Wilson 1995, 169–80.

Huygens, Christiaan. (1901). *Oeuvres Complètes de Christiaan Huygens*. Vol. 9. The Hague: Martinus Nijhoff. Translation adapted from Koyré 1968, 117.

———. (1937). *Traité de la Lumière*. In *Oeuvres Complètes de Christiaan Huygens*. Vol. 19, 450–537. The Hague: Martinus Nijhoff. Translation from Matthews 1989, 126.

———. (1944a). *Oeuvres Complètes de Christiaan Huygens*. Vol. 21. The Hague: Martinus Nijhoff. Translation adapted from Koyré 1968, 116.

———. (1944b). *Discours de la cause de la pesanteur*. In ibid., 443–88. Translations by Karen Bailey.

Kepler, Johannes. (1995). *Epitome of Copernican Astronomy* and *Harmonies of the World*. Translated by Charles Glenn Wallis. Buffalo, N.Y.: Prometheus Books.

Koyré, Alexandre. (1968). *Newtonian Studies*. Chicago: University of Chicago Press.

Lakatos, Imre. (1978). "Newton's Effect on Scientific Standards." In *The Methodology of Scientific Research Programmes, Philosophical Papers*, vol. 2, edited by John Worrall and Gregory Currie. Cambridge: Cambridge University Press, 193–222.

Lambeck, Kurt. (1980). *The Earth's Variable Rotation: Geophysical Causes and Consequences*. Cambridge: Cambridge University Press.

———. (1988). *Geophysical Geodesy: The Slow Deformations of the Earth*. Oxford: Oxford University Press.

Laplace, Pierre Simon, Marquis de. (1966). *Celestial Mechanics*. Vol. 1. Translated by N. Bowditch. Bronx, N.Y.: Chelsea Publishing Company.

Leibniz, Gottfried Wilhelm. (1860). "Illustratio Tentaminis de Motuum Coelestium Causis." *Leibnizens Mathematische Schriffen*. Vol. 6, edited by C. I.

Gerhardt. Halle: Druck und Verlag von H. W. Schmidt, 254–76. Translation of 1689 version in Bertoloni Meli 1993, 126–42.

Machin, John. (1729). "The Laws of the Moon's Motion according to Gravity." Appendix to Isaac Newton, *The Mathematical Principles of Natural Philosophy*, translated by Andrew Motte. London: Benjamin Motte. Reprinted in 1968, London: Dawsons of Pall Mall.

Maclaurin, Colin. (1748). *Account of Sir Isaac Newton's Philosophical Discoveries.* London: Johnson Reprint (1968).

Matthews, Michael R. (1989). *Scientific Background to Modern Philosophy*. Indianapolis: Hackett.

Morando, Bruno. (1995). "The Golden Age of Celestial Mechanics." In Taton and Wilson 1995, 211–39).

Nauenberg, Michael. (1994). "Newton's Early Computational Method for Dynamics." *Archive for History of Exact Sciences* 46, 212–52.

Newton, Isaac. (1962). *Unpublished Scientific Papers of Isaac Newton*. Edited by A. Rupert Hall and Marie Boas Hall. Cambridge: Cambridge University Press.

———. (1974a). "De Motu Corporum in Gyrum." In *The Mathematical Papers of Isaac Newton*. Vol. 6, edited by D. T. Whiteside. Cambridge: Cambridge University Press, 30–80. The augmented version of "De Motu" is also in Newton 1962, 239–92. The translation of the excerpt from the "Copernican Scholium" of the augmented version is by Curtis Wilson in Wilson 1989b, 253.

———. (1974b). Ibid., 569–99.

———. (1999). *The Principia: Mathematical Principles of Natural Philosophy*. Translated by I. Bernard Cohen and Anne Whitman. Berkeley, Calif.: University of California Press.

———. (1960). *The Correspondence of Isaac Newton*. Vol. 2, edited by H. W. Turnbull. Cambridge: Cambridge University Press.

———. (1975). *The Correspondence of Isaac Newton*. Vol. 5, edited by A. Rupert Hall and Laura Tilling. Cambridge: Cambridge University Press.

Palter, Robert. (1972). "Early Measurements of Magnetic Force." *Isis* 63: 544–58.

Pemberton, Henry. (1728). *A View of Sir Isaac Newton's Philosophy.* London: S. Palmer.

Popper, Karl. (1972). "The Aim of Science." In *Objective Knowledge: An Evolutionary Approach*. Oxford: Oxford University Press.

Schliesser, Eric, and George E. Smith. (Forthcoming). "Huygens's 1688 Report to the Directors of the Dutch East India Company on the Measurement of Longitude at Sea and the Evidence It Offered Against Universal Gravity." *Archive for History of Exact Sciences.*

Smith, George E. (2000a). "The Newtonian Style in Book 2 of the *Principia*." In *Newton's Natural Philosophy*, edited by Jed Z. Buchwald and I. Bernard Cohen. Cambridge, Mass.: MIT Press, 249–313.

———. (2000b). "Fluid Resistance: Why Did Newton Change His Mind?" In *Foundations of Newtonian Scholarship*, edited by Richard Dalitz and Michael Nauenberg. Singapore: World Scientific, 3–34.

———. (Forthcoming). "A Note on Newton's Use of His 'Precession Theorem'."

Stein, Howard. (1970a). "Newtonian Space-Time." In *The Annus Mirabilis of Sir Isaac Newton*, edited by Robert Palter. Cambridge, Mass.: MIT Press, 258–84.

————. (1970b). "On the Notion of Field in Newton, Maxwell, and Beyond." In *Historical and Philosophical Perspectives on Science*, edited by Roger B. Stuewer. Vol. 5 of *Minnesota Studies in the Philosophy of Science*. Minneapolis: University of Minnesota Press, 264–87 (followed by comments, 287–99, and replies, 299–310).

————. (1977). "Some Philosophical Prehistory of General Relativity." In *Foundations of Space-Time Theories*, edited by John Earman, Clark Glymour, and John Stachel. Vol. 8 of *Minnesota Studies in the Philosophy of Science*. Minneapolis: University of Minnesota Press, 3–49.

————. (1990a). "On Locke, 'the Great Huygenius, and the Incomparable Mr. Newton'." In *Philosophical Perspectives on Newtonian Science*, edited by Phillip Bricker and R. I. G. Hughes. Cambridge, Mass.: MIT Press, 17–47.

————. (1990b). "'From the Phenomena of Motions to the Forces of Nature': Hypothesis or Deduction?" *PSA 1990* 2: 209–22.

————. (1994). "Some Reflections on the Structure of our Knowledge in Physics." In *Logic, Methodology and Philosophy of Science* 9, edited by D. Prawitz, B. Skyrms, and D. Westerståhl. Proceedings of the Ninth International Congress of Logic, Methodology, and Philosophy of Science. New York: Elsevier Science B.V., 633–55.

Stephenson, F. Robert. (1997). *Historical Eclipses and Earth's Rotation*. Cambridge: Cambridge University Press.

Swerdlow, Noel M. (Manuscript). "The Origin of the Ptolemaic System."

Taton, René, and Wilson, Curtis. (1989). *Planetary Astronomy from the Renaissance to the Rise of Astrophysics. Part A: Tycho Brahe to Newton*. In *The General History of Astronomy*, vol. 2. Cambridge: Cambridge University Press.

————. (1995). *Planetary Astronomy from the Renaissance to the Rise of Astrophysics, Part B: The Eighteenth and Nineteenth Centuries*. In *The General History of Astronomy*, vol. 2. Cambridge: Cambridge University Press.

Torge, Wolfgang. (1991). *Geodesy*. Berlin: Walter de Gruyter.

Valluri, Sree Ram, Curtis Wilson, and Wiliam Harper. (1997). "Newton's Apsidal Precession Theorem and Eccentric Orbits." *Journal for the History of Astronomy* 28: 13–27.

Voltaire. (1967). *The Elements of Sir Isaac Newton's Philosophy*. Translated by John Hanna. London: Frank Cass and Company.

Westfall, Richard S. (1973). "Newton and the Fudge Factor." *Science* 179: 751–58.

Wilson, Curtis. (1968). "Kepler's Derivation of the Elliptical Path." *Isis* 59: 5–25.

————. (1980). Perturbations and Solar Tables from Lacaille to Delambre: The Rapprochement of Observation and Theory." Part 1. *Archive for History of Exact Sciences* 22: 53–188.

————. (1985). "The Great Inequality of Jupiter and Saturn: From Kepler to Laplace." *Archive for History of Exact Sciences* 33: 15–290.

————. (1987). "D'Alembert *versus* Euler on the Precession of the Equinoxes and the Mechanics of Rigid Bodies." *Archive for History of Exact Sciences*. 233–73.

————. (1989a). "Predictive Astronomy in the Century after Kepler." In Taton and Wilson 1989, 161–206.

————. (1989b). "The Newtonian Achievement in Astronomy." In Taton and Wilson 1989, 233–74.

Wilson, Kenneth G. and Constance K. Barsky. (2001). "From Social Construction to Questions for Research: The Promise of the Sociology of Science." In *The One Culture? A Conversation about Science,* edited by Jay A. Labinger and Harry M. Collins. Chicago: University of Chicago Press.

[4]

Howard Stein on Isaac Newton: Beyond Hypotheses?

WILLIAM HARPER

This essay is my tribute to Howard Stein. A main focus will be on what I take to be deep and important issues raised by his 1991 paper, titled "'From the Phenomena of Motions to the Forces of Nature': Hypothesis or Deduction?" Stein's essay, though forced by editorial constraints to be shorter than he would have liked, is the most insightful analysis anyone has ever provided of the extraordinary argument for universal gravity which opens book 3 of Isaac Newton's *Principia*.

My efforts to use Newton's argument as a source of insight into the practice of natural science have been very much enriched by it. Those of us who have seriously grappled with the details of this argument all share a profound and admiring appreciation for Stein's analysis. Especially noteworthy are his unparalleled accounts of the innovations in Newton's conception of a centripetal force, and of Newton's conception of a force of nature as a force of interaction between bodies characterized by laws of interaction. This work builds on his seminal earlier treatments in "On the Notion of Field in Newton, Maxwell, and Beyond" and "Some Philosophical Prehistory of General Relativity" as well as his celebrated "Newtonian Space-time." It is further extended in his forthcoming "Newton's Metaphysics." Stein's work in these papers and others,[1] has done much to rescue philosophy of science from what he has characterized (1977, 13–14) as an "abusive empiricism."

Over and over again my work has benefited by attempts to answer challenges raised by Stein to assumptions I, and others, had been making

This essay has benefited from discussions with Wayne Myrvold, Howard Stein, Abner Shimony, and Corlis Swain, and also from discussions at presentations I have given at the Chicago Steinfest Conference and at Western, Cal Tech, Irvine, and Concordia. I have also benifited from Curtis Wilson's kindness of proofreading a version. These critics, Howard especially, are not responsible for any errors I have continued to make.

without having sufficiently thought them through. At my first meeting
with Stein, when he came to a paper I was giving on Newton's argument
at Chicago, he asked me how the evidence I was citing from Newton for
the inverse-square law for gravitation toward the sun—the absence of significant orbital precession for each orbit and Kepler's harmonic law for the
system of those six orbits—gave evidence against a hypothesis for variation
of force with distance that agreed with the inverse-square in the distances
explored by each orbit and in the inverse-square relation among the forces
at those six small distance ranges, but differed wildly from the inverse-
square power law in the large ranges of distance not explored by the
motions of those planets. My attempts to come to grips with this Humean
challenge have led me to a much deeper appreciation of the role, in
Newton's methodology, of empirical success as convergent agreeing measurements of causal parameters by the phenomena explained by them, of
the significance of Stein's account of the link between Newton's three
measures of centripetal force and acceleration fields, and of Newton's
appeal to what he describes as Rules for Philosophizing to generalize such
measured parameter values.

In this paper I will attempt to respond to Stein's challenge that
Newton's crucial application of the third law of motion to construe gravity as a force of interaction between bodies is an assumed hypothesis rather
than what, according to Newtonian method, would be called a "deduction
from phenomena." My attempt to answer this challenge will appeal to passages from some of Stein's other papers, which outline the argument in the
context of its role in Newton's solution to the two chief worlds systems
problem and which discuss reactions to it by Huygens and Leibniz. I will
argue that the richer notion of empirical success that guides Newton's
method offers more resources for empirically discriminating between theories than are provided by limiting empirical success to prediction alone.
My most important conclusion will be that it is appropriate to make the
finer empirical discriminations recommended by Newton's method.

I hope that the extent to which my effort to answer Stein's challenge
reveals some of the great depth of his insight into Newton and science will
make it count as an appropriate tribute to him.

I. Howard Stein's Challenge

1. Background

I want to use passages from Stein's beautiful 1967 essay, "Newtonian
Space-Time," to set the stage for our consideration of his challenge to
Newton's argument. The first of these passages articulates the role of the

argument in Newton's surprising resolution of the two chief worlds systems problem.

> The central argument of the *Principia* is to be found in the first part of Book III, on the System of the World—that is, as we should say, on the solar system. The dominant question of natural philosophy in the seventeenth century was the question of the structure of this system, and in particular the question of whether it is the earth or the sun that occupies its true fixed center. Newton succeeded in giving a very solid and decisive answer to the general question; and to the particular question, whether the earth or the sun is at the center, he gave an answer that was quite surprising: Neither. The argument upon which this answer was based is a most beautiful one, and repays careful study. (1967, 177)

Stein offers a summary of the salient points of the argument that begins with this description of the astronomical phenomena that Newton takes as premises.

> The premises of the argument, in Newton's formulation, are statements about astronomical phenomena, summing up the best data on the motions of the planets and their satellites essentially in the form of what we know as Kepler's laws (taken as characterizing the *relative* motions). (1967, 177)

Stein emphasizes that the phenomena are relative motions. His later paper expands on this.

> These latter [the phenomena] are actually formulations of astronomical regularities, as regularities of the motions of the heavenly bodies (planets or their satellites), each referred to a suitable frame of reference: for each system of satellites of a central body, the motions are described from a perspective in which the fixed stars and the central body in question are taken to be at rest. (1991, 210)

The inverse-square centripetal acceleration fields Newton infers from these phenomena for each orbital system, such as the sun, Jupiter, Saturn, and Earth, leave open the problem of their consistent combination into a single system. Stein makes the following comment

> Thus we have, at this point in the argument, a picture of several distinct fields of gravitational force, not only directed at as many distinct central bodies, but also characterized with respect to as many distinct frames of kinematical reference; we have so far no account of how this whole ensemble is to be organized into a single representation of the motions and the forces. (1991, 214)

Newton's solution to the two chief worlds systems problem is his solution to this problem of combining these separate inverse-square acceleration fields.

Stein's summary continues with this elegant reference to the Laws of Motion and the theorems Newton derives from them.

> Along side these premises are the general principles of force and motion: the Laws of Motion that stand as "Axioms" at the beginning of the *Principia*, and a formidable battery of theorems in the preceding Books, especially Book I. (It is, by the way, these theorems that Newton specifically refers to as the "mathematical principles of philosophy.") (Stein 1967, 177)

Stein also singles out for comment Newton's rules for philosophizing.

> And the argument is conducted in accordance with yet another class of premises or principles, which Newton in later editions calls "Rules of Philosophizing," although in the first edition they are simply listed as "Hypotheses" together with the astronomical premises. The presence of this third class of principles shows that Newton does not present his argument as simply a mathematical deduction from astronomical premises and principles of mechanics. It is also not a deduction from these together with the Rules of Philosophizing—the later function as guiding principles rather than as premises or precise rules of inference. (Stein 1967, 177)

He makes clear that Newton's appeal to these rules of philosophizing is not to be understood as an attempt to appeal to premises from which conclusions follow by logic alone. An important part of my response to Stein will by my contention that Newton's applications of these rules in his argument are informed by his commitment to an ideal of empirical success as convergent accurate measurement of causal parameters by the phenomena explained by them.

Stein's summary goes on to point out that a chief subtlety of Newton's argument is the fact that its conclusion, universal gravitation, requires correcting the premises, Kepler's orbital laws, that it begins from.

> But in saying that the argument *conceals* subtleties I have in mind more than this: the reasoning has this extraordinary character, that *its conclusions* (so far from being logical consequences) *stand in formal contradiction to its premises*. For according to the theory of universal gravitation, which is of course the product of this argument, Kepler's laws (which were its premises) cannot give an exact representation of the motions of the planets. (Stein 1967, 177)

Duhem (1906; trans. 1991, 193), and many philosophers of science such as Feyerabend (1970, 165), take this as grounds to dismiss the evidential force of Newton's argument. Stein's reaction is far more insightful.

Stein, first, points out that the initial stages of Newton's argument led Huygens and Leibniz to accept the inverse-square law in astronomy.

The astronomical phenomena that Newton starts from can be represented, according to his analysis, by supposing the major astronomical bodies to be surrounded by central acceleration-fields whose intensities vary inversely with the square of the distance from the center; this yields Kepler's laws exactly. Newton's demonstration of this fact led his greatest contemporaries, Huygens and Leibniz, to accept the inverse square law in astronomy; indeed Huygens, who had previously doubted that Kepler's laws were more than an empirical approximation to the planetary motions, was actually convinced by Newton's theory that Kepler's laws hold exactly! (1967, 177–78)

He, then, goes on to provide the following comment on their failure to follow Newton to universal gravitation.

Huygens and Leibniz both, however, *rejected* the theory of universal gravitation, as theoretically objectionable and empirically unproved. Since Huygens and Leibniz were men of formidable intellect—and Huygens in particular a skillful and profound investigator of nature; and since on the other hand Newton's fantastic conclusion that *all bodies attract one another*—and how really extraordinary a conclusion this was, and even (compared with our ordinary experience of bodies) still is, only the dulling effects of what we call "education" can have succeeded in obscuring—has proved to be entirely correct; there is *prima facie* reason to consider that there may be something both sound and deep in the method that led Newton along this path where his great contemporaries could not follow. (1967, 178)

We are encouraged to open our minds to the prospect of learning something both sound and deep from our exploration of Newton's method.

2. The Challenge

Stein argues that Newton's appeal to the third law of motion to construe the gravitation of a primary towards its satellite as the equal and opposite reaction of the gravitation of that satellite towards the primary is an assumed *hypothesis.*[2]

The third law of motion does not tell us that whenever one body is urged by a force directed towards a second, the second body experiences an equal force towards the first; it tells us, rather, that whenever one body is acted upon *by* a second, the second body is subject to a force of equal magnitude and opposite direction. Therefore—putting the point in proper generality—what we may legitimately conclude, from the proposition that each planet is a center of gravitational force acting upon all bodies, is that for each body *B* there must be some body (or system of bodies) *B'* which, exerting this force on *B* is subject to the required equal and opposite reaction. (1991, 217)

Imagine, in analogy with the example used by Cotes in a passage we shall presently quote, that the moon is maintained in its orbit of the earth by invisible hands pushing it toward the earth's center. In this case it would be the pushing hands, rather than the earth, which would be subject to the equal and opposite reaction to the force accelerating the moon towards the center of the earth.

Stein goes on to point out that the leeway implied by the above formulation would not have counted as far-fetched to Newton's readers.

> It must not be thought that the leeway implied by this formulation is one merely of far-fetched possibilities—that the only *plausible* subject of the reaction to gravitational force towards a planet is the planet itself. On the contrary, the very widespread view of Newton's time that one body can act on another only by contact—a view that is known to have had a powerful influence on Newton himself—makes for precisely the opposite assessment: that it is far-fetched to apply the third law in the way Newton does. (1991, 217)

The widespread view was a fundamental commitment of the mechanical philosophy—that to make motion phenomena intelligible one had to produce a hypothesis that would show how such motions could result from contact pushes among bodies. On vortex theories, which were directly motivated by this commitment, the subject of the equal and opposite reaction to the gravitation of the moon toward the earth would be the vortical particles—invisible hands pushing it towards the earth.

Stein's objection was anticipated by Roger Cotes, the wonderful editor who did so much to improve the second (1713) edition of the *Principia*. The following passage is from a letter Cotes sent to Newton on March 18, 1713.

> But in the first Corollary of the 5th I meet with a difficulty, it lyes in these words *Et cum Attractio omnis mutua sit.* I am persuaded that they are then true when the Attraction may properly be so call'd, otherwise they may be false. You will understand my meaning by an Example. Suppose two Globes *A* & *B* placed at a distance from each other upon a Table, & that whilst *A* remains at rest *B* is moved towards it by an invisible Hand. A by-stander who observes this motion but not the cause of it, will say that *B* does certainly tend to the centre of *A*, & thereupon he may call the force of the invisible Hand the Centripetal force of *B*, or the Attraction of *A* since ye effect appears the same as if it did truly proceed from a proper & real Attraction of *A*. But then I think he cannot by virtue of the Axiom [Attractio omnis mutua est] conclude contrary to his Sense & Observation, that the Globe *A* does also move towards the Globe *B* & will meet it at the common center of Gravity of both Bodies. (Corresp., vol. 5, 392)

Cotes goes on to suggest that Newton respond to this objection, either on the last sheet which was not yet printed out or in an errata table.

> This is what stops me in the train of reasoning by which as I said I would make out in a popular way the 7th Prop. Lib. III. I shall be glad to have your resolution of the difficulty, for such I take it to be. If it appeares so to You also; I think it should be obviated in the last sheet of Your Book which is not yet printed off, or by an Addendum to be printed with ye Errata Table. For 'till this Objection be cleared I would not undertake to answer anyone who should assert You do *Hypothesim fingere* I think You seem tacitly to make this Supposition that the Attractive force resides in the Central Body. (Corresp., vol. 5, 392)

He points out that what seems to be Newton's tacit assumption that the attractive force resides in the central body appears to count as an assumed hypothesis on which the argument is based, rather than as a conclusion supported by the evidence adduced.

II. Newton's Initial Response

1. Hypotheses vs. Deductions from Phenomena

Newton responded to this challenge in a letter sent to Cotes on Saturday 28 March 1713.

> the Difficulty you mention wch lies in these words [Et cum Attractio omnis mutua sit] is removed by considering that in Geometry the word Hypothesis is not taken in so large a sense as to include the Axiomes & Postulates, so in experimental Philosophy it is not to be taken in so large a sense as to include the first Principles or Axiomes wch I call the laws of motion. These Principles are deduced from Phaenomena & made general by Induction: wch is the highest evidence that a Proposition can have in this philosophy. And the word Hypothesis is here used by me to signify only such a Proposition as is not a Phaenomenon nor deduced from any Phaenomenon but assumed or supposed without any experimental proof. (Corresp., vol. 5, 396–97)

Newton points out that in his experimental philosophy he does not use the word "hypothesis" in so large a sense as to include those first principles he calls the laws of motion. These laws of motion, he tells Cotes, have the highest evidence a proposition can have in his philosophy. Such evidence, he claims, results from deducing propositions from phenomena and making them general by induction. In contrast, he says that he is using "hypothesis" neither for phenomena nor for propositions deduced from phenomena but only for assumptions supposed without any experimental proof.

The letter we are examining instructs Cotes to add remarks to the new edition to clarify these points for readers of the *Principia*. These instructions are to follow up the famous *hypotheses non fingo* passage,

> Indeed, I have not yet been able to deduce the reason [or cause] of these properties of gravity from phenomena, and I do not feign hypotheses. (Cohen and Whitman [C & W], 943)

in the general scholium being added to book 3, with the following remarks.

> For whatever is not deduced from the phenomena must be called a *hypothesis*; and hypotheses, whether metaphysical or physical, or based on occult qualities, or mechanical, have no place in *experimental philosophy*. In this experimental philosophy, propositions are deduced from the phenomena and are made general by induction. The impenetrability, mobility, and impetus of bodies, and the laws of motion and the law of gravity have been found by this method. And it is enough that gravity really exists and acts according to the laws that we have set forth and is sufficient to explain all the motions of the heavenly bodies and of our sea. (C & W, 943)[3]

These remarks characterize his experimental philosophy as one in which propositions are deduced from the phenomena and made general by induction. *Hypotheses* have no place in experimental philosophy and are explicitly identified as "*whatever is not deduced from the phenomena.*" So, *deductions from phenomena* are to be construed widely enough to include not just *propositions deduced from the phenomena*, directly, but, also, propositions resulting from *making general such propositions by induction*.[4]

Newton's basic inferences from phenomena are backed up by systematic dependencies that make the propositions inferred count as parameter values measured by the phenomena from which they are inferred.[5] Such measurements by phenomena are good examples of what we might consider *direct* deductions from phenomena, according to Newton's methodology. What Newton counts as making general such propositions by induction is to infer the extension of such parameters, with values found constant on all bodies within the reach of our experiments, to bodies beyond the direct reach of the phenomena accessible to us. Stein's challenge forces us to investigate the extent to which Newton's appeal to law 3 to count gravity as an interaction between bodies can be supported by such reasoning from phenomena.

2. *Appeal to Law One*

After his initial remarks to Cotes, contrasting deductions from phenomena from hypotheses, Newton went on to appeal to his general defense, in the scholium to the laws, of applying the third law of motion to attractions.

> Now the mutual & mutually equal attraction of bodies is a branch of the third Law of motion & how this branch is deduced from Phaenomena you may see in the end of the Corollaries of ye Laws of Motion, pag. 22. If a body attracts another body contiguous to it & is not mutually attracted by the other: the attracted body will drive the other before it & both will go away together wth an accelerated motion in infinitum, as it were by a self moving principle, contrary to ye first law of motion, whereas there is no such phaenomenon in all nature. (Corresp., vol. 5, 397)

He argues that a failure of this application of the third law of motion to attractions would violate the first law of motion. This argument, he claims, counts as a deduction from phenomena of the challenged application of the third law of motion to construe attractions as mutual interactions between bodies.

Let us see how this argument is developed in the passage referred to.

> I demonstrate the third law of motion for attractions briefly as follows. Suppose that between any two bodies that attract each other any obstacle is interposed so as to impede their coming together. If one body A is more attracted toward the other body B than that other body B is attracted toward the first body A, then the obstacle will be more strongly pressed by body A than by body B and accordingly will not remain in equilibrium. The stronger pressure will prevail and will make the system of the two bodies and the obstacle move straight forward in the direction from A toward B and, in empty space, go on indefinitely with a motion that is always accelerated, which is absurd and contrary to the first law of motion. For according to the first law, the system will have to persevere in its state of resting or of moving uniformly straight forward, and accordingly the bodies will urge the obstacle equally and on that account will be equally attracted to each other. (C & W, 1999, 427–28)

This argument appeals to the first law of motion, together with the (already established[6]) application of the third law of motion to contact pushes,[7] to extend application of the third law to attractions. Having the obstacle and bodies meet in empty space insures that, to the extent that the system consisting of the two bodies and the obstacle can be treated as a body,[8] the first law of motion would be violated if the pressures on the obstacle are not equal and opposite.

Newton goes on to outline an actual experiment in which there is such an attraction between two bodies.

> I have tested this with a lodestone and iron. If these are placed in separate ves-
> sels that touch each other and float side by side in still water, neither one will
> drive the other forward, but because of the equality of the attraction in both
> directions they will sustain their mutual endeavors toward each other, and at
> last, having attained equilibrium, they will be at rest. (C & W, 428)

Here, in reasonably close approximation to the thought experiment in
empty space, even small differences in attraction would generate motion
with respect to the water in the direction of the stronger. Still water provides no directionally specific resistances to the motion of vessels floating
in it; therefore, any invisible hands preventing such motion would have to
be coordinated with the direction of the attraction.

Magnetic attraction exhibits general regularities—phenomena—that
make it count as an attraction. You can make either lodestone or iron move
towards the other by holding the other still. You can feel the pull on the
lodestone towards the iron, just as you can feel the pull on the iron
towards the lodestone. Moreover, the directions of these pulls towards one
another are independent of orientation with respect to the still water on
which the vessels containing the lodestone and iron float. In 1666 John
Wallis appealed to the following comment on the phenomena of magnetic
attraction at a distance

> it is harder to shew How they have, than That they have it. That the Load-
> stone and Iron have somewhat equivalent to a Tye; though we see it not, yet
> by the effects we know. (*Philosophical Transactions* [*Phil Trans.*] vol. 1, 282)

to support his conjecture of a similar attraction between the earth and
moon.[9]

Even when they are touching one another, the lodestone and iron
attract with no visible mechanism pushing them together. This makes
such attraction, like the Aristotelian account of gravity as the endeavor of
heavy bodies to seek the center of the earth as their natural place, hard
to reconcile with the commitment of the mechanical philosophy to make
motion phenomena intelligible by showing how they could be produced
by having bodies push on one another[10]. Newton, himself, as late as
1675, proposed that the gravitating attraction of the earth, and a similar
attraction towards the sun, may be caused by some "aethereal spirit"
streaming downward to condense within them "[i]n which descent it
may bear down with it the bodies it pervades with force proportional to
the superficies of all their parts it acts upon"[11] (Cohen 1958, 181). This
account is in line with a definition of force offered by Newton in his very
early Waste Book manuscript:[12] "Force is the pressure or crouding of one
body upon another" (Herivel, 138). The aim of the mechanical philoso-

phy, to make motion phenomena intelligible by showing how they could be produced by such contact pushes among bodies, would motivate attempting to suppose some sort of ethereal or other mechanism whereby the iron and lodestone would be pushed toward each other by invisible particles.

In his scholium to proposition 69, Newton makes clear that his use of attraction is not intended to exclude such mechanisms.

> I use the word "attraction" here in a general sense for any endeavor whatever of bodies to approach one another, whether that endeavor occurs as a result of the actions of the bodies either drawn toward one another or acting on one another by means of spirits emitted or whether it arises from the action of aether or of air or of any medium whatsoever—whether corporeal or incorporeal—in any way impelling toward one another the bodies floating therein. I use the word "impulse" in the same general sense, considering in this treatise not the species of forces and their physical qualities but their quantities and mathematical proportions, as I have explained in the definitions. (C & W, 588)

Newton is, thus, using "attraction" in a wider sense than is customary. The generality in this usage of "attraction," however, does not make every example of a force directing one body toward another count as an "attraction" toward that other body. Cotes's example with the invisible hand pushing ball *B* towards ball *A*, is not an endeavor of these bodies to approach one another.

In contrast to Cotes's example, Newton's thought experiment and his actual experiment with lodestone and iron are cases in which there is such a mutual endeavor. For these cases, Newton argues that unless the endeavors of those bodies to approach one another are equal and opposite they would lead to violation of the first law of motion for the system construed as a body. This makes satisfaction of equal quantity, needed to count these oppositely directed endeavors as action and reaction according to the third law of motion, a mathematical proportion that must be satisfied by the quantities of motion toward one another produced by the cause of such an attraction between bodies—whatever that cause might be.

However implausible they may seem, the mechanical philosophy must be able to provide for coordinated pushes that conserve momentum separately amongst the visible bodies, if it is to accommodate the phenomena of magnetic attraction. Stein has suggested to me, in conversation, that one way to have particles pushing bodies toward one another produce equal and opposite endeavors of those bodies to approach one another would be to have them act something like the two ends of a pair of tweezers pushing the bodies together. To account for the attractions in Newton's experiment by particles acting as invisible pushing hands would require two invisible hands coordinated like tweezers ends pushing the bodies together.

Such coordination of invisible hands—like a contracting string drawing two bodies together—would make the resulting attraction fit Newton's concept of a force of interaction in which both bodies enter symmetrically. Here is a passage from Newton's earlier version of book 3 [13] to which Stein has called attention.[14]

> And though the mutual actions of two planets may be distinguished and considered as two, by which each attracts the other, yet as those actions are between both, they do not make two but one operation between two terms. Two bodies may be attracted each to the other by the contraction of a cord interposed. There is a double cause of action, namely, the disposition of both bodies, as well as a double action in so far as the interaction is considered as upon two bodies; but as between two bodies, it is but a single one. It is not one action by which the sun attracts Jupiter, and another by which Jupiter attracts the sun; but it is one action by which the sun and Jupiter mutually endeavor to approach each other. By the action with which the sun attracts Jupiter, Jupiter and the sun endeavor to approach each other (by the third Law of Motion); and by the action with which Jupiter attracts the sun, likewise Jupiter and the sun endeavor to come nearer together. But the sun is not attracted towards Jupiter by a twofold action, nor Jupiter by a two fold action towards the sun; but it is one single intermediate action, by which both approach nearer together. (Cajori, 569)

If Newton's defence of his application of the third law of motion to attractions succeeds in supporting extension to gravitation between the sun and planets, then any mechanical account of the cause of gravity will have to be compatible with having gravitation between Jupiter and the sun satisfy his concept of a force of interaction between bodies.

3. Gravity as Attraction

In the following corollary to proposition 6, book 3, Newton points out that the force of gravity differs from magnetic force.

> Corol. 5, prop. 6, bk. 3 The force of gravity is of a different kind from the magnetic force. For magnetic attraction is not proportional to the [quantity of] matter attracted. Some bodies are attracted [by a magnet] more than [in proportion to their quantity of matter], and others less, while most bodies are not attracted [by a magnet at all]. And the magnetic force in one and the same body can be intended and remitted [i.e. increased and decreased] and is some times far greater in proportion to the quantity of matter than the force of gravity; and this force, in receding from the magnet, decreases not as the square but almost as the cube of the distance, as far as I have been able to tell from certain rough observations. (C & W, 810)

Magnetic attraction does not accelerate bodies equally. So, unlike gravitation, a centripetal force toward a magnet does not count as an acceleration field. It also, so far as Newton's rough observations reveal, decreases much faster than the square with distance.

Newton does not take these differences to undercut his argument to extend application of the third law of motion to construe gravity as an attraction between bodies. He takes the appeal to law one in his thought experiment, and the outcome of his experiments with magnet and iron, to establish that the equality of the endeavors to approach one another is a mathematical proportion holding generally for what he takes to be attractions between bodies.

Newton follows up his discussion of the experiment with lodestone and iron with the following argument for the claim that gravity is such a mutual attraction between the earth and its outer parts.

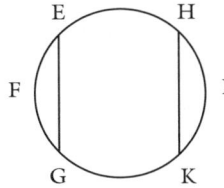

In the same way gravity is mutual between the earth and its parts. Let the earth FI be cut by any plane EG into two parts EGF and EGI; then their weights toward each other will be equal. For if the greater part EGI is cut into two parts EGKH and HKI by another plane HK parallel to the first plane EG, in such a way that HKI is equal to the part EFG that has been cut off earlier, it is manifest that the middle part EGKH will not preponderate toward either of the outer parts but will, so to speak, be suspended in equilibrium between both and will be at rest. Moreover, the outer part HKI will press upon the middle with all its weight and will urge it toward the other outer part EGF, and therefore the force by which EGI, the sum of the parts HKI and EGKH, tends toward the third part EGF is equal to the weight of the part HKI, that is, equal to the weight of the third part EGF. And therefore the weights of the two parts EGI and EGF toward each other are equal, as I set out to demonstrate. And if these weights were not equal, the whole earth, floating in an aether free of resistance, would yield to the greater weight and in receding from it would go off indefinitely. (C & W, 428)

The last part, corresponding to the above appeal to the first law of motion, was added in the second edition (C & W, 428; Koyré and Cohen, 71). Given that the parts EGF and EGI have weights towards one another, one might appeal to law 1 to infer that the earth is not being accelerated by this interaction among its parts, and so conclude that the weights of the parts EGF and EGI toward each other will be equal. This is how the argument is made in Newton's earlier version of book 3, where the diagram has the earth cut into two parts by a single plane and where the analogy with the lodestone and iron is stressed (Cajori, 570).

Such an argument assumes that there is attraction between the parts
EFG and EGI. This assumption of mutual attraction between these parts
of the earth might seem to be exactly the sort of mutual attraction of bodies toward one another that Huygens would dismiss out of hand. The middle of the above passage, however, is a separate argument, which generates
the pressing of the parts EGF and EGI upon one another—as well as the
equality of those oppositely directed pressings—from the assumption that
weights of outer parts of the earth are directed towards its center of gravity and distributed about it so as to be in equilibrium. This assumption is
one that mechanical hypotheses about the cause of terrestrial gravity,
including Huygens's own proposal,[15] were designed to accommodate.

The outer part EGF will be made up of terrestrial material bodies
which are impenetrable to one another and to the central parts of the
earth. The weights toward the center of the earth of all these smaller parts
will make the outer piece EGF singled out by Newton's diagram press
upon the rest of the earth EGI. In the present argument, the larger part
EGI is cut by an additional plane HK parallel to the original plane EG,
which cuts off another outer part HKI on the opposite side of the earth
that is equal to the original outer part EGF. By Newton's construction, the
total force by which this outer part HKI presses the middle part EGKH is
equal to the oppositely directed total force by which the original outer part
EGF presses that middle part from the other side.[16] The middle part
EGKH—being, so to speak, suspended in equilibrium between both outer
parts—will not exert a net force toward either.[17] This makes the total force
of the larger part EGI pressing on the first outer part EGF equal the total
force exerted by the second outer part HKI to drive the middle part
EGKH toward that first part. By the construction of HKI, this total force
by which the larger part EGI presses the outer part EGF equals the oppositely directed force by which that outer part presses it.

The third law of motion, thus, applies to gravitation between the earth
and its outer parts. As any body lying on the earth can count as an outer
part, this argument shows that a large class of cases of gravity as weight
toward the center of the earth count as attractions, in Newton's sense,
between the earth and terrestrial bodies.

III. Gravitation as Attraction between
Solar System Bodies?

Let us examine the extent to which arguments, such as the foregoing gravitational equilibrium argument or Newton's appeal to law 1, can be applied
to extend law 3 to count gravitation of Jupiter toward the sun as an attraction between them. To this end we will be considering interactions among
visible solar system bodies as though they were isolated systems.

1. Measuring Inverse-Square Centripetal Acceleration Fields

In his very interesting essay "On the Notion of Field in Newton, Maxwell, and Beyond," Stein argues that the initial inferences in Newton's argument, to inverse-square centripetal forces directed to Jupiter, Saturn, the sun, and the earth, are inferences to inverse-square centripetal acceleration fields. In the definitions which open the *Principia* Newton characterizes centripetal force, with its three distinct measures or quantities: absolute, accelerative, and motive. His discussion includes the following remarks.

> These quantities of forces, for the sake of brevity, may be called motive, accelerative, and absolute forces, and, for the sake of differentiation, may be referred to bodies seeking a center, to the places of the bodies, and to the center of the forces: that is, motive force may be referred to a body as an endeavor of the whole directed toward a center and compounded of the endeavors of all the parts; accelerative force to the place of the body as a certain efficacy diffused from the center through each of the surrounding places in order to move the bodies that are in those places; and absolute force to the center as having some cause without which the motive forces are not propagated through the surrounding regions. (C & W, 407)

Stein argues that Newton's three measures or quantities of a centripetal force—absolute, accelerative, and motive—are respectively the strength of a centripetal acceleration field, the equal accelerations it would produce on all bodies at any given distance from the center, and the product of this acceleration with the mass of each body accelerated by it.[18] That there should be an accelerative measure—that the field intensity at each place around the center is measured by the equal accelerations that the field would produce on bodies at that distance—is what makes a force field count as an acceleration field.

Here is a comment by Stein on Newton's argument to an inverse-square centripetal force toward the sun.

> That the accelerations of the planets severally and collectively, are inversely as the squares of their distances from the sun is not the conclusion of Newton's induction; that is his deductive inference from the laws established by Kepler. Newton's inductive conclusion is that the accelerations toward the sun are *everywhere*—i.e. even where there are no planets—determined by the position relative to the sun; namely, directed toward that body, and in magnitude inversely proportional to the square of the distance from it. And although the inductive argument is very straightforward—certainly not dependent upon any tortuous constructs—that argument cannot be made, because its conclusion cannot even be sensibly formulated, without the notion of a field. From a mathematical point of view, the idea of an acceleration attached to each point in space is the idea of a function on space, hence a field; from the physical and

methodological point of view, the idea of an *acceleration* characterizing *a point where there happens to be no body* makes no sense at all, unless one accepts the notion of a disposition or tendency; subject to probing, but not necessarily probed. (Stein 1970, 267–68)

Stein preceded this comment with the following assessment of Newton's induction.

The induction is very convincing. The fact that the acceleration is the field intensity is critical, for the evidence comes entirely from six bodies, each exploring the field in a fixed and severely restricted range; the inductive basis would therefore be rather weak if we were not, by good luck, able to relate directly to one another purely *kinematical*—and, thus, ascertainable—parameters of the several bodies motions. This lucky fact is not the work of Newton's definitions, but of nature. Newton's merit was to know how to use what he was lucky enough to find. (Stein 1970, 267)

Newton backs up his inference to the inverse-square centripetal acceleration field surrounding the sun by systematic dependencies that make orbital phenomena measure its parameters. Kepler's areal law measures the sun-centered direction of the forces by which the planets are maintained in their orbits.[19] That the orbits of the planets satisfy Kepler's harmonic law measures the inverse-square variation of a centripetal acceleration field directed towards the sun.[20] This measurement of the inverse-square variation of this field is backed up by measurements of inverse-square variation provided by absence of significant orbital precession for each planet.[21] For any given distance from the center of the sun, the inverse-square adjusted centripetal accelerations exhibited by the planets count as agreeing measurements of the acceleration towards the sun that the sun-centered inverse-square acceleration field would produce on bodies at that distance. Similarly, the harmonic law ratios for each of the planets count as agreeing measurements of the strength—what Newton calls the "absolute quantity"—of this inverse-square centripetal acceleration field directed toward the sun.[22]

Stein has characterized Newton's inductive conclusion:

that the accelerations toward the sun are *everywhere*—i.e. even where there are no planets—determined by the position relative to the sun; namely, directed toward that body, and in magnitude inversely proportional to the square of the distance from it.

Newton's inference to this inductive conclusion is backed up by these agreeing measurements from the six separate orbits of planets of the centripetal direction, the inverse-square accelerative quantity, and the absolute quantity or strength of this single sun-centered inverse-square centripetal acceleration field. I contend that Newton would be correct to count the extension to other distances of dispositions to accelerate bodies toward the sun corresponding to the agreeing measurements of these field parameters as *making general by induction* propositions *deduced from the phenomena*. In this case the propositions being made general are the values of the acceleration field parameters measured by the cited orbital phenomena.

The extension of such inverse-square accelerations toward the sun to bodies at other distances than those explored by the planets is endorsed by Newton's third rule for Natural Philosophizing.

Rule 3. Those qualities of bodies that cannot be intended and remitted [that is, qualities that cannot be increased and diminished] and that belong to all bodies on which experiments can be made should be taken as qualities of all bodies universally. (C & W, 795)

Those qualities of bodies that cannot be intended or remitted are those that count as constant parameter values. This rule, therefore, endorses counting such parameter values found to be constant on all bodies within the reach of experiments as constant for all bodies universally. The inverse-square variation with distance from the center of the sun and the corresponding field intensities of this sun-centered acceleration field are found to be constant for the earth, and with it all terrestrial bodies, as well as for all the planets. These parameter values, therefore, hold for all bodies within the reach of our experiments.[23]

In proposition six Newton offers additional measurements to back up his assumption that the inverse-square centripetal forces directed toward the sun, Jupiter, Saturn, and earth—as well as to the other planets—are acceleration fields. These include pendulum experiments and the equality of the acceleration of terrestrial gravity at the surface of the earth with the inverse-square adjusted centripetal acceleration of the lunar orbit exhibited in the moon test. These put bounds on a parameter Δ_e representing differences in ratios of inertial mass to inverse-square adjusted weight toward the earth for bodies.[24] Newton's data for Jupiter's moons put bounds on the corresponding parameter Δ_j, while his data for the orbits of the planets about the sun put bounds on Δ_h representing differences between ratios of inertial mass to inverse-square adjusted weight of bodies toward the sun. Additional bounds on Δ_h are provided by the absence of polarization with respect to the sun of the orbits of Jupiter's satellites, Saturn's satellites, and the orbit of the moon about the earth.[25] The measurements directly bounding Δ_h are backed up by the phenomena measuring bounds on Δ_e and Δ_j. All these phenomena count as agreeing measurements bounding toward zero a single universal parameter Δ representing differences in ratios of inertial mass to inverse-square adjusted weight toward planets.

Comets provide inverse-square orbits that explore considerably more distances from the sun than the six small ranges explored by the primary planets known to Newton. As comets are not appealed to in the basic argument and are discussed later in book 3, they provide examples of the additional support Stein refers to when he points out that Newton's argument for universal gravity is the whole of book 3. In the general scholium added to the second edition of 1713 Newton includes the following remarks

> The motions of comets are extremely regular, observe the same laws as the motions of planets, and cannot be explained by vortices. Comets go with very eccentric motions into all parts of the heavens, which cannot happen unless vortices are eliminated. (C & W, 939)

In 1759 a particularly striking later vindication of extending the inverse-square law for an acceleration field toward the sun to distances not explored by planetary orbits was provided by Clairaut's celebrated success in predicting the return of Halley's comet.[26]

2. Measuring Acceleration Fields for the Sun and Jupiter

For Jupiter's orbit let the semimajor axis a be 5.21 au, the mean between the value Newton cites from Kepler (5.1965 au) and the value he cites from Boulliau (5.2252 au), and the period t be 4332.514/365.2565 or 11.8616 in units equal to the period t_e (365.2565 sidereal days) Newton cites for the earth.[27] These give a centripetal acceleration[28] of 1.462 au/t_e^2. The ten distinct distance estimates Newton cites for the planets yield ten inverse-square adjusted estimates of the acceleration toward the sun that would be produced by the inverse-square centripetal acceleration field directed towards the sun on any body at the distance 5.21 au we are assuming for Jupiter. These yield

$$1.456 \pm .007 \ au/t_e^2$$

as their combined estimate of this centripetal acceleration towards the sun at distance 5.21 au from its center.[29]

Similarly, the inverse-square adjusted centripetal accelerations of Jupiter's moons measure the strength of an inverse-square centripetal acceleration field centered on Jupiter. By the third edition of *Principia* Newton had available quite good data on Jupiter's moons provided by Pound.[30] This data provided estimates for the strength of the inverse-square acceleration field directed towards Jupiter.

Consider Callisto, Jupiter's outermost Galilean satellite. Newton cites $\theta = 8' \; 16''$ as the maximum elongation at the mean distance of Jupiter from the earth, which equals Jupiter's mean distance from the sun. This makes the radius $r_c = 5.21 \sin \theta$ au $= .01253$ au, for Callisto's nearly circular orbit. He cites 16.689 days equal .04569 t_e for its period[31]. The data cited for Callisto yields a centripetal acceleration toward Jupiter of $4\pi^2 r/t^2 = 236.96$ au/t_e^2.

If, in accordance with rule 3, we extend the inverse-square acceleration field towards Jupiter to the distance of the sun we can extend our inverse-square adjusted estimates from Pound's data for the four moons to 5.21 au, which we are assuming as the length of the semimajor axis of Jupiter's orbit. This leads to an estimate

$$.00137 \pm .00003 \text{ au}/t_e^2$$

of the acceleration toward Jupiter the sun should have at the distance of 5.21 au from Jupiter.[32]

3. Combining these Acceleration Fields

Applying rule three to the inverse-square centripetal acceleration field toward Jupiter revealed by Pound's data on Jupiter's satellite orbits yields an acceleration of the sun towards Jupiter of $.00137 \pm .00003$ au/t_e^2. We cannot use the sun-centered reference frame to represent this, since the sun is motionless in that frame. We also cannot use a Jupiter-centered frame to represent this. In a Jupiter-centered frame, Jupiter is at rest and the centripetal acceleration of the sun toward Jupiter is 1.456 au/t_e^2, not .00137 au/t_e^2. We need to combine the inverse-square centripetal acceleration field toward Jupiter with the inverse-square centripetal acceleration field towards the sun. This requires finding a reference frame that can assign appropriate accelerations to each body.

The ratio $1.456/.00137 = 1063/1$ of the acceleration toward the sun—according to the sun's centripetal acceleration field at the distance we are assuming for Jupiter from the center of the sun—to the acceleration toward Jupiter—at the same distance according to Jupiter's centripetal acceleration field—is a measure of the strengths (the absolute quantities) of these two acceleration fields.[33]

George Smith (1999, 46–49; also this volume)[34] has suggested that Newton could have developed his solution to the problem of combining the acceleration fields of the sun and Jupiter before he had developed the concept of mass. The basic harmonic law ratios $[a^3/t^2]_j$ and $[a^3/t^2]_h$ giving the strengths of the two acceleration fields can do the job. Consider a point c along the line connecting Jupiter with the sun, where r_j and r_h are the respective distances from c to Jupiter and the sun. Let us suppose $r_j/r_h = [a^3/t^2]_h/[a^3/t^2]_j = 1063/1$ so that the sun and Jupiter orbit about c as a common center with circular orbits of respective radii $r_j = 5.21(1063/1064) = 5.205$ au and $r_h = 5.21(1/1064) = .0049$ au with a common period $t = 11.8616\ t_e$. The respective orbits are described by the two ends of the line connecting Jupiter and the sun which rotates about center c, as in the following diagram from Smith.[35]

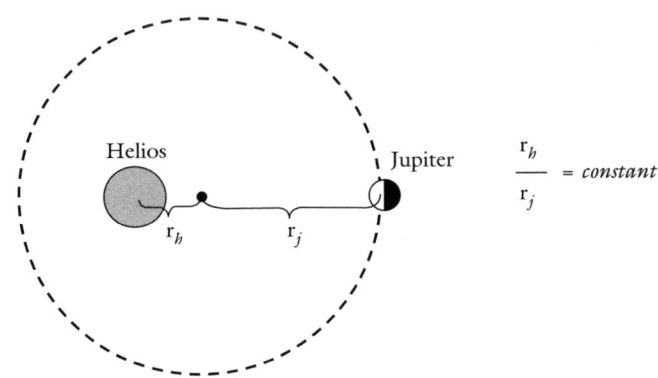

The respective centripetal accelerations of these two circular orbits about c will be $(4\pi^2/t^2)r_j = 1.46$ au$/t_e^2$ for Jupiter and $(4\pi^2/t^2)r_h = .00137$ au$/t_e^2$ for the sun. They will, thus, be in the ratio $r_j/r_h = 1063/1 = a^3/t^2]_h/[a^3/t^2]_j$ needed to combine these two inverse-square acceleration fields.

These orbits will satisfy the two-body correction of the harmonic law, where the radius $r_j = 5.205$ au of the corresponding one-body orbit for Jupiter is $[a^3/t^2]_h/([a^3/t^2]_h + [a^3/t^2]_j) = (1063/1064)$ times the two-body distance 5.21 au between Jupiter and the sun. The two-body correction to the harmonic law is, therefore, recovered by Smith's use of the harmonic law ratios to combine these acceleration fields.

We can think of this as a *kinematical* solution to the problem of combining these two acceleration fields. Newton's definitions of motive measure (def. 8) of a centripetal force and of quantity of motion (def. 2) support an application of the second law of motion in the form $f = ma$ familiar to us. The centripetal acceleration of the sun towards c times the mass of the sun counts as a motive measure $W_j(H)$ of its weight towards Jupiter, while the centripetal acceleration of Jupiter towards c times the mass of Jupiter counts as the motive measure $W_h(J)$ of its weight toward the sun.

One result of combining these acceleration fields is that each has weight towards the other. These weights maintain their oppositely directed orientations toward one another as sun and Jupiter orbit about c. They therefore fulfill one major criterion, distinguishing what Newton counts as attraction from Cotes's invisible hand pushing one body toward another. To have these oppositely directed weights count as a single endeavor of these bodies to approach one another requires, in addition, that they be equal so that they satisfy law 3. To have these oppositely directed endeavors equal is equivalent to having the center c be the center of mass of the sun-Jupiter system. So, this system counts as one where the third law of motion applies to the attraction between the sun and Jupiter just in case c is its center of mass.

4. Laws Three and One for a Sun-Jupiter System?

Do the orbital motions of Jupiter and the sun about c, resulting from combining their acceleration fields into a single system, force Newton's application of the third law to attraction between them? Let us consider this a dynamically isolated system, so that to have the third law apply at all would require the application Newton wants. Will violations of law 3 here lead to absurdities comparable to the resulting violation of law 1 in Newton's thought experiment?

Let us suppose a reference frame with point c as origin and a direction fixed for specifying angles counts as inertial.[36] At time zero let the position vectors r_j and r_h of Jupiter and the sun point in respectively the +y and −y directions along the y-axis. The orbital motion will be in the x–y plane with constant angular rates. Now suppose that the oppositely directed endeavors of Jupiter and the sun toward one another are not equal. That is to suppose that the ratio $W_h(J)/W_j(H)$ of Jupiter's weight toward the sun to that of the sun toward Jupiter is some number k different from 1.

We can still have the ratio $[(4\pi^2/t^2)r_j]/[(4\pi^2/t^2)r_h]$ of Jupiter's acceleration towards the sun to the sun's acceleration towards Jupiter be 1063/1, as required for combining the two acceleration fields. We have $W_h(J)/m(J) = (4\pi^2/t^2)r_j$ and $W_j(H)/m(H) = (4\pi^2/t^2)r_h$. Therefore, having the ratio of Jupiter's mass to the sun's mass be k/1063 instead of 1/1063 will exactly off-set having the force ratio $W_h(J)/W_j(H)$ be k instead of 1.

If k differs from 1, the center of mass c_m will be at a different place than c on the line connecting the centers of Jupiter and the sun. At time zero this will be on the y-axis and as time goes by it will move in a uniform circle around point c. The centripetal acceleration of this orbit of the center of mass c_m is the analogue to the motion of the system as a whole that Newton counts as a violation of the first law of motion in his discussion of the application of the third law of motion to attractions.

Unlike rectilinear acceleration of the center of mass, which (insofar as the system as a whole counts as a body) would clearly violate the first law of motion, the uniform circular motion of the center of mass is not obviously something that ought to count as a violation of the first law of motion. Here is Newton's statement with his elucidating discussion for the first law.

> Law 1. Every body perseveres in its state of being at rest or of moving uniformly straight forward except insofar as it is compelled to change its state by forces impressed.
> Projectiles persevere in their motions, except in so far as they are retarded by the resistance of the air and are impelled downward by the force of gravity. A spinning hoop which has parts that by their cohesion continually draw one another back from rectilinear motions, does not cease to rotate, except insofar as it is retarded by the air. And larger bodies—planets and comets—preserve for a longer time both their progressive and their circular motions, which take place in spaces having less resistance. (C & W, 416)

The elucidating discussion, here, suggests that uniform revolution of the center of mass about c might count as a state of motion that would be preserved according to law 1, rather than as a violation of that Law.[37]

This suggests that, unlike Newton's thought experiment, failure to apply law 3 to Jupiter and Saturn would not result in any comparably clear violation of law 1 for the system of both construed as a body.

Stein's contention that Newton's appeal to law 3 is an assumed hypothesis rather than a deduction from phenomena is supported in so far as the phenomena measuring the acceleration fields directed towards Jupiter and the sun do not measure the equality of their oppositely directed weights toward each other. Without Newton's appeal to law 3, even the two body correction fails to carry the information that these weights are equal.

5. Many Bodies

We can continue this construction adding as many bodies as we want, consistently with our information about the relative strengths of their acceleration fields. For a system of n gravitationally interacting bodies, the acceleration for each body i at time t is given by

$$1 \quad d^2\mathbf{r}_i/dt^2 = \Sigma_{j \sim i} \, GM_j(\mathbf{r}_j - \mathbf{r}_i)/(r_{ij})^3,$$

where the vector's $\mathbf{r}_i$, $\mathbf{r}_j$ are defined relative to some inertial frame, r_{ij}(the scalar length of vector $(\mathbf{r}_j - \mathbf{r}_i)$) is the distance between bodies i and j, M_j

is the mass of body j, and G is the gravitational constant. This is a standard formulation for the accelerations corresponding to the gravitational interactions among n bodies.[38] Now take the same n initial position vectors r_i and r_j, but let the accelerations be given by

$$2 \; d^2 r_i / dt^2 = \Sigma_{j \to i} \; G'_j M'_j (r_j - r_i)/(r_{ij})^2,$$

where each body is allowed to have its own separate gravitational constant G'_j and correspondingly adjusted mass M'_j. So long as for each body j , the product $G'_j M'_j = GM_j$, equation 2 will recover exactly the same accelerations as equation 1, even though it need not count the center of mass of the system as inertial. This shows that law 3, does not follow from law 2. More importantly for our purposes, it shows that, so long as they do not result in collisions, the motion phenomena resulting from gravitational interactions among these bodies will not put any bounds on the ratios among these G',s, nor among the M's, if law 3 does not apply to these interactions.

6a. No Direct Deduction from Phenomena

The foregoing results show that Newton's application of law 3 is not a *direct* deduction from the orbital phenomena of motions of solar system bodies among themselves. Even given that these bodies are interacting according to the result of combining their separate inverse-square acceleration fields into a single system, these phenomena do not measure the equalities of their oppositely directed weights toward one another. As we have seen, unless law 3 is assumed for the pairwise interactions among the visible solar system bodies their orbital phenomena put no bounds on the ratios of the separate G's.

This result strongly supports Stein's challenge to Newton. It is clear that the equality of the weights of the sun and Jupiter toward one another is not directly measured by orbital phenomena. Newton's application of law 3 is, therefore, not what would count according to Newton's methodology as a direct deduction from orbital phenomena.

6b. Making General by Induction?

To count it as, nevertheless, a deduction from phenomena Newton would have to construe it as resulting from a legitimate making general by induction of some measurements by phenomena. Let us review the measurements he could appeal to. Consider his thought experiment. Perhaps it may be counted as, itself, a phenomenon—a general null outcome of many experiments—that "there is no such phaenomenon in all nature" as the violations of law 1 that would result if law 3 did not apply to attractions of the sort considered in Newton's thought experiment. Each null outcome of the many examples of such attractions would, thus, measure the equality of the oppositely directed endeavors. The repeatable outcomes of Newton's actual experiments with lodestone and iron are clear examples of phenomena measuring such equalities. Newton's demonstration of such equal attractions between the earth and its outer parts from the assumption that the weights of outer parts of the earth form an equilibrium about its center may go some way toward supporting extending this measured equality to construe gravity as such an attraction between bodies.

Finally, and perhaps more compelling, we have the requirement that the separate acceleration fields measured by orbital phenomena be combined into a single system. The oppositely directed weights toward one another resulting from combining these acceleration fields is maintained as the bodies move about. This does distinguish gravitation of the sun and Jupiter toward one another from acceleration of a body toward another resulting, merely, from an independent push in that direction. Perhaps Newton regarded fulfilling this criterion for attraction as sufficient to legitimately count the additionally required equality of those weights toward one another as making general by induction the foregoing measured equalities of attractions between bodies.

As we have seen, additional orbital phenomena corresponding to interactions do not measure the equality of the relevant oppositely directed weights. What these do is to provide additional empirical support for the conclusion that gravitation satisfies that one crucial criterion—that these bodies maintain forces toward one another as they move about—which distinguishes what Newton counts as attraction from Cotes's example.

IV. Empirical Success and Rule 4

1. Newton's Second Thought?

Newton's letter to Cotes ends with the remark, "I have not time to finish this Letter but intend to write to you again on Tuesday" (Corresp., vol. 5, 397). His Tuesday letter, dated 31 March 1713, opens with the following remarks.

> Sr
> On saturday last I wrote to you representing that Experimental philosophy proceeds only upon Phenomena & deduces general Propositions from them only by Induction. And such is the proof of mutual attraction. And the arguments for ye impenetrability, mobility & force of all bodies & for the laws of motion are no better. And he that in experimental philosophy would except

against any of these must draw his objection from some experiment or phaenomenon & not from a mere Hypothesis, if the Induction be of any force.[39] (Corresp., vol. 5, 400)

The last sentence, which rejects objections from mere hypotheses while endorsing objections drawn from phenomena, is a clear anticipation of the method advocated in the fourth rule of philosophizing that was introduced in the third edition of 1726.

> Rule 4. In experimental philosophy, propositions gathered from phenomena by induction should be considered either exactly or very nearly true notwith-standing any contrary hypotheses, until yet other phenomena make such propositions either more exact or liable to exceptions.
>
> This rule should be followed so that arguments based on induction may not be nullified by hypotheses. (C & W, 796)

Newton's comment suggests that the point of this rule is to defend arguments based on induction from being undercut by hypotheses. The rule tells us to consider propositions gathered from phenomena by induction as "either exactly or very nearly true" and tells us to maintain this in the face of any contrary hypotheses.

The provision "either exactly or very nearly true" explicitly makes room for acceptance of propositions gathered from phenomena as approximations. This, together with the explicit provision that other phenomena could make such propositions either more exact or liable to exceptions, makes this rule especially appropriate for what George Smith (this volume) has identified as Newton's methodology of science as the development of successive approximations. These features, also, are those appropriate for construing propositions gathered from phenomena by induction as ones adequately supported from measurements by phenomena.

The provisional nature of the acceptance endorsed by Newton's rule 4 is emphasized by Stein.

> Of course, *when* such proof is sufficient to provide adequate warrant, and so constitute the grounds for a proper deduction, remains a very difficult question indeed; but it is clear that Newton is no Popperian: he believes that this diffi-cult question can be answered in practice, even in the absence of a general prin-ciple for deciding it. On the other hand, although a proposition may qualify as deduced from the phenomena, the issue may always be reopened by the dis-covery of new evidence from phenomena—this is explicit in Rule IV of the Rules of Philosophizing; and in this sense, the process of "proof" is in princi-ple unending. (1991, 219)

Newton's ideal of empirical success—convergent accurate measurement of causal parameters by the phenomena they are taken to explain—offers far more resources for answering this question positively than are provided by the empiricist construal of empirical success as limited to accurate prediction.

2. Empirical Success as Convergent Accurate Measurement

In an isolated two-body system, without the assumption that the third law of motion applies, the forces and masses cannot be disentangled sufficiently to be measured. This would make them behave like gorce and morce in the example that Clark Glymour (1980, 357) used to illustrate the advantage of measurement over mere prediction as a criterion of empirical success.

Without Newton's application of law 3, the assumed orbital phenomena are compatible *even* with having Jupiter's mass very much greater than that of the sun. For example if k is set at $1063(1063) = 1,129,969$, then the very same orbits about c would be predicted even though Jupiter's mass would be 1063 times that of the sun. The assumed orbital phenomena cannot disentangle weights and masses sufficiently to measure either because they put no bounds on the ratios of the separate G's.

Since the visible solar system bodies do approximate an isolated system, Newton's application of the third law clearly outstrips alternative hypotheses[40] of separate G's in realizing his ideal of empirical success. Instead of up to as many differing gravitational constants as there are bodies—resulting in masses too entangled with weights to be measured, Newton's application of the third law to gravitation as interactions among visible bodies in the solar system leads to convergent agreeing measurements of relative masses among them.

The fourth rule of reasoning can, thus, be interpreted to nicely accommodate appeal to such success led to. This supports appeal to a wider range of such successes than those which, according to Rule 3, would underwrite making general by induction the equality of attractions measured directly between the magnet and lodestone or between the earth and its outer parts.

3. Agreeing Measurements Yield Resiliency

Newton's moon test provides an informative example of his ideal of empirical success at work. Its outcome is the agreement between the one-second fall corresponding to the acceleration at the surface of the earth measured by inverse-square adjusting estimates of the centripetal acceleration of the lunar orbit and the one-second fall corresponding to the acceleration of terrestrial gravity measured by Huygens's celebrated use of a seconds pendulum. The six lunar distance estimates and other data Newton cites result in the measurement

15.041 ± .429 Paris feet

of the one second fall corresponding to the acceleration that an inverse-square field maintaining the moon in its orbit would produce at the surface of the earth.[41] Huygens's seconds pendulum yields a more precise measurement

15.096 ± .01 Paris feet

for the one-second fall corresponding to the acceleration of gravity.[42]

From this agreement, Newton appeals to his rules 1 and 2 of philosophizing to conclude:

> And therefore that force by which the moon is kept in its orbit, comes out equal to the force of gravity here on earth, and so (by rules 1 and 2) is that very force which we generally call gravity. (C & W, 804)

This application of rules 1 and 2 is not just an appeal to a general commitment to simplicity. The simplicity involved is that special sort of unification which results in counting different phenomena as agreeing measurements of a single causal parameter which explains them. The two phenomena, here, are the length of a seconds pendulum at the surface of the earth and the centripetal acceleration exhibited by the lunar orbit. The single parameter they measure is the strength—Newton's absolute quantity—of an inverse-square centripetal acceleration field that is the action of terrestrial gravity on bodies above the surface of the earth.

To see that the agreement of these measurements counts as *empirical* success note that the cruder data from the lunar orbit empirically back up Huygens's measurement. They do this by making large deviations from Huygens's value more improbable than they would be on Huygens's data alone. In general, when different phenomena count as agreeing measurements of the same parameter their data reinforce one another by increasing the resiliency—the evidential basis for resisting large changes—of the estimate of the value of that parameter. Resilient estimates are well supported as approximations.[43]

V. Responding to Stein's Challenge

As section III has made clear, the orbital phenomena cited by Newton do not provide direct measurements of the equality of the weights toward one another of Jupiter and the sun. This testifies to the depth of Stein's challenge. The issue he has raised is the problem which has limited the precision to which we have yet been able to measure the gravitational constant.

> The gravitational force between masses of planetary size is not so weak, but this is no help in determining *G* because only the combination *GM* (where *M* is the mass of the attracting body) appears in the equations of motion of bodies with purely gravitational interactions; hence, planetary observations cannot determine separate values for *G* and *M*. (Ohanian and Ruffini 1994, 3)

That orbital phenomena don't separate *G* and *M* limits the precision to which *G* has been measured to experiments using laboratory bodies.

Another deep insight of Stein's is that appeal to further consequences, rather than the phenomena cited in Newton's argument up to proposition 7, is what provides the strongest evidence to back up Newton's appeal to law 3. Stein has suggested that only by appeal to such additional consequences can Newton provide sufficient evidence to counter a charge of wild hypothesis.

> In the first place, it is essential to recognize that Proposition VII implies a vast range of consequences not implied by the propositions antecedent to it—and in part, *contradictory* of the statements of "Phænomena" on which the initial reasoning of book III was based. That Newton understood this cannot possibly be called into question: the entire remainder of Book III of the *Principia* is devoted to the derivation of such consequences, and to *their confrontation*, so far as it was possible at the time, *with actual phenomena*. In short, in the formal terms I have suggested above as characteristic of Newton's usage, the remainder of Book III can be seen as devoted to the "proof by phenomena" of the law of gravitation; and furthermore, the proofs so obtained, in so far as they involve in part new (and confirmed) astronomical discoveries, and a great increase in both the scope and precision of astronomical prediction, provide a kind of warrant for that law that can quite reasonably be seen as drawing the sting from the charge of "wild hypothesis" that could otherwise be levelled at Newton's way of applying the third law of motion. (1991, 220)

This statement, however, suggests that what Stein is counting for Newton as "proof by phenomena" might be just success at accurate prediction of increased scope and accuracy and of new discoveries that become confirmed as phenomena.

Let us consider consequences of limiting empirical success to *confrontation with actual phenomena* for attempts to empirically rule out alternatives with differing *G*'s for different solar system bodies. The striking new phenomena corresponding to increasingly precise successive approximations made possible by taking into account gravitational interactions— such phenomena as the two-body correction to the harmonic law for Jupiter's orbit or corrections of Saturn's motion corresponding to perturbations by Jupiter—are all consequences of any such alternative theory. No such phenomenon, therefore, can support Newton's application of law 3 to gravitation by empirically refuting differing *G*'s.

Such phenomena empirically back up the existence of oppositely directed weights of solar system bodies toward one another that follows from combining their inverse-square acceleration fields. They, thus, add empirical support for attributing to gravitation the first main criterion distinguishing what Newton calls "attraction" from examples like Cotes's ball being pushed toward another. They do not, however, force the equalities of those oppositely directed weights required by Newton's application of law 3 to gravitation as such an attraction between bodies.

Without collisions only *confrontation with actual phenomena* corresponding to nongravitational interactions—such as effects of tidal friction on lunar motion or actions on comet motions due to escaping gasses heated by the sun—could empirically falsify differing G's. To do so, such phenomena would have to be developed with far more precision than was available in Newton's time. I doubt whether even Laplace's treatment of such phenomena would have provided resources sufficient for the job.

My position is similar with regard to Stein's earlier Humean challenge to me. If only confrontation with actual phenomena counts as evidence then the orbital phenomena of planets do not provide any evidence against a conjectured alternative that would agree with a sun-centered inverse-square acceleration field in the ranges explored by the orbits of planets but would differ wildly for distances not so explored. On an account of evidence limited to prediction alone there would be no evidence against such an alternative hypothesis until phenomena such as comet orbits which explored distances not explored by planets became available.

As we have seen above, Newton's application of law 3 to gravitation clearly outstrips alternative hypotheses of separate G's in realizing his ideal of empirical success. Similarly, on an account of evidence informed by this richer ideal of empirical success, the agreeing measurements of its parameters provided by the planetary orbits count as evidence for expecting that comet motions, also, will *approximate* motions in accord with the sun's inverse-square centripetal acceleration field. On this methodology departures from motions in accordance with the sun-centered inverse-square acceleration field are to be sought for. Such departures provide second-order phenomena that testify to additional causal factors and support increasingly precise successive approximations.[44]

The following passage is the first part of the paragraph in the general scholium containing the famous *hypotheses non fingo* passage and the explicit changes introduced by Newton in response to Cotes.

Thus far I have explained the phenomena of the heavens and of our sea by the force of gravity, but I have not yet assigned a cause to gravity. Indeed, this force arises from some cause that penetrates as far as the centers of the sun and planets without any diminution of its power to act, and that acts not in proportion to the *surfaces* of the particles on which it acts (as mechanical causes are wont to do) but in proportion to the quantity of *solid* matter, and whose action is extended everywhere to immense distances, always decreasing as the squares of the distances. Gravity toward the sun is compounded of the gravities toward the individual particles of the sun, and at increasing distances from the sun decreases exactly as the squares of the distances as far out as the orbit of Saturn, as is manifest from the fact that the aphelia of the planets are at rest, and even as far as the farthest aphelia of the comets, provided those aphelia are at rest. (C & W, 943)

These consequences that Newton takes as support are not just accurate predictions. These appeals to phenomena explained are appeals to explanations realizing his stronger ideal of empirical success as accurate measurement of causal parameters by the phenomena they explain.

Newton's second corollary to proposition 7—an important part of his statement of the law of universal gravitation—provides another realization of this stronger ideal of empirical success.

Corollary 2 (prop. 7, bk. 3) The gravitation toward each of the individual equal particles of a body is inversely as the square of the distance of places from those particles. This is evident by book 1, prop. 74, corol. 3.

Corollary 3 (prop. 74, bk. 1) If a corpuscle placed outside a homogeneous sphere is attracted by a force proportional to the square of the distance of the corpuscle from the center of the sphere, and the sphere consists of attracting particles, the force of each particle will decrease in the squared ratio of the distance from the particle.

Just as is the case with Newton's classic inferences from phenomena, the inference from (a) inverse-square variation of the total force on a corpuscle outside a sphere with respect to distance from the center to (b) the inverse-square variation of the component attractions toward particles, is backed up by systematic dependencies. Any difference from the inverse-square law for attraction toward the particles would produce a corresponding difference from the inverse-square for the law of attraction toward the center resulting from summing the attractions toward the particles.[45] These dependencies make phenomena measuring inverse-square variation of attraction toward the whole count as measurements of inverse-square variation of the law of attraction toward the particles.

Stein's contention that Newton explicitly regarded his argument up to proposition 7 as in need of supplementation is supported by Newton's discussion of proposition 8, book 3.

After I had found that the gravity toward a whole planet arises from and is compounded of the gravities toward the parts and that toward each of the individual parts it is inversely proportional to the squares of the distances from the parts. I was still not certain whether that proportion of the inverse square obtained exactly in a total force compounded of a number of forces, or only nearly so. For it could happen that a proportion which holds exactly enough at very great distances might be markedly in error near the surface of the planet, because there the distances of the particles may be unequal and their situations dissimilar. But at length, by means of book 1, props. 75 and 76 and their corollaries, I discovered the truth of the proposition dealt with here.

In the corollaries to proposition 8 Newton applies proposition 7 to use harmonic law ratios to measure the masses of the sun and planets with moons (C & W, 812–13). These measurements of relative masses of bodies with satellites are clear examples of a realization of Newton's stronger ideal of empirical success that flows from his challenged application of the third law of motion to gravitation.[46]

Unlike the isolated systems we have been considering, Stein's challenge to Newton's application of the third law to construe gravitation as a pairwise attraction between visible bodies in the solar system does not deny that the third law applies. On the vortex theory it would be the changes in motion of invisible vortical particles resulting from their pushing the planets into orbital motion—rather than the gravitation of the sun towards the planet—that count as the equal and opposite reaction to the weight of a planet towards the sun. As Stein makes clear, failure to grant Newton's application of the third law of motion to construe gravitation as a pairwise interaction among visible solar system bodies is compatible with the orbital phenomena together with all three laws of motion.

Appeal to invisible vortical particles pushing the planets into their orbits—like the outright violations of the third law in the isolated systems we considered above—fails to realize Newton's ideal of empirical success as well as Newton's application of the third law of motion to construe gravitation toward visible solar system bodies as a pairwise interaction among them. The agreement among the measurements of relative inertial masses among solar system bodies provided by orbital phenomena counts as evidence supporting Newton's application of the third law of motion to construe gravitation among solar system bodies as pairwise attractions between them. Without some comparably accurate way of measuring masses and motions of vortical particles, the alternative application of the third law of motion to construe gravitation towards the sun as an interaction between planets and vortical particles pushing them towards the sun will not do nearly as well.

This suggests that perhaps, already by the corollaries of proposition 8, Newton had provided enough evidence supporting his application of the third law to gravitation between bodies for his Rule 4 to endorse regarding vortical alternatives as mere contrary hypotheses—no longer to be counted as rivals to be taken seriously.

This comparative advantage of being able to approximate the motions of visible solar system bodies as those in an isolated system is not just a policy to avoid commitment to invisible causes. Perrin's 1909 comment on implications of the many agreeing measurements of Avogadro's number from diverse phenomena illustrates the central role played by what I am identifying as Newton's ideal of empirical success in development of support for the existence of invisible molecular particles.

> I think it is impossible that a mind free from all preconception can reflect upon the extreme diversity of the phenomena which converge upon the same result without experiencing a strong impression, and I think that it will henceforth be difficult to defend by rational arguments a hostile attitude to molecular hypotheses. (quoted in Pais, 95)

On Newton's methodology, the convergent agreeing measurements of Avogadro's number from diverse phenomena contributes significantly toward raising the status of molecular theory from that of a mere hypothesis to that of a proposition gathered from phenomena by induction.

NOTES

1. The admiration of our small band of Newton scholars is matched by the admiration of the much wider group of philosophers who have had their work on space-time informed by "Newtonian Space-Time" and "Some Philosophical Prehistory of General Relativity," as well as Howard's seminal 1968 correction to Hilary Putnam and C. W. Rietdijk in "On Einstein-Minkowski Space-Time" and his related 1991 correction to Nicholas Maxwell in "On Relativity Theory and Openness of the Future."

2. In earlier essays (1967, 179–80 and 1970, 269) Howard, also, mentions difficulties with Newton's appeal to law 3 to argue that gravitation is an interaction. Dana Densmore (1999, 104–11) independently raises this objection.

3. Here is Newton's instruction along with the latin of his proposed revision.

> And for preventing exceptions against the use of the word Hypothesis I desire you to conclude the next paragraph in this manner

> Quicquid enim ex phaenomenis non deducitor Hypothesis vocanda est, et ejusmodi Hypotheses seu Metaphysicae seu Physicae seu Qualitatum occultarum seu Mechanicae in Philosophia experimentali locum non habent. In hac Philosophia Propositions deducunter ex phaenomenis & reddunter generales per Inductionem. Sic impenetrabilitas mobilitas & impetus corporum & leges motuum & gravitis innotuere. Et satis est quod Gravitas corporum revera existat & agat secundum leges a nobis expositas & ad corporum caelestium et maris nostri motis omnes sufficiat. (Corresp., vol. 5, 397)

The printed Latin replaces [, et ejusmodi] in line 2 with [; &]. It also italicizes *hypothesis* in line 2 and *philosophia experimentali* in line 4. (Koré and Cohen, 764)
Here is a translation of the original passage Newton had earlier sent to Cotes.

> Indeed, I have not yet been able to deduce the reason [or cause] of these properties of gravity from phenomena, and I do not feign hypotheses. For whatever is not deduced from phenomena is to be called a hypothesis; and I do not follow *hypotheses*, whether metaphysical or physical, whether of occult qualities or mechanical. It is enough that gravity should really exist and act according to the laws expounded by us, and should suffice for all the motions of the celestial bodies and of our sea. (C & W, 276)

4. Those of us who associate "deduction" with strictly logical or mathematical inference may be somewhat surprised to find that Newton's deductions from phenomena include inductions. Stein has provided a useful clarification of Newton's usage, which points out that Newton uses "deduction" widely enough to encompass any reasoning competent to establish a conclusion as warranted.

> Baldly, then: In Newton's terminology, three terms describing kinds of arguments are used in sharply distinguished fashion: namely, *demonstration, deduction*, and *proof*. The first of these is Newton's characteristic term for *purely mathematical reasoning*. The second—"deduction"—is used by him in a quite wide sense, for reasoning *competent to establish a conclusion as warranted* (in general, on the basis of *available evidence*). As for "proof," Newton typically means by it *the subjection of a proposition to test by experiment or observation* (with a successful outcome). (1991, 219)

Though Newton's usage may not have as sharply limited "demonstration" to purely mathematical reasoning as is suggested, it is clear that Newton used "deduction" in the quite wide sense Stein has described.
5. See the systematic dependencies backing up Newton's inferences to inverse-square centripetal acceleration fields in notes 20 to 22 below. For a more general exploration of this point see Harper 1991.
6. Newton had just cited work by Wren, Wallis, Huygens, and Marriot together with a description of his own careful pendulum experiments supporting the application of law 3 to collisions (C & W, 424–27).
7. The immediate applications of law 3 to pushes between body A and the obstacle and body B and the obstacle make the equilibrium of the system support the further application of law 3 to the attraction between A and B themselves.
8. Stein pointed out to me the need to specify this reservation. This is just one of many examples where his insight has led me to a deeper understanding of issues raised by Newton's argument.
   Another version of the experiment with magnet and iron would have both fastened to a single floating block. (See Knudsen and Hjorth 1996, 29.) Here the relations among the magnet, block, and iron are, rather obviously, sufficiently isolated from disturbance for it to be intuitive to count the system as a body. In Newton's thought experiment the system, though not actually fastened together, is sufficiently isolated from disturbance of relations among its parts to count as a body for purposes of the experiment. Indeed, part of Newton's evidence for law 1 is that "there is no such phaenomenon in all nature" as having unequal attraction

between bodies make the system of both "go away together wth an accelerated motion in infinitum, as it were by a self moving principle."

In an interesting manuscript paper Dana Jalobeanu and Katherine Brading (1999) have argued that Newton provided resources which allowed the concept of an isolated system, to which conservation principles can be applied, to contribute significantly toward the solution to the metaphysical and epistemological problems for the concept of body arising from the destruction of the Aristotelian cosmos.

9. Wallis's appeal to magnetic attraction was addressed to the following objection to his conjectured motion of earth and moon about a common center of gravity: "it appears not how two bodies, that have no tye, can have one common center of gravity" (*Phil. Trans.*1: 282). Gilbert's 1600 book *De magnete* inspired Kepler's proposals for a celestial physics based on forces modeled on magnetic attraction and repulsion, as well as later proposals such as Wallis's appeal to a "tye" between the earth and moon to explain the tides (*The General History of Astronomy*, Taton and Wilson, eds., vol. 2A, chaps. 4 and 5).

10. In his new paper on Newton's metaphysics, Stein offers the following description of such commitment to contact action in the mechanical philosophy of Huygens which departed from Descartes by allowing void and atoms.

> Within this (as it were) "revisionist" conception—also known as the "corpuscular philosophy"—it was still maintained that all the processes in nature consist in the motions of bodies, and that all natural *changes* of motion are occasioned by direct actions of one body pushing on another. (2000, 7)

11. This conjecture, as well as Huygens's hypothesis which we shall consider below, is compatible with having gravitation be an acceleration field. One needs the superficies of parts so distributed that the total force exerted on a body by the downward streaming ethereal spirit is proportional to its quantity of matter.

12. Stein calls attention to this early passage in his forthcoming essay on Newton's metaphysics.

13. In his preface to book 3, Newton describes this earlier version as "composed in popular form" (C & W, 793). A translation into English is available in Cajori 1962, 549–626).

14. Stein offers an excellent discussion of this passage together with a preceding related passage (1991, 218). This discussion is expanded upon in his forthcoming essay on Newton's metaphysics.

15. On Huygens's hypothesis, gravity is the centripetal force generated on ordinary bodies by many very thin spherical shells of tiny vortical particles whirling in arbitrarily large numbers of different directions around the center of the earth. This whirling "matter that causes gravity passes through the pores of all known bodies" and, by causing gravity through its impact on the particles composing ordinary matter makes the gravities on ordinary bodies "maintain the same proportion as the quantities of matter that compose them." The quotations are respectively from the citations for pages 139 and 140 in Huygens's *Discourse on the Cause of Gravity* (1690).

16. As long as the weights of the outer parts form an equilibrium with respect to the center, this construction can be carried out—even if those weights are not uniformly distributed.

17. This is in accord with Huygens's hypothesis in which the only force directed by the middle part on either outer part would be that resulting from its resistance to being pushed away from its centered location with respect to the center of the whirling spherical shells.

The middle part, EGKH, of the earth plays the same role in this equilibrium argument as the obstacle plays in Newton's basic equilibrium thought experiment.

18. See Stein 1970, 265–66 and his forthcoming paper on Newton's metaphysics.

19. Proposition 1 and its corollary 6 together with proposition 2 from book 1 make a constant areal rate by radii to an inertial center equivalent to the centripetal direction of forces maintaining a body in orbit about that center (C & W, 444–47). Corollary 1 of proposition 2, together with these propositions, makes an increasing areal rate equivalent to having the force deviate forward (in the direction of tangential motion) and, for orbits in non-resisting spaces, a decreasing areal rate equivalent to having the force deviate backwards (C & W, 447). These dependencies make the constancy of the rate at which areas are being swept out by radii to a center measure the centripetal direction of the force maintaining a body in an orbit about that center, provided the center can be treated as inertial. Proposition 3 extends approximations of such results to noninertial centers (see, e.g., Harper 1998, 274–75).

20. The appeals to the harmonic law are backed up by corollaries of proposition 4 of book 1, which are proved for concentric circular orbits but can be straightforwardly extended to elliptical orbits. (See note 28 below; and also Harper 2000, 84–87).

$$\text{Cor. 7} \quad t \propto R^n \text{ iff } f_{ac} \propto R^{1-2n}$$

To have the periods t be as some power n of the distances R is to have the result of plotting $\mathrm{Log}\, t$ against $\mathrm{Log}\, R$ be fit by a straight line of slope n. This is equivalent to having the centripetal accelerations f_{ac} produced by the centripetal forces maintaining those bodies in their orbits be as the 1-$2n$ power of their distances. To have the harmonic law is to have the result of plotting $\mathrm{Log}\, t$ against $\mathrm{Log}\, R$ be fit by a straight line of slope $3/2 = 1.5$. It follows from cor. 7, therefore, that

$$\text{Cor. 6} \quad t \propto R^{3/2} \text{ iff } f_{ac} \propto R^{-2}$$

the harmonic law is equivalent to having the centripetal accelerations be as the -2 power of the distances. The systematic dependencies of cor. 7, also, would make alternatives to the harmonic law phenomena measure alternative power laws. Having n greater than $3/2$ would be equivalent to having the accelerations fall off faster than the -2 power with distance, while having n less than $3/2$ would be equivalent to having the accelerations fall off less fast than the -2 power.

21. Newton's appeal to the absence of significant orbital precession for planetary orbits about the sun is backed up by corollary 1 of proposition 45, book 1, according to which

Precession is		Power law is
p degrees per	iff	$f_{ac} \propto R^x$, where
revolution		$x = (360/360 + p)^2 - 3$

This makes zero orbital precession measure the inverse-square law for distances explored by an orbit. These systematic dependencies would also make, otherwise unaccounted for, positive (or negative) orbital precessions measure power laws falling off faster than (or less fast than) the –2 power of distance from the center.

This result applies to orbits of small eccentricity. It can be extended to orbits of arbitrary eccentricity, with the result that increased eccentricity increases the sensitivity with which departures from the inverse-square would generate orbital precession (Valluri et al. 1997).

22. Given two acceleration fields with the same law relating accelerations to distance from their respective centers, the ratio of the accelerations they would produce on bodies at equal distances is constant for all distances. This makes the ratio of their accelerative measures—for any given distance—equal the ratio of their strengths.

23. In his earlier version of book 3, Newton calls his phenomena "astronomical experiments" (Cajori, 595).

24. Newton offers the first part of corollary 2 of proposition 6 to back up the argument for extending gravitation towards the earth to all bodies universally.

> Corollary 2. All bodies universally that are on or near the earth are heavy [or gravitate] toward the earth, and the weights of all bodies that are equally distant from the center of the earth are as the quantities of matter in them. This is a quality of all bodies on which experiments can be performed and therefore by Rule 3 is to be affirmed of all bodies universally. (C & W, 809)

The quality of bodies which is generalized is weight toward the earth. Newton explicitly tells us that weights for bodies at equal distances from the center of the earth have equal ratios to the quantities of matter in them. Rule 3 tells us to conclude that the ratio of mass to inverse-square adjusted weight toward the earth is equal for all bodies universally—that this ratio is a constant parameter for all bodies—if its equality holds for all the bodies in reach of our experiments.

25. See Harper 1999, 91–93 for discussion and references.

26. As early as 1705 Halley had proposed elements for this retrograde orbit with a perihelion distance of about .58 au and a period of on average about 75.5 years, corresponding to a semi-major axis of about 17.86 au and an eccentricity of about .97 (See Halley 1705; Hughes p. 359 in Thrower; and Wilson p. 83 in *GHA*, vol. 2b).

For general background see Thrower 1990, part 4 and *GHA*, vol. 2b, chap. 19 with appendix.

27. Newton points out that there is agreement among astronomers about the periodic times of the orbits of the planets and cites 4332.514 decimal days as Jupiter's period (C & W, 800).

28. Consider an elliptical orbit with period t, semimajor axis a, and the force maintaining a body in it directed toward a focus. The magnitude f_{ac} of the acceleration toward the focus in an elliptical orbit with the force to the focus where r is the location vector from the focus to the orbiting body is

$$f_{ac} = 4\pi^2 a^3 / t^2 r^2$$

(French 1971, 588). When the length of vector r equals the length of the semimajor axis a (i.e., when the body is at either end of the minor axis) then the acceleration toward the focus

$$f_{ac} = 4\pi^2 a/t^2$$

is exactly the centripetal acceleration of a circular orbit concentric to the focus with radius equal in length to the semimajor axis a of the ellipse and the same period t as that of the elliptical orbit. Kepler's value for the semimajor axis of Jupiter's orbit $a_j = 5.1965$ au, yields $4\pi^2 a/t^2 = 1.4581$ au$/t_e^2$ as the centripetal acceleration of a uniform motion concentric circular orbit of radius $= a_j$ and period $t_j = 11.8616 t_e$. Boulliau's estimate $a_j = 5.2252$ yield's 1.4661 au$/t_e^2$ as the centripetal acceleration at distance a_j of Jupiter towards the sun. The mean of these estimates is 1.462 au$/t_e^2$.

29. Boulliau and Kepler agree on periods. Here they are in units $t_e = 365.2565$ days for

	Mercury,	Venus,	Earth,	Mars,	Jupiter,	Saturn.
t	.2408	.6150	1.0000	1.8808	11.8616	29.4568

For Kepler, the semimajor axes a in units au equal to the semimajor axis of the orbit of the earth are

a	.38806	.724	1.0000	1.5235	5.1965	9.51.

For Boulliau, the estimates for Earth and Mars are the same as Kepler's. For the others, Boulliau gives

a	.38585	.72398			5.2252	9.54198.

These give the following inverse-square adjusted $4\pi^2 a/t^2$ $(a/5.21)^2$ estimates for the centripetal acceleration toward the sun at the distance 5.21au we are assuming for Jupiter when its distance from the sun equals the semimajor axis of its orbit.

1.466	1.459	1.454	1.454	1.451	1.442
1.441	1.459			1.475	1.456

These yield a mean 1.456 $sd = (\text{avg?}^2)^{1/2} = .00965$

$$sd^+ = (10/9)^{1/2} sd = .01018 \quad SE = ((10)^{1/2}/10)\, sd^+ = .0032$$

95% t-confidence bound $2.26 SE = .007$

30. Pound's observations, including ones using a 123-foot focal-length telescope with a micrometer were made between the second and third editions of the *Principia*.

According to King (1955, 63), the long-focus lens Pound used was presented by Christian Huygens's brother, Constantine, to the Royal Society in 1692. The Huygens brothers were pioneers in telescopes with focal length too long to easily accommodate a tube that would not bend. Newton paid to have a very long maypole made available so that Pound could erect the lens on it at Wansted Park. The eyepiece and micrometer set up could then be positioned on the ground with respect to the lens on the maypole.

Pound's nephew James Bradley, then just past his mid-twenties, assisted on the project. Bradley, who became famous with his discovery of aberration in 1729, succeeded Halley as astronomer Royal in 1742. Bradley, also, was the first to discover and accurately characterize nutation which, along with abberation, had been a major obstacle preventing astronomers from being able to exploit the precision made possible by telescopes with micrometers.

Newton cites 8'16" for the fourth and 4'42" for the third satellite as Pound's angular measures of maximum elongations at the mean Earth-Jupiter distance (C & W, 798). He gives greatest elongations in semidiameters of Jupiter, with one-half of 37.25 assumed as the angular measure at the mean earth-Jupiter distance of the semidiameter of Jupiter.

> Let us assume that this diameter is very nearly 37-1/4" ; then the greatest elongations of the first, second, third and fourth satellites will be equal respectively to 5.965, 9.494, 15.141, and 26.63 semidiameters of Jupiter. (C & W, 798)

These yield the following angular measures from Pound of maximum elongations at the mean earth-Jupiter distance, which may be compared with the corresponding angular measures calculated from the semimajor axes cited in the 1992 *Explanatory Supplement to the Astronomical Almanac*, 807.

	Io	Europa	Ganymede	Callisto
Pound	1.85'	2.95'	4.7'	8.267'
ESAA	1.864'	2.96'	4.726'	8.317'

31. Astronomers in Newton's day had very good data on periods of Jupiter's Galilian satellites. Here in decimal days are the periods cited by Newton.

Io	Europa	Ganymede	Callisto
1.76914	3.55118	7.15458	16.68899

The differences from mean periods cited in the 1992 *ESAA* (708) are,respectively, .0000022, -.000001, .000027, -.000028 decimal days. In the worst case, that of the 4th moon Callisto, the period in Newton's table is 2.42 seconds less than that cited in *ESAA*.

Here are these periods converted to units 365.2565 days that Newton cites for the period t_e of the earth.

.00484	.00972	.01959	.04569

32. Pound's angular measures Φ of maximum elongations of Jupiter's Galilean satellite orbits, at the mean earth-Jupiter distance 5.21 au = the mean Jupiter-sun distance, lead to estimates $5.21 \sin \Phi$ of orbital radii in au of

r	.00280	.00447	.00712	.01253

Combining them with the periods

t	.00484	.00972	.01959	.04569

yields the following results

.001363	.001375	.00138	.00137

for $4\pi^2 r/t^2$ $(r/5.21)^2$ in au/t_e^2—inverse-square adjusting Pound's data for the accelerations of Jupiter's moons to the assumed distance, 5.21 au, of the sun from Jupiter.

These, in turn, lead to a mean of .001373, an sd of .000015, an sd^+ of $(4/3)^{.5} sd = .000018$, an SE of the mean of $(4^{.5}/4) sd^+$ of .000009, for a 95% t-confidence bound for 4–1 = 3 degrees of freedom of 3.18 SE = .00003.

33. Using the orbit of Venus for the harmonic law ratio for orbits about the sun and that of Callisto for the harmonic law ratio for orbits about Jupiter Newton estimates this ratio as 1067/1 (C & W, 812–13).

34. See pp. 46–49 of Smith (1999) "How Did Newton Discover Universal Gravity." This is Smith's impressive contribution to *The St .John's Review* 45, no. 2 (1999), which contains the proceedings of an excellent conference entitled "Beyond Hypothesis: Newton's Experimental Philosophy" held at St. John's College, Annapolis, March 19–21, 1999.

35. These motions are described from a reference frame with center at c and fixed directions toward the stars.

36. We are explicitly allowing for violations of law 3, so that we can explore whether or not the orbits of Jupiter and the sun about c measure the equality of their weights toward one another.

37. Martin Cernohorsky (Kaminski 1988, 28–46) offers a discussion of rotation in Newton's wording of the first law of motion.

38. This is the Newtonian part of the n-body point mass equation used as a basis for the orbital ephemerides for the sun, moon and planets (Standish et al., in Seidelmann 1992, 281)

39. Newton had included a somewhat longer statement in an unsent draft of this letter (Corresp., vol. 5, 401)

> Sr
> On Saturday last I wrote to you representing that Experimental philosophy proce[e]ds only upon Phenomena & makes Propositions general by Induction from them. In this Philosophy neither Explications nor Objections are to be heard unless taken from phaenomena. Nor are Propositions here made general by arguments a priori by [*read* but] only by Induction without exception. And upon such an Induction the mutuall and mutually equal Attraction is founded. One may suppose that there may be bodies penetrable or immoveable or destitute of force, or with attraction mutually unequal, but such suppositions without any instance in Phaenomena are mere hypotheses & have no place in experi[ment]al Philosophy: & to introduce them into it would be to overthrow the Arguments from Induction upon wch all the general Propositions in this Philosophy are built.

This somewhat more extended discussion anticipates Newton's comment on rule 4, as well as the method advocated in that rule.

Newton's long draft of his Saturday letter to Cotes (Corresp., vol. 5, 398–99), also, includes significant anticipation of rule 4 and of connections between such considerations and his third rule of philosophizing.

40. Any way of fixing the ratio G_h/G_j will allow the relative acceleration phenomena to measure the weights and masses of these bodies. Moreover, any way of fixing these ratios will predict exactly the same relative accelerations among these

bodies. This might lead one to say that differing specifications of the ratios among the G's are just alternative specifications of the same theory.

Even though these alternatives would predict the same actual history of relative accelerations among these bodies they would not agree on all possible predictions. For example, suppose the tangential velocities of the sun and Jupiter with respect to c were destroyed so that they fell together. Before their collision the motions toward one another would be the same in all these alternative theories, but the different theories would give differing predictions about the motions after they collided. Theories where G_h/G_j is set at 1,129,969 so that the inertial mass of Jupiter is 1063 times that of the sun would make very different post-collision predictions from those where G_h/G_j is set at 1 so that it is the sun's inertial mass which is more than a thousand times that of Jupiter.

41. The six lunar distance estimates are respectively 59, 60, 60, 60 1/3, 60 2/5, and 60 1/2 terrestrial semidiameters, while his cited circumference of the earth is 123,249,600 Paris feet and his lunar period is 27 days 7 hours, 43 minutes or 39,343 minutes (C & W, 803–4). The one-minute fall for a concentric circular lunar orbit is arc^2/D, where arc is the orbital arc length traversed in one minute and D is the diameter of that orbit. Where x is a lunar distance estimate in earth radii we have

$$arc^2/D = (x^2)(123,249,600/39,343)^2 / x \, (123249600/\pi) = .25015x$$

Therefore, for each lunar distance estimate x the corresponding inverse-square adjusted one-second fall at the surface of the earth is

$$d = .25015 \; x \; (x/60)^2.$$

The six cited distance estimates yield a mean of 15.041 Paris feet, with $sd = (avg\delta^2)^{1/2} = .373$, $sd^+ = (n/n-1)^{1/2}sd = .409$, $SE = (n^{1/2}/n)sd^+ = .167$, for a t-confidence bound for 6−1=5 degrees of freedom of 2.57 $SE = .429$ Paris feet.

42. Huygens experimentally established 3 Paris feet and 8 1/2 lines, which is 3.059 Paris feet, as the length l of a seconds pendulum at Paris. As Newton points out, Huygens also showed that $1/2$ of l times π^2 equals the one-second fall corresponding to the acceleration of gravity. This gives $.5l\pi^2 = 15.096$ Paris feet.

George Smith (2000a) argues for an upper bound of 1 part in 1520 on the error of Huygens's measurement of the one-second fall for gravity. 15.096 Paris feet/1520 = .0099 or .01 Paris feet.

43. Measurements by phenomena are backed up by the open-ended—and often quite large—bodies of data fit by each phenomenon. For an example where large bodies of data for the orbit of Mars back up estimates of the mass of the sun see Harper, Bennett, and Valluri 1994.

44. Mercury's celebrated 43" per century of perihelion precession is exactly such a second-order phenomenon. It is 43" per century beyond the 530" per century perihelion precession attributable to perturbation by other planets (Will 1993, 181). As George Smith (conversation) has pointed out to me, this makes the evidence for General Relativity provided by its accounting for these 43" per century of Mercury's perihelion precession depend on supporting Newtonian perturbations as approximations good enough to continue to account for the other 530" per century.

I have argued (Harper 1997, 67–70) that Kuhn's own discussion (1970, 101–3) of the Newtonian limit results of General Relativity presupposes that General Relativity can take over for its parameters the measurement successes that perturbation theory made available to Newtonian gravitation.

45. Chandrasekhar (formula 9, p. 289) provides an integral formulating such dependencies Newton provides in lemma 29 and propositions 79–81, bk. 1.

According to proposition 74, bk. 1, inverse-square attraction toward the center of a uniform sphere on corpuscles outside, right down to the surface, result from summing the inverse-square attractions on the corpuscle toward the particles making up the sphere. This proposition follows from Chandrasekhar's integral when the law of attraction toward particles is the –2 power of distance.

Wayne Myrvold generated a computer graph from this integral which shows that a power law differing even slightly from the inverse square, e.g., a –2.01 power law, for the particles will approach that same alternative power law for attractions to the whole at great distances, but will yield attractions to the whole corresponding to differing nonuniform relations to distance for locations close to the surface of the sphere.

The inverse-square case, and the simple harmonic oscillator case where attraction is directly as the distance, are special in that the law of attraction toward particles yields the same law of attraction toward the whole all the way down to the surface of the sphere. These are the two cases Newton singles out for detailed treatment.

46. The resolution of the problem of combining the separate inverse-square acceleration fields is, also, an important achievement that goes beyond successful prediction of the separate phenomena. This achievement is so important that, according to Michael Friedman (1992, 155) Kant appealed to it as a transcendental deduction of Newton's application of the third law to construe gravitation as a universal force of interaction between bodies.

As we have seen, however, combining the separate acceleration fields into a single system can be achieved without Newton's application of law 3. Any of the alternatives with distinct separate *G*'s will do the job. What these do not do is to realize Newton's stronger ideal of empirical success well enough to count as serious rivals.

REFERENCES

Cajori F., ed., trans. (1934, 1962). *Newton's "Principia," Motte's translation revised.* Los Angeles: University of California Press.

Chandrasekhar, S. (1995). *Newton's "Principia" for the Common Reader.* Oxford: Clarendon Press.

Cohen, I. B., ed. (1958). *Isaac Newton's Papers and Letters on Natural Philosophy.* Cambridge: Harvard University Press.

Cohen, I. B., and Whitman, A., trans. (1999). *Isaac Newton, the Principia.* Los Angeles: University of California Press.

Densmore, D. (1999). "Cause and Hypothesis: Newton's Speculation about the Cause of Universal Gravitation." *The St. John's Review* 45, no. 2: 94–111.

Duhem, P., trans. (1906). *The Aim and Structure of Physical Theory.* Princeton: Princeton University Press, 1991.

Feyerabend, P. (1970). "Classic Empiricism." In *The Methodological Heritage of Newton,* edited by R. E. Butts and J. W. Davis. Toronto: University of Toronto Press.

French, A. P. (1971). *Newtonian Mechanics.* New York: W.W. Norton & Company.

Friedman, M. (1992). *Kant and the Exact Sciences.* Cambridge: Harvard University Press.

Glymour, C. (1980). *Theory and Evidence.* Princeton: Princeton University Press.

Halley, E. (1705). "Astronomiae Cometicae Synopsis." *Philosophical Transactions* 24.

Harper, W. L. (1991). "Newton's Classic Deductions from Phenomena." *Philosophy of Science Association* 2 (1990): 183–96.

———. (1997). "Isaac Newton on Empirical Success and Scientific Method." In *The Cosmos of Science*, edited by L. Earman and J. D. Norton. Pittsburgh: University of Pittsburgh Press, 55–86.

———. (1998). "Measurement and Approximation: Newton's Inferences from Phenomena verses Glymour's Bootstrap Confirmation." In *The Role of Pragmatics in Contemporary Philosophy*, edited by G. Weingartner, G. Schurz, and G. Dorn. Vienna: Hölder-Picher-Tempsky.

———. (1999). "The First Six Propositions in Newton's Argument for Universal Gravitation." *The St. John's Review* 45, no. 2: 74–93.

Harper, W. L., B. H. Bennett, and S. R. Valluri. (1995). "Unification and Support: Harmonic Law Ratios Measure the Mass of the Sun." In *Logic and Philosophy of Science in Uppsala*, edited by D. Prawitz and D. Westerståhl. Dordrecht: Kluwer Academic Publishers, 131–46.

Herivel, J. (1965). *The Background to Newton's "Principia."* Oxford: The Clarendon Press.

Huygens, C. (1690). *Discourse on the Cause of Gravity.* Trans. K. Bailey with comments by K. Bailey and G. Smith. Forthcoming.

Jalobeanu, D., and K. Brading. (1999). "All Alone in the Universe: Descartes, Newton and the Introduction of Isolated Systems into Natural Philosophy." Manuscript.

Kaminski, W. A., ed. (1988). *Isaac Newton's "Philosophiae Naturalis Principia Mathematica."* Lublin, Poland and Singapore: World Scientific.

Knudsen, J. M., and P. G. Hjorth. (1966). *Elements of Newtonian Mechanics.* Berlin: Springer.

Koyré, A., and I. B. Cohen, eds. (1972) *Isaac Newton's "Philosophiae Naturalis Principia Mathematica."* Cambridge: Harvard University Press.

Kuhn, T. S. (1970). *The Structure of Scientific Revolutions.* 2nd ed., enlarged. Chicago: University of Chicago Press.

Newton, I. (1959–1977). *The Correspondence of Isaac Newton.* Cambridge: Cambridge University Press.

Pais, A. (1982). *'Subtle is the Lord . . .'.* Oxford: Oxford University Press.

Seidelmann, P. K., ed. (1992). *Explanatory Supplement to the Astronomical Almanac.* Mill Valley: University Science Books.

Smith, G. (1999). "How Did Newton Discover Universal Gravity?" *The St. John's Review* 45, no. 2: 32–63.

———. (2000). "From the Phenomenon of the Ellipse to an Inverse-Square Force: Why Not?" (This volume.)

———. (2000a). "Huygens' Empirical Challenge to Universal Gravity." (Forthcoming.)

Stein, H. (1967). "Newtonian Space-Time." *The Texas Quarterly* 10, no. 3: 174–200.

———. (1968). "On Einstein-Minkowski Space-Time." *The Journal of Philosophy* 65, no. 1: 5–23.

———. (1970). "On the Notion of Field in Newton, Maxwell, and Beyond." In *Historical and Philosophical Perspectives of Science*, edited by R. H. Stuewer. Minneapolis: University of Minnesota Press, 264–87.

———. (1970a). Reprint of Stein 1967 with revision. In *The Annus Mirabilis of Sir Isaac Newton 1666-1966*, edited by R. Palter. Cambridge, Mass: The M.I.T. Press, 258–84.

———. (1977). "Some Philosophical Prehistory of General Relativity." In vol. 8 of *Minnesota Studies*, edited by Earman, Glymour, and Stachel. Minneapolis: University of Minnesota Press, 3–49.

———. (1991) "'From the Phenomena of Motions to the Forces of Nature': Hypothesis or Deduction?" *Philosophy of Science Association* 2 (1990): 209–22.

———. (1991b). "On Relativity Theory and Openness of the Future." *Philosophy of Science* 58: 147–67.

———. (2000). "Newton's Metaphysics." In *Cambridge Companion to Newton*, edited by I. B. Cohen and G. Smith. Forthcoming.

Taton, R., and C. Wilson, eds. (1989, vol. 2A; 1995, vol. 2B). *The General History of Astronomy.* Vol. 2. Cambridge: Cambridge University Press.

Valluri, S. R., C. Wilson, and W. L. Harper. (1997). "Newton's Apsidal Precession Theorem and Eccentric Orbits." *Journal for the History of Astronomy* 27: 13–27.

Wallis, J. (1666). "An Essay Exhibiting His Hypothesis about the Flux and Reflux of the Sea, with appendix." *Philosophical Transactions* 1: 263–89.

Will, C. M. (1993). *Theory and Experiment in Gravitational Physics.* 2nd ed. Cambridge: Cambridge University Press, 1993.

[5]

Some Fruit for Howard: Descartes's Melon and Newton's Apples

ROBERT PALTER

Introduction

In the course of my research for a book entitled *The Duchess of Malfi's Apricots, and Other Literary Fruits*, I came across certain—as I call them— "philosophical fruits"; two of these were apples and pomegranates. The very banality of ordinary apples may be precisely what permits them to enter that class of commonsensical pedagogical paradigms (including white envelopes, red tomatoes, and brown sticks) so much favored in recent epistemological exercises designed to distinguish "the real" from mere simulacra, illusions, or—a late-twentieth-century technological wrinkle— holographs. It may occasion some surprise, though, that the notion of such a "philosophical" apple goes back at least as far as the Roman writer Macrobius in the early fifth century C.E., in whose *Saturnalia*, written for the edification of his son, Eustachius, Macrobius concludes the fifteenth and penultimate chapter of the seventh and last book with an analysis of reason and the senses as sources of knowledge:

> the reason does not always find the evidence of a single sense enough to establish the identity of an object; for, if I see from afar an object with the shape of the fruit called an apple, it does not necessarily follow that the object is an apple—it might have been made from some material to resemble an apple. I must therefore call for the advice of a second sense and let smell judge. But, if the object had been placed in a heap of apples, it could have acquired the smell of an apple, and so at this point I must consult my sense of touch, which enables me to judge by the weight. But there is a risk that this sense too may itself be deceived, should a cunning craftsman have chosen a material equal in weight to an apple's. I must therefore have recourse to my sense of taste, and, if the taste of the object agrees with its appearance, then I have no hesitation in regarding the object as an apple. (Davies 1969, 505–6)

Whether philosophically sound or not, it should be deeply satisfying to all pomophiles that the ultimate criterion of the apple's reality is said to be its *taste*.

Like the apple, the pomegranate has occasionally figured in arguments about the nature of knowledge, for no other reason, I suppose, than its ready recognizability. In the life of the Stoic philosopher, Sphaerus, by Diogenes Laertius (early third century C.E.), we learn that Sphaerus once engaged in a philosophical discussion with King Ptolemy Philopator (222–205 B.C.E.). Sphaerus having maintained that a wise man could never permit himself to hold mere opinions, the king attempts a refutation by deceiving Sphaerus with some waxen pomegranates. Diogenes then reports the following exchange:

> Sphaerus was taken in and the king cried out, "You have given your assent to a presentation which is false." But Sphaerus was ready with a neat answer. "I assented not to the proposition that they are pomegranates, but to another, that there are good grounds for thinking them to be pomegranates. Certainty of presentation and reasonable probability are two totally different things." (Hicks 1925, vol. 2, 285)

"Are there philosophical melons?" at one point I found myself asking.

Descartes's Melon

There is a mysterious melon which occurs in the first of the three extraordinary dreams of René Descartes (1596–1650) on the night of November 10 to 11, 1619. Descartes thought the dreams were prophetic about his future philosophical career. In the earliest biography of Descartes, by Adrien Baillet, Descartes's interpretation of the melon episode is reported as follows: "The melon, which someone had wished to give him in the first dream, signified, he said, the charms of solitude, but presented by purely human solicitations" [Le melon, dont on vouloit luy faire présent dans le prémier songe, signifoit, distoit il, le charmes de la solitude, mais présentez par des sollicitations purement humaines] (Browne 1977, 266).

If no one has yet provided convincing interpretations of Descartes's dreams, it has not been for want of trying. Perhaps the lengthiest and most detailed set of interpretations to date are in a book by John R. Cole, *The Olympian Dreams and Youthful Rebellion of René Descartes* (1992). Specifically in connection with Descartes's melon—which has been interpreted as anything from the apple of Paradise to the forbidden fruit of Genesis—Cole cites French and Dutch texts which might have been known to Descartes and which assert a parallel between the difficulty of selecting a good melon and the difficulty of selecting a good friend. (Descartes had met his most important intellectual mentor, the Dutch Isaac Beeckman, exactly one year prior to the dream day.) Thus, the entry for *melon* in Randle Cotgrave's French-English dictionary cites as a proverb the saying: "A peine connoist on la femme, & le melon" (Cotgrave 1611); or—as translated by Cole—"a good woman is as hard to pick as a good melon" (Cole 1992, 264, n. 16). And, he continues, a book of verse by the sixteenth-century author Claude Mermet (ca. 1550–ca. 1601), *Le temps passé* (Paris, 1585), contains the following quatrain:

> Les amis de l'heure *présente*
> Ont le naturel du *melon*;
> Il en faut essayer cinquante,
> Avant qu'en rencountrer un bon.

> [Friends in the present day
> Have this in common with the melon,
> You've got to try fifty
> Before you get a good one.] (Cole 1992, 143)

Finally, Cole cites a collection of versified Dutch proverbs by Jacob Cats (1577–1660), *Spiegel van den Ouden ende Nieuwen Tijdt* (1632), containing "a handsome plate of a buxom melon vendor and her perplexed customer, who is trying to select a good piece of fruit by its aroma" (see figure 1), as well as a Dutch poem and two French poems on the same theme. Here are translations of the Dutch and French poems from Cats's book, together with the original French texts:

> In choosing Friends, it's requisite to use
> The self-same care as when we Melons choose:
> No one in haste a Melon ever buys,
> Nor makes his choice till three or four he tries;
> And oft indeed when purchasing this fruit,
> Before the buyer can find one to suit,
> He's e'en obliged t'examine half a score,
> And p'rhaps not find one when his search is o'er.
> Be cautious how you choose a friend;
> For Friendships that are lightly made,
> Have seldom any other end
> Than grief to see one's trust betray'd! (Pigot 1860, 225–26)

> Amys sont comme le melon;
> De dix souvent pas un est bon.

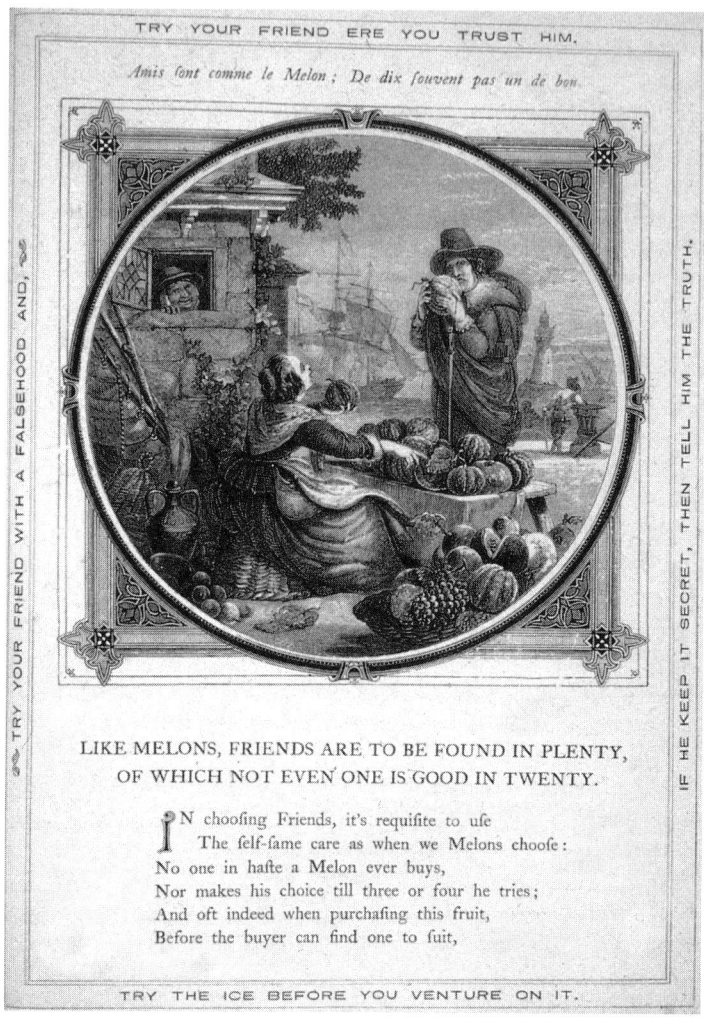

FIGURE 1
Moral Emblems. . . from Jacob Cats and Robert Earlie, illustrated by John Leighton, ed. Richard Pigot (1860).

[Friends are like melons.
In ten, you might not find a good one.]

Les amys sont comme le melon;
Il faut essayer plusieurs pour rancontrer un bon.

[Friends are like melons;
You've got to try many to get a good one.] (Cole 1992, 143)

The relevance of all this to Descartes's dream-life, according to Cole, is "a line of associations extending back from the *présent melon* in the dream through Descartes's memory of this proverbial maxim in poetic form to the painful infidelity of Isaac Beeckman, who had written so as to hurt, when his friend had needed help" (Cole 1992, 143). (Descartes was to break with Beeckman some ten years later but at the time of the dreams they were still friends.) What Cole is suggesting is Descartes's ambivalent feelings toward the older Dutch thinker, exacerbated perhaps by a letter from Beeckman of 6 May 1619 (some six months before the dreams) which might have offended Descartes—if he ever received it, which is not certain (and if he did not, that too would have been an affront, since Descartes had written asking Beeckman for assistance).

The most recent discussion of Descartes's dreams I have come across, by Alan Gabbey and Robert E. Hall, proposes an interpretation of the melon as "representing a complete world of knowledge, an integrated science of the whole, an entire science of the cosmos, a *sphérique encyclopédie*" (Gabbey 1998, 662), citing in support the following passage from the Italian philosopher, Francesco Patrizi (1529–1597), whose work "Descartes surely read or knew about" (Gabbey 1998, 661):

> But if from the beginning and end of each sign lines are prolonged through the surface of heaven all the way to each pole, they will divide it into twelve equal parts, like a melon. These parts must each have just as many parts of that empty space contiguous to them. And so the whole heaven and the whole world must be surrounded by an outermost space. (Gabbey 1998, 661)

Newton's Falling Apple

The rotundity and gravity of apples are invoked in this slight poem by the Spanish poet, Jorge Teillier (b. 1935), here translated by Mary Crow:

GIFT

A friend from the South
has sent me an apple
too beautiful
to eat right away.
I hold it in my hands:
It is heavy and round
like the Earth.

REGALO

Un amigo del sur
me ha enviado una manzana
demasiado hermosa
para comerla de inmediato.
La tengo en mis manos:
es pesada y redonda
como la Tierra. (Teillier 1990, 60–61)

Teillier's heavy, round apple, "like the earth," must remind us of the second most famous apple in Euro-American cultural history: the apple in Newton's (1642–1727) garden, whose fall—so the legend goes—somehow inspired him to formulate his law of gravitation. Four different individuals claimed to have heard the story from Isaac Newton himself near the end of his life. One account, by William Stukeley (1687–1765), goes like this:

> After dinner, the weather being warm, we went into the garden and drank thea, under the shade of some appletrees, only he and myself. Amidst other discourse, he told me, he was just in the same situation, as when formerly, the notion of gravitation came into his mind. It was occasion'd by the fall of an apple, as he sat in contemplative mood. Why should that apple always descend perpendicularly to the ground, thought he to him self. Why should it not go sideways or upwards, but constantly to the earths centre? Assuredly, the reason is, that the earth draws it. There must be a drawing power in matter: and the sum of the drawing power in the matter of the earth must be in the earths center, not in any side of the earth. Therefore dos this apple fall perpendicularly, or towards the center. If matter thus draws matter, it must be in proportion of its quantity. Therefore the apple draws the earth, as well as the earth draws the apple. That there is a power, like that we here call gravity, which extends its self thro' the universe. (White 1936, 19–20)

Newton scholars today are disposed to believe the anecdote of Newton's falling apple but to doubt that the incident made any substantial contribution to his discovery or demonstration of the law of universal gravitation.

The first publication of the apple incident was in several works by Voltaire (1694–1778), who claimed to have heard the story, while visiting London, from Newton's niece Catherine Barton Conduitt. Here is a modern translation of the anecdote from Voltaire's *Lettres Philosophiques* (published in 1734, a year after the publication of the English version, "Letters Concerning the English Nation"): "Having withdrawn in 1666 into the country near Cambridge, one day as he walked in his garden and noticed fruit falling from a tree [*fruits tomber d'un arbre*] he drifted off into deep meditation on that problem of gravity" (Voltaire 1961, 68; Voltaire 1986, 104). In Voltaire's first published version of the story in *Epick Poetry of the European Nations from Homer Down to Milton* (1727) and also in the version published in his *Elements of the Philosophy of Newton* (1738), the fruit is an apple [*pomme*].

The apple incident is memorialized in the form of a spray of apple blossoms on a British one-pound note and in the form of a bright red apple on an eighteen pence British postage stamp, both issued in 1987 to commemorate the three-hundredth anniversary of the publication of Newton's *Principia* (see figure 2). The actual tree from which the apple is imagined to have fallen has also not been neglected: scions from the supposed tree (which was taken down in 1814) have produced apples in Britain, the U.S., and New Zealand, recognizable as Flower of Kent (a variety known in Britain since the seventeenth century). It is even possible that a descendant of the original tree still survives in Newton's garden at Woolsthorpe Manor (near Grantham, in Lincolnshire); at least, so I am informed by Peter Joyce, the present Custodian of the Manor, now a National Trust.

A notable poetic rendition of Newton's falling apple occurs in the first stanza of the tenth canto of *Don Juan* (1823) by Lord Byron (1788–1824):

FIGURE 2
British Postage Stamp (1987).

When Newton saw an apple fall, he found
    In that slight startle from his contemplation—
'Tis *said* (for I'll not answer above ground
    For any sage's creed or calculation)—
A mode of proving that the Earth turned round
    In a most natural whirl, called "Gravitation";
And this is the sole mortal who could grapple,
Since Adam—with a fall—or with an apple. (McGann 1986,
    vol. 5, 37)

Since Newton's law of gravitation accounted for the orbital motion of the
Earth about the Sun—rather than for the daily rotational motion of the
Earth suggested by Byron's "turn[ing] round/In a most natural whirl"—
one might question the poet's physics. Byron's very next stanza, however,
contains an unmistakable reference to the Earth's orbit, in the course of
reassuring us that man's Biblical fall is redeemed by his evolving knowledge (and Byron even hints that applied science—in the form of the steam-
engine—may soon get man into the heavens):

Man fell with apples, and with apples rose,
    If this be true; for we must deem the mode
In which Sir Isaac Newton could disclose
    Through the then unpaved stars the turnpike road,
A thing to counterbalance human woes;
    For ever since immortal man hath glowed
With all kinds of mechanics, and full soon
Steam-engines will conduct him to the Moon. (McGann 1986,
    vol. 5, 37)

At the time of composition of *Don Juan,* by the way, turnpikes had not yet
begun to be replaced by railways but Byron was evidently alert to the
potentialities of early nineteenth century technological advance.
    The Russian poet Vladimir Soloukhin (b. 1924) accepts the legend of
Newton's apple but, unlike Byron, Soloukhin professes to favor the apple's
horticultural and alimentary roles as compared with its intellectual role
(quoting Daniel Weissbort's translation):

THE APPLE

I am convinced that finally
Isaac Newton ate
The apple that taught him
The law of gravity.

The apple, born of Earth and Sun,
Came into being,
Sprang from the seed,
Ripened
(And before this bees flew to it,
Rain fell and a warm wind blew),
Not so much that it might drop
And by its direct motion demonstrate
That gravity exists,
But to become
   heavy and sweet,
Beautiful, juicy,
To be admired and picked,
Its scent enjoyed—
And with its sweetness
To delight a Man. (Glad 1992, 204–5)

Soloukhin subscribes to the ideology of Pamyat'—a Russian political party with a seemingly contradictory mixture of slavophilic, culturally preservationist, and socially reactionary impulses—though whether (and if so how) his politics influences his view of Newton's apple is unclear to me.

Another contemporary Russian poet, the Jewish emigré Lev Mak (b. 1939), refers to "Newton-apples" in his "Eden" (1979). Mak was trained as an engineer but expelled from Russia in 1974 after the confiscation of all his writings; he now lives in the U.S. His political point of view is obviously drastically different from Soloukhin's, and includes deep feelings for his lost Eden: "What have they done with your peoples, garden,/My homeland, where have they been hidden?" (Schwartz 1980, 1106)—lines from the second half of the poem, which, in a way, brings the ancient myth of the first half up to date. Thus, the first half of the poem describes the apple tree of Eden:

Do not forget, Temptation has its price,
And the fruit on the apple tree is from the Serpent:
On the graftings of knowledge and retribution,
Newton-apples hang from the Tree of Evil. (Schwartz 1980, 1105)

The American poet John Hollander (b. 1929) contrasts Granny Smiths with American apple varieties, while alluding in passing in his "Granny Smith" (1971) to the fall of Newton's apple: the principal action in the poem is the falling of a Granny Smith apple, "green levin, to her grave/From Newton's skyward tree" (Hollander 1971, 72). Granny Smith apples—and ripe green apples ("greenings") in general—were rare in the U.S. in 1971 (less so in England, where Granny Smiths were introduced from Australia in the 1930s), and the archaic term "levin" (meaning lightning) strikes an appropriately exotic note. It is curious that Hollander refers, like Soloukhin, to "Apples of the earth and sun," which he takes to include all the more ordinary apple varieties: "Bright Americans fallen or/Plucked" (Hollander 1971, 72).

Another American poet, Charles Simic (b. 1938), has written a poem "Dear Isaac Newton" (1982), explicitly about Newton's apple, which is from the start identified with the Edenic apple:

> Your famous apple
> Is still falling.
>
> Your red, ripe,
> Properly notarized
> Old Testament apple. (Brown 1998, 297)

Simic's main conceit is that Newton's apple—identified as "(The famous *malus pumila*)" (that is, crab, or dwarf, apple)—has never stopped falling despite all our attempts "To cause her to stay up there." I cannot say I find the chattiness or affected jauntiness of some of the lines very attractive: "(Is she suffering for us, Isaac,/In some still incomprehensible way?)"; "And wasn't that one of her/Prize worms/We saw crawling off/Into the unthinkable?" One other interrogatory stanza—the penultimate one—incorporates the pun on "fall" which seems to be the point of the poem:

> O she's falling lawfully,
> But isn't she now
> Perhaps even more mysterious
> Than when she first started? (Brown 1998, 298)

Newton and Cider

Cider has long occupied a prominent place in the English potatory imagination, patriotic imbibers having no alternative but to reconcile themselves to the sobering fact that the English environment supports excellent apples but only mediocre wine grapes. No less a figure than Isaac Newton was concerned with the quality of the cider served at Cambridge University (where he had been appointed Lucasian Professor of Mathematics in 1669). In a letter of 2 September 1676 to the Secretary of the Royal Society, Henry Oldenburg, Newton discusses at some length the question of which are the best cider apples for growing in Cambridge. It seems that "Red Streaks (the famous fruit for cyder in other parts) will not succeed in this country. The tree thrives well here, and bears as much fruit, and as good to look as in other countries; but the cyder made of it they find harsh and churlish, and so this fruit begins here to be generally neglected" (Turnbull 1960, vol. 2, 93). (A late-seventeenth-century text, cited by Stuart Peachey (1995, vol. 1, 16), characterizes "Red Strake" apples as follows: "greenish, striped all over with red; this is a good Sider Apple.") Newton then continues: "The ill success of Red Streaks here, I perceive, is generally imputed to the soil; but since the tree thrives, and bears as well here as in other parts, I am apt to think it is in the manner of making the cyder" (Turnbull 1960, vol. 2, 93). In particular, Newton suggests that the mixing of Red Streaks with other apple varieties *in the proper proportions* may be what is required for making the best cider:

> What sort of fruit are best to be used, and in what proportion they are to be mixed, and what degree of ripeness they ought to have? Whether it be material to press them as soon as gathered, or to pare them? Whether there be any circumstances to be observed in pressing them? or what is the best way to do it? If you can direct us to, or procure for us a short narrative of the way of making and ordering cyder in the cyder countries, which takes in a resolution of these, or the most material of these queries, you will oblige your humble servant. (Turnbull 1960, vol. 2, 94)

Newton's concern for mathematical exactitude was, of course, a dominating feature of his approach to nature. It may be worth mentioning that in the months just before and just after writing the above letter, Newton was engaged in research on mathematics, on experiments regarding a mathematical theory of spectral colors, and on a quantitative approach to alchemy. Indeed, in a letter of 14 November 1676 to Oldenburg, Newton's opening paragraph is, once again, about cider, after which he turns to mathematics. As to the cider, Newton is eager to obtain some grafts from the apple trees of "Mr Austin ye Oxonian planter" (Turnbull 1960, vol. 2, 181). This was Ralph Austen (d. 1676), a famous Oxford horticulturist, who, unfortunately, had just died, without an heir to take over his horticulture business; the garden historian Mavis Batey tells us this about him:

> Profits and Pleasures was stamped on the title page of Ralph Austen's *The Spiritual Use of an Orchard or Garden of Fruit Trees*, published in Oxford in 1653. The first part of the book gave clear instructions for planting based on experimental methods and the second part of the book was aimed at promoting Puritanism. Austen had his own nursery in Oxford and academics started their own orchards. So great was his trade that he estimated that he could sell 20,000 plants a year from his seedlings and grafts. (Batey 1982, 37)

REFERENCES

Batey, Mavis. 1982. *Oxford Gardens: The University's Influence on Garden History*. Amersham, England: Avebury.

Brown, Kurt, ed. 1998. *Verse and Universe*. Minneapolis, Minn.: Milkweed Editions.

Browne, Alice. 1977. "Descartes's Dreams." *Journal of the Warburg and Courtauld Institutes*, 40: 256–73.

Cole, John R. 1992. *The Olympian Dreams and Youthful Rebellion of René Descartes*. Urbana and Chicago: University of Illinois Press.

Cotgrave, Randle. 1611. *A Dictionarie of the French and English Tongues*. London: Adam Islip; Columbia, S.C.: University of South Carolina Press (1950).

Cummins, Walter, ed. 1993. *Shifting Borders: East European Poetries of the Eighties*. London and Toronto: Associated University Presses.

Davies, Percival Vaughan, trans. 1969. *Macrobius, the Saturnalia*. New York: Columbia University Press.

Gabbey, Alan and Robert E. Hall. 1998. "The Melon and the Dictionary: Reflections on Descartes's Dreams." *Journal of the History of Ideas* 59: 651–68.

Glad, John, and Daniel Weissbort, eds. 1992. *Twentieth-Century Russian Poetry*. Iowa City: University of Iowa Press.

Hicks, R. D., trans. 1925. *Diogenes Laertius, Lives of Eminent Philosophers*. 2 vols. New York: G. P. Putnam.

Hollander, John. 1971. *The Night Mirror*. New York: Atheneum.

McGann, Jerome J., ed. 1986. *Lord Byron, the Complete Poetical Works*. 6 vols. Oxford: Clarendon Press.

Peachey, Stuart. 1995. *Fruit Variety Register, 1580–1660*. 2 vols. Bristol: Stuart Press.

Schwartz, Howard, and Anthony Rudolf, eds. 1980. *Voices within the Ark: The Modern Jewish Poets*. New York: Avon.

Teillier, Jorge. 1990. *From the Country of Nevermore*. Translated by Mary Crow. Hanover and London: University Press of New England.

Turnbull, H. W. et al., eds. 1960. *The Correspondence of Isaac Newton*. 7 vols. (1960–1977). Cambridge: Cambridge University Press.

Voltaire. 1986. *Lettres philosophiques*. Edited by Frédéric Deloffre. Paris: Gallimard.

———. 1961. *Philosophical Letters*. Translated by Ernest Dilworth. Indianapolis: Bobbs-Merrill.

White, A. Hastings, ed. (1936). *Memoirs of Sir Isaac Newton's Life*. London: Taylor and Francis.

ACKNOWLEDGMENTS

Figure 1 is reproduced with the permission of Trinity College Library (Hartford, Connecticut) from a book in its collection: Richard Pigot, trans. and ed., *Moral Emblems with Aphorisms, Adages, and Proverbs, of All Ages and Nations, from Jacob Cats and Robert Farlie* (New York: D. Appleton, 1860).

For permission to quote, grateful acknowledgment is made to the publishers listed below.

The University of Illinois Press: for lines from "Dream I and Dream II" from *The Olympian Dreams and Youthful Revellion of René Descartes* by John R. Cole. Copyright 1993 by the Board of Trustees of the University of Illinois Press.

The University of Iowa Press: for "The Apple" by Vladimir Soloukhin, trans. Daniel Weissbort, from *Twentieth-Century Russian Poetry*, ed. John Glad and Daniel Weissbort. Copyright 1992 by the University of Iowa Press.

The University Press of New England: for "Gift" from *From the Country of Nevermore: Selected Poems of Jorge Teillier*, trans. Mary Crow. Copyright 1990 by Mary Crow and Wesleyan University Press.

PART II

Nineteenth- and Twentieth-Century Science

[6]

Maxwell and "the Method of Physical Analogy": Model-based Reasoning, Generic Abstraction, and Conceptual Change

NANCY J. NERSESSIAN

1. Introduction

As my teacher at Case Western Reserve University, Howard Stein gave me a piece of advice—one of many for which I will always be grateful—that I recall went something like this: if you want to understand the nature of science, read the scientists not just what philosophers have to say about science. To a young student who had just switched from physics to philosophy of science, this was a revelation. One could actually read the words of Newton, Maxwell, and Einstein and not just science textbooks! His advice was all the more significant because I found what Carnap, Nagel, Hempel, and the others we were assigned to read in classes had to say about science did not address the problems that had led me from physics to philosophy. Reading the scientists was the start of developing my own analysis of these problems, which is a task that still occupies me all these many years later. I started with Einstein, since we were working on the general theory of relativity, but interest in the origins of the field concept led me back to Faraday and Maxwell. In this paper I want to return to Maxwell for several reasons, not the least of which is Howard's often expressed admiration for his acuity and insight into scientific method and his encouragement to explore the nature of Maxwell's "method of physical analogy" and its role in his discovery of the electromagnetic field equations.

When I first read Maxwell I found it surprising how many commentators on his work failed to take seriously what seemed to me to be the generative role of the analogy developed in the 1861–62 paper (Maxwell 1861–62).[1] Maxwell's own comments on analogy as a method of discovery—in letters, publications, and lectures—were largely dismissed with his analogies characterized as at best "merely suggestive" (Heimann 1970), offering "slight" value as a heuristic guide (Chalmers 1973, 137),[2] and at worst as dishonest post-hoc fabrications (Duhem 1914, 98). In this last case, Duhem claimed that Maxwell had cooked up the analogy after the fact and even falsified an equation (see also Duhem 1902) while "the results he obtained were known to him by other means" (1914, 98).

The "known by other means" claim is one I frequently encounter from philosophers in response to presentations of my interpretation of Maxwell. When pressed as to what other means, the response is usually that the equations were derived "by induction from the experiments," with the nature of the inductive process left mysterious. Certainly experimental results played a key role in Maxwell's analysis, but the process of deriving the equations was not what philosophers usually call "induction." Another frequent response is the "symmetry argument," presented in figure 1. For example, Steiner states "Once the phenomenological laws of Faraday, Coulomb, and Ampère had been given differential form, Maxwell noted that they contradicted the conservation of electrical charge. . . . Yet, by tinkering with Ampère's law adding to it the 'displacement current,' Maxwell succeeded in getting the laws actually to imply charge conservation" (Steiner 1989, 458). The "tinkering" is usually interpreted as Maxwell's having noticed that equation for the magnetic field did not include a contribution from the electric field. This account is often given by physicists as well (see, e.g., Jackson 1962, 177).[3] But what is left out of the symmetry account is the central problem of how Maxwell gave the phenomenological laws differential form, which leads to a study of "the method of physical analogy" and ultimately to a quite different interpretation of the "tinkering" process. Another common move made by philosophers is to claim that since, in fact, the aether can be eliminated from Maxwell's laws,

Coulomb Law: div $\mathbf{D}$ = $4\pi\rho$

Ampère Law: **curl $\mathbf{H}$** = $4\pi\mathbf{J}$

Faraday Law: **curl $\mathbf{E}$** = $-\dfrac{\partial \mathbf{B}}{\partial t}$

Absence of free magnetic poles: div $\mathbf{B}$ = 0

Conservation of charge requires

Equation of continuity: div $\mathbf{J}$ + $\dfrac{\partial \rho}{\partial t}$ = 0

Considerations of consistency and symmetry lead to alteration of Faraday Law

curl $\mathbf{E}$ = $-\dfrac{\partial \mathbf{B}}{\partial t}$ + $\dfrac{1}{c^2}\dfrac{\partial \mathbf{E}}{\partial t}$

FIGURE 1
The Symmetry Account

its role in the development of the theory is insignificant, thus reducing the generative role of physical analogy. For example, Kitcher does not discuss Maxwell's analogies but indirectly eliminates them in claiming that the aether was not a "working posit" involved in problem solving, explanation, and prediction, but a "presuppositional posit," which was thought to be required to make the claims of the theory true (Kitcher 1993, 174). It can simply be removed from Maxwell's theory. Yet without the working posit of the existence of the aether as a continuum mechanical medium Maxwell could not have derived the equations at all.[4]

Failing to follow the well-known dictum of Einstein,[5] Maxwell's deeds are also dismissed, for in dismissing the role of analogy, we are left with no evidence in Maxwell's papers, drafts, and correspondence as to what "other means" he employed. I contend that in examining Maxwell's "deeds" we see they exhibit a remarkable harmony between word and deed. And, as with Einstein and Newton, Maxwell is one scientist from whom we can learn a great deal about the nature of scientific practice by listening to his words.

Finally, in coming to grips with how Maxwell employed models in constructing the electromagnetic field representation we gain significant insight into a central problem of creativity: Given that we must start from existing representations, how is it possible that we ever create anything genuinely novel? In this case, the problem is how, starting from Newtonian systems, did Maxwell derive the laws of a non-Newtonian dynamical system? What is needed in order to resolve both the general and the specific problem is an explanatory account of how analogy, and more generally what I call "model-based reasoning," functions to generate new representations—representations that ultimately come to transcend the specific models that were employed in their generation.

2. Cognitive-Historical Analysis

My research into conceptual change in several episodes in physics led to my characterizing specific concept formation practices as forms of "model-based reasoning" (Nersessian 1984a; Nersessian 1984b; Nersessian 1988; Nersessian 1992a; Nersessian 1992b; Nersessian 1995; Nersessian 1999). Here I will present just the main features of the case that model-based reasoning is generative of conceptual change in science, by focusing on three central forms: analogical modeling, visual modeling, and thought experimenting (simulative modeling). I will then provide an interpretation of the role of physical analogy in the development of Maxwell's theory. I present a unified analysis of them because they are often employed together in reasoning episodes. For example, as will be discussed in section 4, the idle

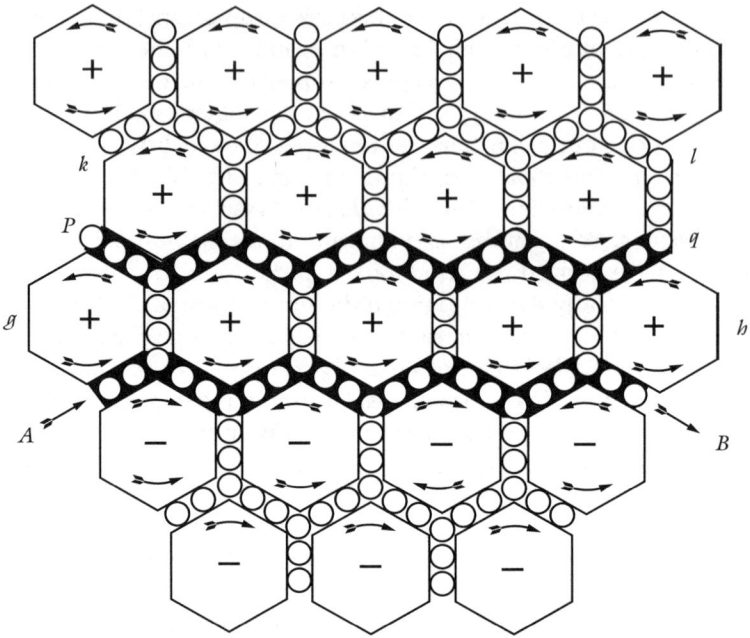

FIGURE 2
Maxwell's drawing of the vortex-idle wheel medium (Maxwell 1890, vol. 1, plate 7)

wheel–vortex model employed by Maxwell in his derivation of the electromagnetic field equations and illustrated by him in figure 2 exemplifies why a unified account is needed. On my interpretation this is a *visual* representation of an *analogical* model that is accompanied with instructions for *simulating* it correctly in thought: "Let the current from left to right commence in AB. The row of vortices *gh* above AB will be set in motion in the opposite direction to a watch. . . . We shall suppose the row of vortices *kl* still at rest, then the layer of particles between these rows will be acted on by the row *gh* on their lower sides and will be at rest above. If they are free to move, they will rotate in the negative direction, and will at the same time move from right to left, or in the opposite direction from the current, and so form an *induced* electric current" (Maxwell 1890b, vol. 1, 477, italics in original).

My analysis draws from practices employed in physics, but investigations of other sciences by philosophers, historians, and cognitive scientists establish that these practices are employed across the sciences (see, e.g., Darden 1980; Darden 1991; Gentner et al. 1997; Giere 1994; Giere

1988; Giere 1992; Gooding 1990; Griesemer 1991a; Griesemer 1991b; Griesemer and Wimsatt 1989; Holmes 1981; Holmes 1985; Latour 1986; Latour 1987; Lynch and Woolgar 1990; Rudwick 1976; Shelley 1996; Thagard 1991; Trumpler 1997; Tweney 1987; Tweney 1992). Further, although these practices are ubiquitous and significant they are, of course, not exhaustive of the practices that generate new conceptual structures.

The account of model-based reasoning developed here stems from an interdisciplinary method that I have called "cognitive-historical" analysis. The objective of that method is to create accounts of the nature and development of science that are informed by studies of historical and contemporary scientific practices and cognitive science investigations of aspects of human cognition pertinent to these practices. When used in analyzing conceptual change, the "historical" dimension of the method is required to uncover the practices scientists employ and to examine these over extended periods of time and as embedded within local communities and wider cultural contexts. The "cognitive" dimension assumes the need to factor into the analysis how human cognitive capacities and limitations could produce and constrain the practices of scientists. Neither the practices nor the cognitive factors can be known *a priori*, empirical research is needed in both cases. Thus cognitive-historical analyses make use of the customary range of historical records for gaining access to practices and draw on and conduct cognitive science investigations into how humans reason, represent, and learn.

In contemporary philosophical parlance, the cognitive-historical method is a "naturalistic" method of analysis. Underlying the method is a "continuum hypothesis": the cognitive practices of scientists are extensions of the kinds of practices humans employ in coping with their physical and social environments and in problem solving of the more ordinary kind. Scientists extend and refine basic cognitive strategies in explicit and critically reflective attempts to devise methods for understanding nature. From this perspective, scientific cognition is shaped by the evolutionary history of the human species, by the developmental processes of the human child, and by the cultural development of human societies. Biological and sociocultural factors co-determine human cognitive development and the various expressions of that development, such as science. The representational and reasoning practices of scientists are analyzed as bearing both the imprint of human cognitive development and the imprint of the sociocultural histories of the communities, internal and external to science, in which it has developed and has come to be practiced. Placing the scientific practices within the broader framework of human cognitive abilities and limitations provides a basis from which to develop an epistemological account that moves beyond the specific case study to more general conclusions about the nature and function of the scientific practices. Such placement aids in establishing that the fragments of scientific research and discovery investigated are more widely representative of scientific practices and thus acts to support drawing more general conclusions from specific aspects of case studies. That scientific cognition is contextual does not preclude developing generalizations from specific cases. As the cognitive anthropologist Edwin Hutchins has argued, "There are powerful regularities to be described at the level of analysis that transcends the details of the specific domain [case]. It is not possible to discover these regularities without understanding the details of the domain [case], but the regularities are not about the domain [case] specific details, they are about the nature of cognition in human activity" (Woods 1997, 15; see also Hutchins 1995).

That there is a continuum, however, does not rule out the possibility that there are salient differences between the scientific and ordinary cognition. Since most of the research in cognitive science has been conducted on ordinary cognition, the cognitive-historical method has to be reflexive in application. Cognitive theories and methods are drawn upon insofar as they help interpret the historical and contemporary practices, while at the same time cognitive theories are evaluated as to the extent to which they can be applied to scientific practices. The assumptions, methods, and results from both sides are subjected to critical evaluation, with corrective insights moving in both directions. One major impact the issues posed by this research into model-based reasoning can have on the field of cognitive science is to push the field toward integration and unification of phenomena largely treated in isolation. Accounting for scientific cognitive practices requires an analysis that integrates research in what are customarily separate research areas in cognitive science, such as analogy, imagery, conceptual change, categorization, problem solving, and decision making.

Although it is not possible to go into the details or give extensive references within the confines of this paper, my account of model-based reasoning derives from extensive historical and cognitive research. The historical research includes my own studies, mainly of but not limited to, nineteenth- and early-twentieth-century field physicists and pertinent research by historians and philosophers of science into other scientific domains. The cognitive research includes the literatures on analogy, mental modeling, mental simulation, mental imagery, imagistic and diagrammatic reasoning, expert/novice problem solving, and conceptual change. Further, my AI collaborators and I have developed and implemented a computational model, the ToRQUE system, that is a model-based reasoner derived from experimental problem-solving protocols that exhibit the kinds of abstraction and constraint satisfaction processes discussed in the next section (Griffith, Nersessian, and Goel 1996; Griffith 1999; Griffith, Nersessian, and Goel 2000).

3. Model-based Reasoning

Within philosophy the identification of reasoning with argument and logic is deeply ingrained. Traditional accounts of scientific reasoning have restricted the notion of reasoning primarily to deductive and inductive arguments. Embracing modeling practices as "methods" of conceptual change in science requires expanding philosophical notions of scientific reasoning to encompass forms of creative reasoning, many of which cannot be reduced to an algorithm in application, are not always productive of solutions, and where good usage can lead to incorrect solutions. Some accounts have proposed abduction as a form of creative reasoning, but the nature of the processes underlying abductive inference and hypothesis generation are left largely unspecified. Examining the modeling practices of scientists as forms of reasoning generative of conceptual change provides a means of specifying the nature of some forms of abductive inference.

The notion of model-based reasoning opens a set of issues about the role of representational format (internal and external) in the reasoning. Different kinds of representations such as linguistic, formulaic, imagistic, and analog/iconic enable different kinds of operations. Operations on linguistic and formulaic representations, for example, include the familiar operations of logic and mathematics. These representations are interpreted as referring to physical objects, structures, processes, or events descriptively. Customarily, the relationship between this kind of representation and what it refers to is "truth" and thus the representation is evaluated as being true or false. Operations on such expressions are rule based and truth preserving if the symbols are interpreted in a consistent manner and the properties they refer to are stable in that environment. Additional operations can be defined in limited domains provided they are consistent with the constraints that hold in that domain. On the other hand, analog models, diagrams, and imagistic representations are interpreted as representing demonstratively. The relationship between this kind of representation, which I will call "iconic," and what it represents is "similarity" or "goodness of fit." Iconic representations are similar in degrees and aspects to what they represent, and are thus evaluated as accurate or inaccurate. Operations on iconic representations involve transformations of the representations that change their properties and relations in ways consistent with the constraints of the domain. Significantly, transformational constraints represented in iconic representations can be implicit, for example, a person can do simple reasoning about what happens when a rod is bent without having an explicit rule such as "given the same force a longer rod will bend farther." The form of representation is such as to enable simulations in which the model behaves in accord with constraints that need not be stated explicitly during this process.

In cognitive psychology there is an ongoing controversy about the nature of human reasoning that parallels the issues raised about reasoning in philosophy. This is not surprising since many philosophers who adhere to the traditional view have played a significant role in shaping this debate. On the traditional psychological view, the mental operations underlying reasoning consist of applying a mental logic to proposition-like representations. For some time critics of this view have contended that a purely syntactical account of reasoning cannot account for significant effects of semantic information exhibited in experimental studies of reasoning (see, e.g., Johnson-Laird 1982; Johnson-Laird 1983; Mani and Johnson-Laird 1982; McNamara and Sternberg 1983; Oakhill and Garnham 1996; Perrig and Kintsch 1985; Wason 1960; Wason 1968). Instead, they propose adopting a hypothesis, first proposed by Craik (1943), that in many instances people reason by carrying out thought experiments on internal models. In its contemporary instantiation, the "mental modeling" hypothesis has been investigated for numerous domains, including: reasoning about causality in physical systems (see, e.g., DeKleer and Brown 1983); the role of representations of domain knowledge in reasoning (see, e.g., Gentner and Gentner 1983); logical reasoning (see, e.g., Johnson-Laird 1983); narrative comprehension (see, e.g., Perrig 1985); and induction (see, e.g., Holland et al. 1986). Additionally, there is considerable experimental protocol evidence collected by cognitive psychologists to support mental modeling as a fundamental form of problem solving employed by contemporary scientists (see, e.g., Chi, Feltovich, and Glaser 1981; Clement 1989; Dunbar 1995; Dunbar 1999; Griffith, Nersessian, and Goel 1996). These studies of reasoning processes provide further support for interpreting the modeling practices exhibited in the historical records of conceptual change as indicating that mental modeling has played a role in the past episodes.

Though "the" mental modeling hypothesis is far from unitary and in need of critical examination, my analysis of model-based reasoning has required adopting only a "minimalist" hypothesis: that in certain problem-solving tasks humans reason by constructing an internal iconic model of the situations, events, and processes that in dynamic cases can be manipulated through simulation. In constructing a model, information in various formats, including linguistic, formulaic, and imagistic, where the latter is taken here to include various perceptual modalities, can be used. In mundane cases the reasoning performed via mental modeling is usually successful because the models and manipulative processes embody largely correct constraints governing everyday real-world events. Think, for example, of how people often reason about how to get an awkward piece of furniture through a door. The reasoner usually figures out how to get a large chair through the door by mentally simulating turning over a geometrical structure approximating the configuration of the chair through various rotations. The task is made easier when the physical chair is in front of the reasoner acting to support the structure in imagination. In the case of science where the situations are more removed from human sensory experience and the assumptions more imbued with theory, there is less assurance that a simulative reasoning process, even if carried out correctly, will yield success. Clearly scientists create erroneous models—revision and evaluation are crucial components of model-based reasoning. In the evaluation process, a major criterion is goodness of fit to the constraints of the target phenomena, but success can also include such factors as enabling the generation of a viable mathematical representation that can push the science along while other details of representing the phenomena are still to be worked out, as Newton did with the concept of gravitation and Maxwell did with the concept of electromagnetic field.

To explain how model-based reasoning could be generative of conceptual change in science requires a fundamental revision of the understandings of concepts, conceptual structures, conceptual change, and reasoning customarily employed explicitly in philosophy and at least tacitly in the other science studies fields. It is not possible to provide all the necessary details and arguments in the confines of this paper. Only an outline of my account will be developed here. Hopefully it will be sufficient to enable the reader to understand the interpretation provided of Maxwell's "method of physical analogy." A basic ingredient of the revision is to view the representation of a concept as providing sets of constraints for generating members of classes of models. Concept formation and change is a process of generating new, and modifying existing, constraints. This is accomplished through iteratively constructing models embodying specific constraints until a model of the *same type* with respect to the salient constraints of the phenomena under investigation, the "target" phenomena, is achieved. My hypothesis is that the prevalence of analogies, visual representations, and thought experiments in periods of radical conceptual change indicates that model-based reasoning is a highly effective means of examining, revising, and abstracting constraints of existing representational systems and, in light of constraints provided by the target problem, an effective means of generating new sets of constraints that the new representational structures come to embody. I will now provide brief encapsulations of how.

To engage in analogical modeling one calls on knowledge of the generative principles and constraints for models in a known "source" domain. These constraints and principles can be represented mentally and externally in different informational formats and knowledge structures that act as explicit or tacit assumptions employed in constructing and transforming models during problem solving. Inter- or intra-domain models can be retrieved directly from the source domain and applied with suitable adaptation, but often, and especially in cases of conceptual change, no direct analogy exists and construction of an initial model itself is required. In these cases the source domain(s) provides constraints that are used together with those provided by the target problem to create the initial as well as subsequent models (Nersessian 1992a; Nersessian 1999; Nersessian 2000). Evaluation of the analogical modeling process is in terms of how well the salient constraints of a model fit the salient constraints of a target problem, with key differences playing a significant role in further model generation (Griffith 1999; Griffith, Nersessian, and Goel 1996). There is an extensive cognitive science literature on analogy with much empirical evidence that substantiates the claim that it is generative in instances of conceptual change. This literature provides theories of the processes of retrieval, mapping, transfer, elaboration, and learning employed in analogy and the syntactic, semantic, and pragmatic constraints operating on these processes (see, e.g., Gentner 1983; Gentner 1989; Gentner et al. 1997; Gick and Holyoak 1980; Gick and Holyoak 1983; Holyoak and Thagard 1989; Holyoak and Thagard 1996; Thagard et al. 1990). Although no current cognitive theory is able to handle the complexity of the Maxwell case, the literature does agree with my analysis in that analogies are not "merely" guides to reasoning but form the creative heart of the reasoning processes in which they are employed. There is also widespread agreement on criteria for good analogical reasoning, drawn from psychological studies of productive and nonproductive use of analogy and formulated by Gentner (Gentner 1983; Gentner 1989): 1. "structural focus": preserves relational systems, 2. "structural consistency": isomorphic mapping of objects and relations, and 3. "systematicity": maps systems of interconnected relationships, especially causal and mathematical relationships.

Constraints in both the target and source domains are domain-specific and need to be understood in the reasoning process at a sufficient level of abstraction for retrieval, transfer, and integration to occur. I call this level of abstraction "generic." That is, the various representations employed have to function with some of their features considered as unspecified. In model-based reasoning processes, a central objective is to create a model that is of the *same kind* with respect to salient dimensions of the target phenomena one is trying to represent. Thus, although an instance of a model is specific, inferences made with it in a reasoning process are generic. In viewing a model generically, one takes it as representing features, such as structure and behaviors, common to members of a class of phenomena. The relation between the generic model and the specific instantiation is similar to the type-token distinction used in logic. Generality in representation is achieved by interpreting the components of the representation as referring to object, property, relation, or behavior types rather than tokens of these. One cannot draw or imagine a "triangle in general" but only some specific instance of a triangle. However, in considering what it has in common with all triangles, humans have the ability to view the specific triangle as lacking specificity in its angles and sides. In considering the behavior of a physical system such as a spring, again one often draws or imagines a specific representation. However, to consider what it has in common with all springs, one needs to reason as though it lacked specificity in length and width and number of coils; to consider what it has in common with all simple harmonic oscillators, one needs to reason as though it lacked specificity in structure and aspects of behavior. That is, the reasoning context demands that the interpretation of the specific spring be generic.

The kind of creative reasoning employed in conceptual innovation involves not only applying generic abstractions but creating and transforming them during the reasoning process. The process of abstracting to the generic level is a significant reasoning process in analogical modeling in conceptual change which often requires recognition of potential similarities across disparate domains, and abstraction and integration of information from these. There are many significant examples of generic abstraction in conceptual change in science. In the domain of classical mechanics, for example, Newton can be interpreted as employing generic abstraction in reasoning about the commonalities among the motions of planets and of projectiles, which enabled his formulating a unified mathematical representation of their motions. The models he employed, understood generically, represent what is common among the members of specific classes of physical systems viewed with respect to a problem context. Newton's inverse-square law of gravitation abstracts what a projectile and a planet have in common in the context of determining motion, for example, that within the context of determining motion, planets and projectiles can both be represented as point masses. After Newton, the inverse-square-law model of gravitational force served as a generic model of action-at-at-distance forces for those who tried to bring all forces into the scope of Newtonian mechanics.

A variety of perceptual resources can be employed in modeling. Here I focus on the use of the visual modality since it figures prominently in cases of conceptual change across the sciences. A possible reason why is that employing the visual modality might enable the reasoner to bypass specific constraints inherent in current linguistic and formulaic representations of conceptual structures. There is a vast cognitive science literature on mental imagery that provides evidence that humans can perform simulative imaginative combinations and transformations that mimic perceptual spatial transformation (Kosslyn 1980; Shepard and Cooper 1982). These simulations are hypothesized to take place using internalized constraints assimilated during perception. Other research indicates that people use various kinds of knowledge of physical situations in imaginary simulations. For example, when objects are imagined as separated by a wall, the spatial transformations exhibit latency time consistent with having simulated moving around the wall rather than through it. There are significant differences between spatial transformations and transformations requiring causal and other knowledge contained in scientific theories. Although the research on imagery in problem solving is scant, recently cognitive scientists have undertaken several investigations examining the role of causal knowledge in mental simulation involving imagery, for example, experiments with problems employing gear rotation provide evidence of knowledge of causal constraints being utilized in imaginative reasoning (Hegarty 1992; Hegarty and Just 1994; Hegarty and Sims 1994; Schwartz and Black 1996).

In model-based reasoning, that the internal representations are iconic does not mean that they need to be picture like in format at all, but can be highly schematic. Thus this modality could be operative even in the reasoning of scientists, such as Bohr, who claim not to experience imagery, in other words pictures, in reasoning. The conflation of mental imagery with pictures-in-the-head stems from the fact that we presently lack an adequate means for expressing the notion of a representational format that is neither picturelike nor linguistic. External visual representations provide support for the processes of constructing and reasoning with a mental model. They aid significantly in organizing cognitive activity during reasoning, such as fixing attention on the salient aspects of a model enabling retrieval and storage of salient information, and exhibiting salient interconnections, such as structural and causal, in appropriate co-location. Further they facilitate construction of shared mental models within a community and transportation of scientific models out of the local milieu of their construction.

As used in model-based reasoning in physics, visual representations participate in modeling phenomena in several ways, including providing abstracted and idealized representations of aspects of phenomena and embodying aspects of theoretical models. For example, early in Faraday's construction of an electromagnetic field concept, the imagistic model he constructed of the lines of force provided an idealized representation of the patterns of iron filings surrounding a magnet (figure 3). But research substantiates that later in his development of the field concept, the imagistic model functioned as the embodiment of a dynamical theoretical model of the transmission and interconversion of forces, generally, through stresses and strains in, and various motions of, the lines (Gooding 1981; Gooding 1990; Nersessian 1984b; Nersessian 1985; Tweney 1985; and Tweney 1992). But, as I have argued, the visual representation Maxwell presented of the idle wheel–vortex model was intended as an embodiment of an imaginary system, displaying a generic dynamical rela-

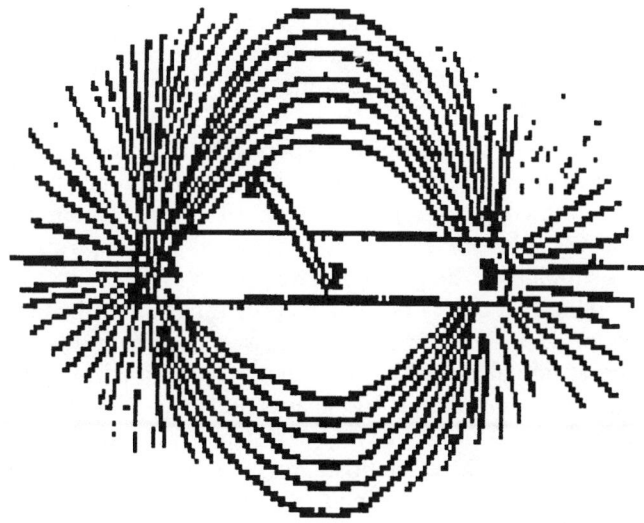

FIGURE 3
Faraday's drawing of the lines of force surrounding a bar magnet (Faraday 1839–55, vol. 1, Plate 1)

tional structure, and not as a representation of the theoretical model of electromagnetic field actions in the aether (figure 2).

As a form of model-based reasoning, thought experimenting can be construed as a specific form of the simulative reasoning that can occur in conjunction with the other kinds of model-based reasoning. Such simulative reasoning would involve constructing a model and using tacit and explicit knowledge to produce new states from it. Constructing a thought-experimental model requires understanding the salient constraints governing the kinds of entities or processes in the model and the possible causal, structural, and functional relations among them. Conducting a simulation requires tacit or explicit understanding of the constraints governing how those kinds of things behave and interact and how the relations can change. A simulation creates new states of a system being modeled, which in turn creates or makes evident new constraints. Changing the conditions of a model enables inferences about differences in the way that a system can behave. Because the simulation complies with the same constraints of the system it represents, performing a simulation with a model enables inferences about real-world phenomena.

In the case of scientific thought experiments implicated in conceptual change, the main historical traces are in the form of narrative reports constructed after the problem solving has taken place. These have often provided a significant means of effecting conceptual change within a scientific community. Accounting for the generative role of this form of model-based reasoning begins with examining how these thought-experimental narratives support modeling processes, and by means of cognitive-histori-cal analysis infers that the original experiment involves a similar form of model-based reasoning. What needs to be determined is: (1) how a narra-tive facilitates the construction of a model of an experimental situation in thought and (2) how one can reach conceptual and empirical conclusions by mentally simulating the experimental processes.

From a mental modeling perspective, the function of the narrative form of presentation of a thought experiment would be to guide the reader in constructing a mental model of the situation described by it and to make inferences through simulating the events and processes depicted in it. A thought-experimental model can be construed as a form of "discourse" model studied by cognitive scientists, for which they argue that the oper-ations and inferences are performed not on propositions but on the con-structed model (see, e.g., Johnson-Laird 1982; Johnson-Laird 1989; Morrow, Bower, and Greenspan 1989; Perrig and Kintsch 1985). Unlike a fictional narrative, however, the context of the scientific thought experi-ment makes the intention clear to the reader that the inferences made per-tain to potential real-world situations. The narrative has already made significant abstractions, which aid in focusing attention on the salient dimensions of the model and in recognizing the situation as prototypical (generic). Thus, the experimental consequences are seen to go beyond the specific situation of the thought experiment. The thought-experimental narrative is presented in a polished form that "works," which should make it an effective means of getting comparable mental models among the members of a community of scientists. Undoubtedly some experimental revision and tweaking goes on in the original reasoning and in the narra-tive construction, although accounts of this process are rarely presented by scientists.

Although some kinds of mental modeling may employ static represen-tations, those derived from thought-experimental narratives are usually dynamic. The narrative delimits the specific transitions that govern what takes place. In constructing and conducting the experiment a scientist makes use of inferencing mechanisms, existing representations, and scien-tific and general world knowledge to make constrained transformations from one possible physical state to the next. Much of the information employed in these transformations is tacit. Thus, expertise and learning play a crucial role in the practice. The thought-experimental reasoning processes link the conceptual and the experiential dimensions of human cognitive processing (see also Gooding 1992). Thus, the constructed sit-uation inherits empirical force by being abstracted both from our experiences and activities in the world and from our knowledge, conceptualizations, and assumptions of it. In this way, the data that derive from thought experimenting have empirical consequences and at the same time pinpoint the locus of the needed conceptual reform. The derived understanding forms the basis of further problem-solving efforts to construct an empirically adequate conceptualization.

In summation, there are several key ingredients common to the various forms of model-based reasoning considered in this section. The models are intended as interpretations of target physical systems, processes, phenomena, or situations. The models are retrieved or constructed on the basis of potentially satisfying salient constraints of the target domain. In the modeling process, various forms of abstraction, such as limiting case, idealization, generalization, generic modeling, are utilized, with generic modeling playing a highly significant role in the abstraction and integration of constraints. Evaluation and adaptation take place in light of structural, causal, and/or functional constraint satisfaction and enhanced understanding of the target problem that has been obtained through the modeling process. Simulation can be used to produce new states and enable evaluation of behaviors, constraint satisfaction, and other factors.

4. Maxwell's Use of "the Method of Physical Analogy": A Cognitive-Historical Interpretation

4.1 The "Method of Physical Analogy"

Maxwell's writings are peppered with talk of "mental operations" in physical and mathematical reasoning. One particularly nice expression of such concerns appears in an article he wrote for *Nature* about the new mathematical formalism, the method of quaternions, that he employed in the *Treatise:* "It does not . . . encourage the hope that mathematicians may give their minds a holiday, by transferring all their work to their pens. It calls upon us at every step to form a mental image of the geometrical features represented by the symbols, so that in studying geometry by this method, we have our minds engaged with geometrical ideas, and are not permitted to fancy ourselves geometers when we are only arithmeticians" (Maxwell 1873a, 137). Although I would not be so foolish as to place Maxwell in the ranks of early cognitive scientists, I nevertheless believe my cognitive-historical analysis of his reasoning to be in the spirit of a man who would make such assertions.

Although ignored by many philosophers and historians, Maxwell's own comments on his method of analysis are most insightful. In investigating a new area in science, Maxwell asserted that one begins with a process of "simplification and reduction of the results of previous investigation to a form in which the mind can grasp them" (Maxwell 1855–56, 155). That process requires a "method of investigation, which allows the mind at every step to lay hold of a clear physical conception, without being committed to any theory founded on they physical science from which that conception is borrowed so that it is neither drawn aside from the subject in pursuit of analytical subtleties, nor carries beyond truth by favourite hypotheses" (ibid., 156). A "physical analogy" is "that partial similarity between the laws of one science and those of another which makes each of them illustrate the other" (ibid.). As Howard has pointed out, "'analogy' in Maxwell's sense is an isomorphism, an equivalence of form" (Stein 1976, 35). This can also be said of the many analogies Thomson constructed, such as between heat and electrostatics, and Maxwell wrote to him about the method that he "intended to borrow it for a season . . . but applying it in a somewhat different way (Larmor 1937, 17–18).[6]

However, Thomson's method was to take an existing mathematical representation of a known physical system, in this case Fourier's analysis of heat, and substitute the parameters for the system under investigation into the equations, in this case electrostatic parameters. What makes Maxwell's modeling process different is that the analogical source to be mapped to the domain of electromagnetism was not ready to hand, but had to be constructed. That is, Maxwell did not know the mathematical structures that could be applied, but through the discovery of an equivalence of form between the dynamical structure of certain mechanical relations and certain electromagnetic relations, he was able to construct the requisite structures. This kind of model-based reasoning process has the potential to lead to genuinely new representational structures, in other words, conceptual change. It does not matter whether the mechanical systems employed in the models do or do not exist in nature; all that matters is that they are "mechanically conceivable." That is, that they supply mechanisms belonging to the classes of phenomena with dynamical relational structure common to mechanics and electromagnetism. Throughout his reasoning processes Maxwell abstracted from the specific mechanism to find the mathematical form of that class of mechanism, in other words, of the generic dynamical structure.

In constructing the mathematical representation of the electromagnetic field concept, Maxwell created several models of an imaginary fluid medium drawing from the source domains of continuum mechanics and of machine mechanics. On my analysis, these analogical domains served as sources for constraints used together with those provided by the target problem to create the imaginary analog models that served as the basis of his reasoning. Maxwell also employed several imagistic representations, such as that in figure 2, which we discussed previously, and those in figure 4, which were

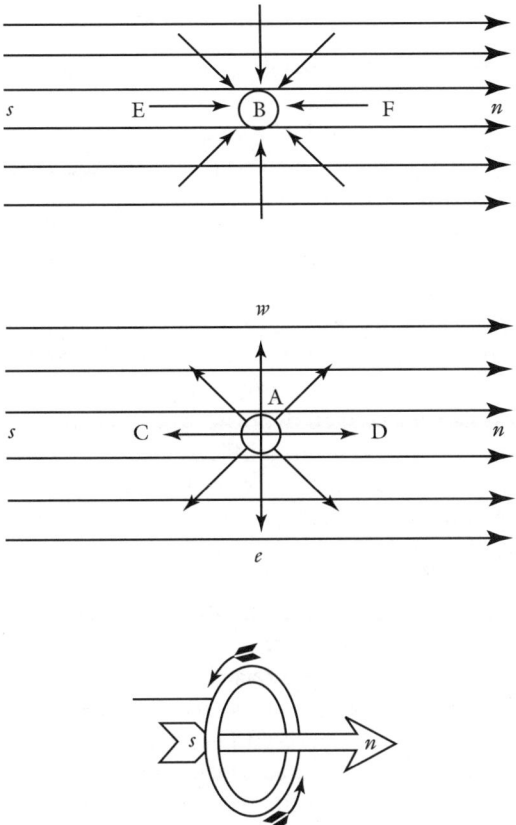

FIGURE 4
Maxwell 1890, vol. 1, p. 460, figs. 1–3

accompanied by text for how to imagine the motion of the vortices in the planes above and below the plane of the paper on which the figures were drawn. The analysis presented below is just of the published work, but my interpretation of Maxwell's reasoning also draws on his letters to Thomson during the period, what little draft material exists in the archives, and other published work. As I have argued elsewhere (Nersessian 1984; Nersessian 1992a) I believe there is sufficient evidence to support my contention that the reasoning in the published papers accurately presents Maxwell's own reasoning processes. We can look at him as attempting to lead his audience through his own reasoning processes as a rhetorical move to help his colleagues to understand the new field representation.

4.2 The Initial Model

Maxwell constructed a mathematical representation for the electromagnetic field concept over the course of several papers. I will focus on "On physical lines of force" (Maxwell 1861–62) and show the relevance of that analysis for "A dynamical theory of the electromagnetic field" (Maxwell 1864). In part 1 of the 1861–62 paper, the mathematical representation of various magnetic phenomena derives from a vortex-fluid model of the aether. Maxwell began with discussing general features of stress in a medium. Stress results from action and reaction of the contiguous parts of a medium and "consists in general of pressures or tensions different in different directions at the same point in the medium" (1861–62, 454). The force is a pressure along the axis of greatest pressure and a tension along the axis of least pressure. Maxwell hypothesized that the causes of electromagnetic phenomena are stresses in a mechanical continuum, the electromagnetic aether, which transmits electromagnetic actions continuously through the space surrounding bodies and charges. Given this hypothesis, one can assume a "resemblance in form" between dynamical relations that hold in the domains of continuum mechanics, such as fluids and elastic solids, and those that hold in the domain of electromagnetism. Thus, continuum mechanics can serve as a source domain for constructing models. The problem is to determine which relations, and determination of these was guided by constraints drawn from the domain of electromagnetism. As specific constraints on stresses that could account for magnetism, Maxwell considered (1) a tension in the direction of the lines of force and (2) a pressure greater in the equatorial than in the axial direction. These constraints are consistent with the geometrical configurations of the magnetic lines of force and Faraday's interpretation of them as resulting from lateral repulsion and longitudinal attraction. The magnetic constraints specify a configuration of forces in the medium and this configuration, in turn, is explained as resulting from the centrifugal forces of vortices in the medium with axes parallel to the lines of force. The centrifugal force of a vortex would cause it to expand equatorially and contract longitudinally. Each vortex is dipolar in that it rotates in opposite directions at its extremities. Further, the geometric constraints on the lines of force are satisfied by the shape of the vortex, which is wider the farther it is from its origin, so the lines become farther apart as they approach their midpoints. Thus, the vortex motion supplies a causal process that is capable of producing the configuration of the lines of force and the stresses in and among them. Figure 5 is my representation of such a vortex drawn from Maxwell's description. The vortex-fluid model is also consistent with constraints derived from known experimental results: (1) electric and magnetic actions are at right angles to each other; (2) magnetism is dipolar; and (3) the plane of polar-

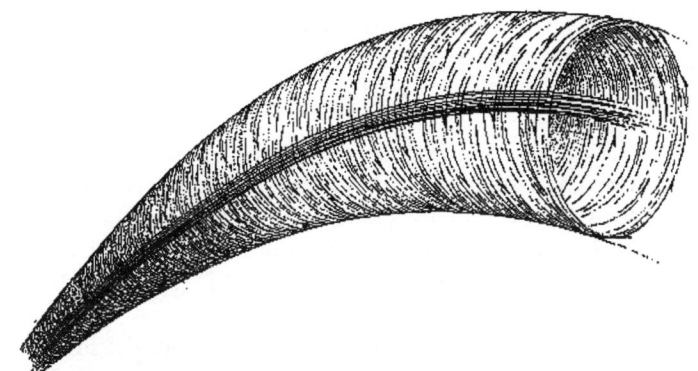

FIGURE 5
Vortex segment

ized light passed through a diamagnetic substance is rotated by magnetic action.

The system of infinitesimal vortices does not correspond to any known physical system. Maxwell constructed it to serve as the basis for deriving mappings between known dynamical relations in continuum mechanics and those thought to produce electromagnetic phenomena. The mathematical expressions for the magnetic phenomena are derived from the mathematical formula for the stresses in the vortex-fluid model by substitution. The vortex-fluid model is "generic" in that it is to be understood as satisfying constraints that apply to the *types* of entities and processes that can be considered as constituting either domain. The model represents the class of phenomena in each domain that are capable of producing the specific configurations of stresses. The modeling process Maxwell used throughout the analysis went as follows. First he constructed a model representing a specific mechanism. Then he treated the dynamical properties and relations generically by abstracting features common to the mechanical and the electromagnetic classes of phenomena. He proceeded to formulate the mathematical equations of the generic model and substituted in the electromagnetic variables.

In part 1 of the analysis Maxwell used the mathematical properties of the limiting case of a single vortex to derive formulas for quantitative relations consistent with the known constraints on magnetic systems. I will only give the highlights of that analysis. From the vortex-fluid model he derived an expression for the resultant force on an element of the medium due to variation in internal stress: $\mathbf{F} = [\mathbf{v}((1/4\pi)(\text{div } \mu\mathbf{v}))] + [(1/8\pi) (\mu \textbf{grad } v^2)] + [\mu\mathbf{v} \textbf{ x } (1/4\pi)(\textbf{curl } \mathbf{v})] - \textbf{grad } p_1$ (equation 5, p. 458, where '$\mathbf{x}$' denotes the cross product),[7] which we now call the general mechanical stress tensor). He then constructed the electromagnetic version by mapping quantitative properties as follows. He stipulated that the quantities related to the velocity of the vortex (α = vl, β = vm, and γ = vn, with 1, m, n the direction cosines of the axes of the vortices) be mapped to the components of the force acting on a unit magnetic bar pointing north. So, the magnetic intensity, which in contemporary notation is designated as 'H', is here related to the velocity gradient of the vortex at the surface. The quantity 'μ' is taken to represent the magnetic permeability, thus relating it to the mass of the medium. The quantity 'μH' represents the magnetic induction.

Substituting the magnetic quantities, Maxwell rewrote the first term of the mechanical stress tensor for the magnetic system as $H(1/4\pi \ (\text{div} \ \mu H))$ (equation 7, p. 459). He followed the same procedure of constructing a mapping between the model and the magnetic quantities and relations to rewrite all of the components of the stress tensor for magnetism. The resulting *electromagnetic* stress tensor represents the resultant force on an element of the magnetic medium due to its internal stress. The four components of the mechanical stress tensor, as interpreted for the electromagnetic medium, are $\mathbf{F} = [\mathbf{H}(1/4\pi \ (\text{div} \ \mu H))] + [(1/8\pi)(\mu \ \mathbf{grad} \ H^2)] + [\mu \mathbf{H}$ x $(1/4\pi)(\mathbf{curl} \ \mathbf{H})] - \mathbf{grad} \ p_1$ (equations 12–14, p. 463). By component they are (1) the force acting on magnetic poles, (2) the action of magnetic induction, (3) the force of magnetic action on currents, and (4) the effect of simple pressure. The last component is required by the model—it is the pressure along the axis of a vortex—but had not yet been given an electromagnetic interpretation.

I will not go through the details of additional constructions and mappings, except to point out that he also derived an expression relating current density to the circulation of the magnetic field around the current-carrying wire $\mathbf{j} = 1/4\pi \ (\mathbf{curl} \ \mathbf{H})$ (equation 9, p. 462). This equation agreed with the differential form of Ampère's law he had derived from kinematic considerations in the first paper of this series of three (1855–56, p. 194). The derivation given here still did not provide a mechanism connecting current and magnetism.

4.3 Introducing Idle Wheels

Thus far, then, Maxwell had been able to provide a mathematical formulation for magnetic induction, paramagnetism, and diamagnetism through modeling these phenomena by means of a nonexistent, but mechanically conceivable dynamical system. A mechanical inconsistency in the vortex-fluid model led Maxwell to a means of representing the causal relationships between magnetism and electricity. He began part 2 by stating that his purpose was to inquire into the connection between the magnetic vortices

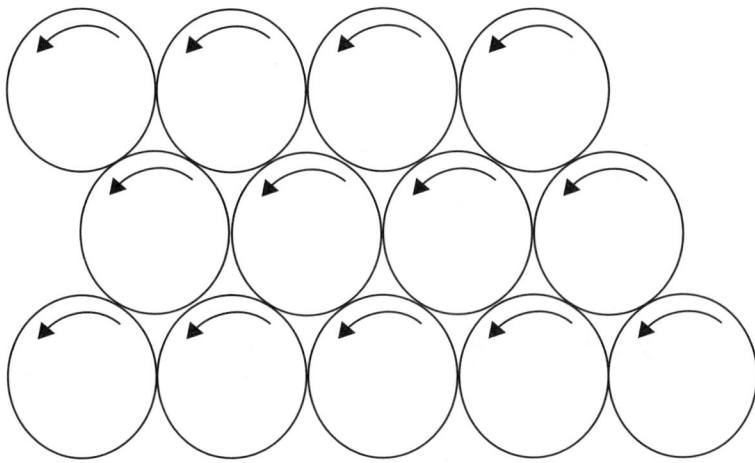

FIGURE 6
Cross section of model of vortex-fluid medium

and current. Thus he could no longer simply consider a single generic vortex in his analysis. He admitted a serious problem with the model in that he "found great difficulty in conceiving of the existence of vortices in a medium, side by side, revolving in the same direction" (468). Figure 6 is my drawing of a cross section of the vortex-fluid model as described by Maxwell. By imagining the motion of the vortices in this figure, it becomes evident that direct contact between consecutive vortices poses a problem in that there will be friction and, thus, jamming. Further, since they are all going in the same direction and, thus, at points of contact they would be going in opposite directions, in the case where they are revolving at the same rate, the whole mechanism should stop. Maxwell noted that in machine mechanics this kind of problem is solved by the introduction of "idle wheels." On that basis he proposed to enhance his imaginary model by supposing that "a layer of particles, acting as idle wheels is interposed between each vortex and the next" (468). In introducing the idle wheels, Maxwell stipulated that the particles would revolve in place in direction opposite to the vortices without slipping or touching. This is consistent with the constraint that the lines of force around a magnetic source can exist for an indefinite period of time, so there can be no loss of energy in the model. He also stipulated that there should be no slipping between the interior and exterior layers of the vortices, making the angular velocity constant. This constraint simplified calculations, but is inconsistent with the mechanical constraint that the vortices have elasticity, and would be eliminated in part 3. Figure 2 is Maxwell's rendering of the idle wheel–vortex model. The diagram shows a cross section of the medium. The vortex cross sections are represented by hexagons rather than circles, presumably to provide a better representation of how the particles are packed around the vortices, with the three-dimensional dodecahedra approximating to spheres in the limit.

   The idle wheel–vortex model is a hybrid constructed from two source domains: fluid dynamics and machine mechanics. As discussed earlier, to combine salient entities and processes from two disparate domains requires abstraction of these to a sufficient level. My explanation of how generic abstraction could have led to the introduction of the idle-wheel particles is illustrated in figure 7. First Maxwell abstracted a generic model of spinning

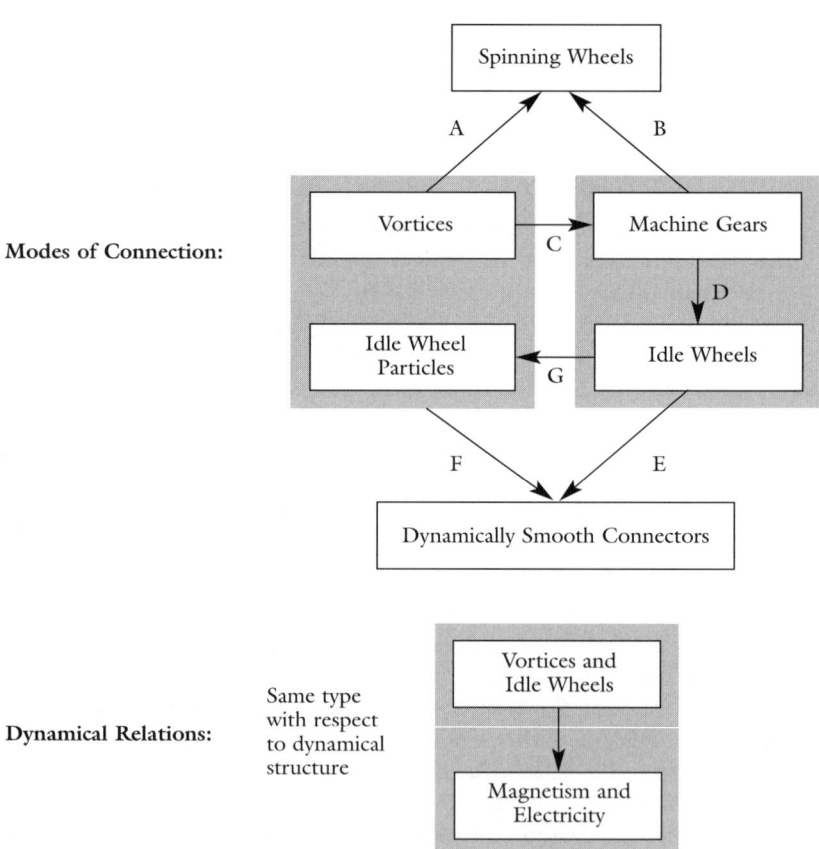

FIGURE 7
Introducing idle wheels via generic modeling

wheels from the vortex-fluid model (A). The generic model of spinning wheels reminded him of specific mechanical systems containing machine gears (B). He noticed an analogy between the vortices and the gears (C), but how this analogy would provide a new mode of connection for the vortices was not immediately evident. Next, from the model of the machine gears he abstracted the generic model of idle wheels (D), and then further abstracted that model into the generic model of dynamically smooth connectors (E). Finally, he instantiated the generic model of dynamically smooth connectors in the vortex-fluid model in the form of idle-wheel particles (F), where the instantiation is guided by both the analogous case of idle wheels (G) and constraints of the continuum mechanical system.

Maxwell, himself, stressed that the idle-wheel mechanism is not to be considered "a mode of connexion existing in nature" (486). Rather it is "a mode of connexion which is mechanically conceivable and easily investigated, and it serves to bring out the actual mechanical connexions between the known electro-magnetic phenomena" (ibid). It does so because the dynamical relations between the idle wheels and vortices are of the same kind as those between electricity and magnetism in the process of induction. That is, although a concrete mechanism is provided, in the reasoning process, the idle wheel–vortex system is taken to represent the class of dynamical systems having certain abstract relations in common. This class includes electric and magnetic interactions on the assumptions of Maxwell's treatment. Thus, in the analysis of electromagnetic induction discussed below, the idle wheel–vortex mechanism is not the *cause* of electromagnetic induction; it represents the *causal structure* of that kind of process. Throughout part 2 Maxwell provided analogies with machinery as specific mechanical interpretations of the relations he had derived between the idle-wheel particles and the fluid vortices to establish for the reader that there are real physical systems that instantiate the generic relations.

In ordinary mechanisms, idle wheels rotate in place. In the model this allows representation of action in a dielectric, or insulating, medium. To represent current, though, the idle wheels need to be capable of translational motion in a conducting medium. Maxwell noted that there are mechanical systems such as the "Siemens governor for steam-engines" which have idle wheels that can translate out of place. The major constraints that need to be satisfied are that (1) a steady current produces magnetic lines of force around it, (2) commencement or cessation of a current produces a current, of opposite orientation, in a nearby conducting wire, and (3) motion of a conductor across the magnetic lines of force induces a current in it. The dynamical relations between the vortices and the idle wheels serve to model the constraints governing the dynamical relations between electric currents and magnetism.

Going through the piece of the analysis in which Maxwell derived the equations for the translational motion of the particles in the imaginary system will help us to understand more fully the relationship between the model and the electromagnetic system. There is a tangential pressure between the surfaces of the spherical particles and the surfaces of the vortices, treated here as approximating rigid pseudospheres. So conceived, these vortices appear to be inconsistent with the geometrical constraints of the vortices in part 1 which should require the vortices to be elastic, but this would not be addressed until the analysis of static electricity in part 3. Maxwell derived the equation for the average flux density of the particles as a function of the circumferential velocity of the vortices $\mathbf{p} = \frac{1}{2} (\rho \ \mathbf{curl} \ \mathbf{v})$ (equation 33, p. 471) and noted that it is of the same form as the equation relating current density and magnetic field intensity $\mathbf{j} = 1/4\pi \ (\mathbf{curl} \ \mathbf{H})$ (equation 9), the form of Ampère's law for closed circuits he had derived in part 1. All that is required to make the equations identical is to set 'ρ', the quantity of particles on a unit of surface, equal to $1/2\pi$. He concluded that "it appears therefore, according to our hypothesis, an electric current is represented by the transference of the moveable particles interposed between the neighboring vortices" (471). That is, the flux density of the particles is taken to represent the electric current density.

We can see how the model provides a mathematical interpretation for constraint (1). Current is represented by translational motion of the particles. In a conductor the particles are free to move but in a dielectric (which the aetherial medium is assumed to be) the particles can only rotate in place. In a nonhomogeneous conducting medium, different vortices would have different velocities and the particles would experience translational motion. They would experience resistance and waste energy by generating heat, as is consistent with current. A continuous flow of particles would thus be needed to maintain the configuration of the magnetic lines of force about a current. The causal relationship between a steady current and magnetic lines of force is captured in the following way. When an electromotive force, such as from a battery, acts on the particles in a conductor it pushes them and starts them rolling. The tangential pressure between them and the vortices sets the neighboring vortices in motion in opposite directions on opposite sides—thus capturing the polarity of magnetism—and this motion is transmitted throughout the medium. The mathematical expression (equation 33) connects current with the rotating torque the vortices exert on the particles. Maxwell went on to show that this equation is consistent with the equations he had derived in part 1 for the distribution and configuration of the magnetic lines of force around a steady current (equations 15–16, p. 464).

Although I won't go through the derivations, Maxwell derived the laws of electromagnetic induction in two parts. Again we see the role of the model in the derivation, since the two cases are different mechanically in the idle wheel–vortex system. In the first case, the mechanism for communicating rotational velocity in the medium accounts for induction of currents by starting or stopping a primary current (constraint [2]), such as switching current off and on in a conducting loop and thereby inducing a current in a nearby conducting loop. A decrease or increase in current will cause a corresponding change in the velocity of the adjacent vortices. The difference in velocity between this row and the next adjacent row will cause the particles surrounding those vortices to speed up or slow down, and this motion will in turn be communicated to the next row and so on until the conducting wire is reached. The particles in the wire will be set in translational motion by the differential electromotive force between the vortices, thus inducing a current oriented in direction opposite to the initial current, which agrees with experimental results. In the second case, in which a current is induced by motion of a conductor across the lines of force (constraint [3]), Maxwell used considerations pertaining to the changing form and position of the medium. Briefly, the portion of the medium in the direction of the conducting wire would become compressed, causing the vortices to elongate and speed up, while vortices behind the wire contract back into place and decrease in velocity. The net force pushes the particles inside of the conductor, producing a current provided there is a circuit connecting the ends of the wire. The case of open circuits is considered in part 3.

4.4 The "Displacement Current" and Inconsistent Signs

By the end of part 2, Maxwell had given mathematical expression to some electromagnetic phenomena in terms of actions in a mechanical medium and had shown the representation coherent and consistent with known phenomena. The full mathematical representation of the electromagnetic field was constructed in part 3 with Maxwell's treatment of static electricity. I will focus on one piece of Maxwell's analysis—the introduction of what he called "the displacement current"—since this feature of the model leads to a formal inconsistency in the equations Maxwell presented in his next paper on the subject (1864). Understanding why he tolerated it and how he eliminated it in 1873 in the *Treatise* (Maxwell 1873b) will provide a deeper appreciation of the role of generic modeling in his analysis.

The fact that Maxwell submitted the part of the analysis pertaining to electrostatics in the 1861–62 paper eight months after the work on magnetism and electromagnetic induction was published indicates that the initial representation of static electricity was difficult for him to work out on the basis of the idle wheel–vortex model. It would seem quite natural to identify charge with the accumulation of idle-wheel particles at the boundary of dielectric and conducting media. Thus, it is puzzling why Maxwell did not immediately proceed to the type of analysis he ultimately presented and he provides no clues of his path to that solution. Siegel (1991) presents a detailed and plausible analysis of the nature of the problems Maxwell would have encountered in trying to construct an account of the interface between conducting and dielectric media given the specific mechanism of the model.[8] The solution to these problems was to make the vortices elastic. So, the idle wheel–vortex model was modified in part 3 by again considering its plausibility as a mechanical system. In part 2, the system contains vortex cells of rotating fluid separated by particles very small in comparison to them. To simplify the calculations for the transmission of rotation from one cell to another via the tangential action between the surface of the vortices and the particles, Maxwell had assumed the vortices to be rigid. But he now noted that in order for the rotation to be transmitted from the exterior to the interior parts of the cells, the cell material needs to be elastic. And, although he does not comment on it, conceiving of the molecular vortices as pseudospherical blobs of elastic material would also give them the right configuration on rotation (figure 5), and thus eliminate the inconsistency of the rigid vortices with the geometrical constraints of part 1 for magnetism.

He began by noting the constraint that "electric tension" associated with a charged body is the same, experimentally, whether produced from static or current electricity. If there is a difference in tension in a body, it will produce either current or static charge, depending on whether the substance is a conductor or insulator. He likened a conductor to a "porous membrane which opposes more or less resistance to the passage of a fluid" (490) and a dielectric to an elastic membrane which does not allow passage of a fluid, but "transmits the pressure of the fluid on one side to that on the other" (491). In the process of electrostatic induction, electricity can be viewed as "displaced" within a molecule of a dielectric, so that one side becomes positive and the other negative, but does not pass from molecule to molecule. Although Maxwell did not immediately link his discussion of the different manifestations of electric tension to the hybrid model of part 2, it is clear that it figures throughout the discussion. This is made explicit in the calculations immediately following the general discussion. I note this because the notion of "displacement current" introduced before these calculations cannot properly be understood without the model. Maxwell claimed that the displacement of electricity in electrostatic induction can be likened to a current in that *change* in displacement is similar to "the commencement of a current" (491). That is, given the model, an electrostatic force produces a slight elastic distortion in the vortices causing a slight translational motion of the idle-wheel particles, which is propagated throughout the dielectric medium.

The mathematical expression relating the electromotive force and the displacement that Maxwell established is: $E = -4\pi k^2 D$, where 'E' is the electromotive force (electric field), 'k' the coefficient for the specific dielectric, and 'D' is the displacement (491). The amount of current due to displacement is $j_{disp} = \partial D/\partial t$. The equation relating the electromotive force and the displacement has the displacement in the direction opposite from that which is customary now and in Maxwell's later work. The orientation given here can be accounted for if we keep in mind that on the model an elastic resorting force is opposite in orientation to the impressed force. Although Maxwell stressed that the relations expressed by the above formula are independent of a specific *theory* about the actual internal mechanisms of a dielectric, they are not independent of the *model*. Without the mechanism of the model, there is no basis on which to call the motion a "current." It is translational motion of the particles which constitutes current. Thus, in its initial derivation, the "displacement current" is modeled on a specific mechanical process. We can see this in the following way.

Recall the difference Maxwell specified between conductors and dielectrics when he first introduced the idle-wheel particles. In a conductor, they are free to move from vortex to vortex. In a dielectric, they can only rotate in place. In electrostatic induction, then, the particles can only be urged forward by the elastic distortion of the vortices, but cannot move out of place. This motion is similar to that of the "commencement of a current." But, their motion "does not amount to a current, because when it has attained a certain value it remains constant" (491). That is, the particles do not actually move out of place by translational motion as in conduction, instead they accumulate, creating regions of stress. Since they are not free to flow, they must react back on the vortices with a force to restore their position. The system reaches a certain level of stress and remains there. 'Charge', then, is interpreted as the excess of tension in the dielectric medium created by the accumulation of displaced particles. Without the model, "displacement current" loses its physical meaning, which is what bothered so many of the readers of the *Treatise*, where the mechanical model is no longer employed. As we will see below, it also created problems for Maxwell in his 1864 analysis.

Since the vortices are now elastic and since in a conductor the particles are free to move, the current produced by the medium (that is, net flow of particle per unit area) must include a factor for their motion due to the elasticity. So Maxwell corrected the equation for Ampère's law (equation 9) to include the total current, $j = 1/4\pi \, (\text{curl } H) - \partial E/\partial t$ (equation 112, p. 496). Since the emf has rotation opposite to the rotation of the vortices, the "displacement current" actually opens the closed current of equation 9, creating a noncircuital current.[9] He coupled this equation with the equation of continuity for charge, which links current and charge, to derive an expression linking charge and the electric field, $e = 1/4\pi$ (k^2 div $\mathbf{E}$) (equation 115, p. 497), which is equivalent to $\rho = -$ div $\mathbf{D}$. This latter expression looks similar to what we now call Coulomb's law except for two features that turn out to be highly salient for understanding Maxwell's reasoning. First, the form of this equation and the modified equation for current (equation 112) again demonstrates Maxwell's field conception of current and charge: interpreted left to right, charge and current arise from the field. Second, the minus sign is not part of the contemporary equation, but arises out of the model because the elastic restoring force exerted on the vortices by the particles and the electromotive force have opposite orientation. Through what can be interpreted simply as a substitution error in equation 104 the equations in this paper are consistent.

Turning to the 1864 paper, we can see how the specific interpretation of the displacement current created problems. In this paper, Maxwell rederived the field equations without explicit reference to the mechanical model. Once he had abstracted the electromagnetic dynamical properties and relations it was possible to derive the electromagnetic equations using generalized dynamics and assuming only that the electromagnetic aether is a "connected system," possessing elasticity and thus having energy. Here Maxwell made the identification of the electromagnetic and light aethers that he failed to do at the end of the 1861–62 analysis, where the close agreement of the velocity of transverse vibrations in these two hypothetical media led him to state "we can scarcely avoid the inference that *light consists in transverse undulations of the same medium which is the cause of electric and magnetic phenomena*" (italics in the original, 500). We can interpret Maxwell's reticence to draw the inference in the earlier analysis as due to the value of the transverse velocity in the electromagnetic medium being derived from specific features of the idle wheel–vortex model. There were no grounds on which to assume vortex motion in the light aether. Note also that Maxwell was not avoiding the inference that light is an electromagnetic phenomenon but only the possible identity of the two media. On the then prevailing view light is a transverse wave in an elastic medium and this is not the same kind of mechanism as that provided by the model for propagating electromagnetic actions. In the 1864 paper Maxwell treated the aether as a generic elastic medium whose constraints could be satisfied by many specific mechanical instantiations (in the *Treatise*, Maxwell says an "infinite" number) and thus saw no reason for multiplying aethers. Elastic systems can receive and store energy in two forms, what Maxwell called the "energy of motion," or kinetic energy, and the "energy of tension," or potential energy. He identified kinetic energy with magnetic polarization and potential energy with electric polarization. Figure 8 illustrates my interpretation of how the 1861–62 analysis enabled him to do this through generic abstraction from the model. Although in 1864

Maxwell is thinking of the aetherial medium in more abstract, general dynamical terms, vestiges of the earlier specific mechanical model can be shown to have remained in his thinking and this created a problem with the current and charge equations.

The infamous inconsistency arises in the 1864 paper because in the absence of the mechanical model there is no basis on which to distinguish conduction current and displacement current. Thus current is treated generically in terms of the stresses in the medium created by the flow of electricity, so $E = kD$ and coupling the equation for the total current $j = 1/4\pi$ (**curl H**) $- \partial D/\partial t$ with the equation of continuity $\partial\rho/\partial t + \text{div } j = 0$, yields ρ - div **D**. However, Maxwell wrote the equivalent of $\rho = -$ div **D** as the "equation of free electricity" (charge)—that is, the equation in the form of the 1861–62 analysis. So the complete set of equations which he gathers together in part 3 of the 1864 paper is formally inconsistent. My interpretation is that Maxwell continued to think of charge as associated

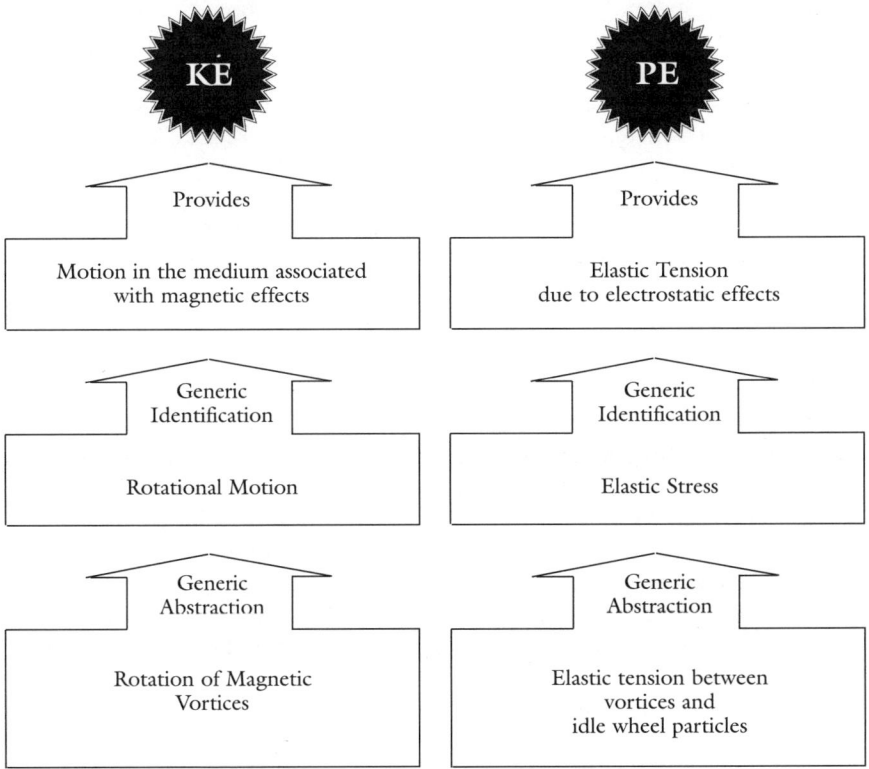

FIGURE 8
Identifying energy components via generic modeling

with the accumulation of the idle-wheel particles through the specific mechanism of "displacement" and thus with the reactive force that is oriented away from the accumulation point. However, the mathematical equations of the generic medium require that it be pointing toward the stress point and so clearly require $\rho = \mathrm{div}\ \mathbf{D}$.

There are few existing drafts of Maxwell's published papers, but fortunately there are a couple of pages of a draft pertaining to this derivation.[10] These reveal that Maxwell had some confusion about how to think of current and charge without the medium. In the draft equation for current, Maxwell wrote the components of "electric resistance," in other words, the electromotive force required to keep the current flowing through a conductor, as pointing in the opposite direction, as would have been the case in the mechanical medium, but in the published paper, these are written correctly. In the draft equation for "statical electricity" the components are written with the negative sign as above. Maxwell's handwriting can be interpreted as indicating that he struggled with this result.[11] In writing the first two components Maxwell made the equals sign three times the length used in other samples of his handwriting and he actually blends the equals sign into the minus sign. Only for the final component is the equals sign of regular length and clearly separated from the minus sign. As noted above, it is also written with the minus sign in the published paper.[12]

Maxwell never discussed this inconsistency and then in the *Treatise*, again without discussion, the inconsistency is gone. Although it is just speculation on my part, given how Maxwell collects all of his equations in part 3 of the published paper, it is hard to imagine that he did not notice the inconsistency. I believe he kept the Coulomb equation $\rho = -\mathrm{div}\ \mathbf{D}$ (equation G, p. 561) in 1864 because he had not figured out how to conceive of charge generically, that is to abstract it from the specific mechanism through which he had been able initially to represent it. In the *Treatise* charge is abstracted from the notion of stress and reactive force due to accumulating particles and treated generically as elastic stress from the flow of electricity through a medium, with orientation in the direction of flow. In conduction the current flow is unimpeded and in electrostatics, stress is created at points of discontinuity, such as where a charging capacitor and a dielectric meet. The generic notion of charge as associated with elastic stress is compatible with Faraday's field notion of charge, but was to cause difficulties in comprehending Maxwell's for those who held the customary action-at-a-distance notion that charge is associated with a particle. As H. A. Lorentz noted in a discussion of the need for a clear separation of field and charge, "Poincaré mentions a physicist who declared that he understood the whole of Maxwell's theory, but that he still had not grasped what an electrified sphere was !" (Lorentz 1891, 95, my translation from Dutch).

5. Further Reflections: Model-based Reasoning and Conceptual Change

My analysis has shown how Maxwell was able to formulate the laws of the electromagnetic field by abstracting from specific mechanical models the dynamical properties and relations that continuum-mechanical systems, certain machine mechanisms, and electromagnetic systems have in common. In their mathematical treatment these common dynamical properties and relations were separated from the specific instantiations provided in the models through which they had been rendered concrete. Thus the underlying inconsistencies in the models could be ignored. The generic mechanical relationships represented by the imaginary systems of the models served as the basis from which he abstracted a mathematical structure of sufficient generality that it represented causal processes in the electromagnetic medium without requiring knowledge of specific causal mechanisms (see also Stein 1970; Stein 1976; Stein 1981). In the final analysis in the *Treatise* even the "theory of molecular vortices" is presented in generic form, with vortex motion in the medium becoming "something belonging to the same mathematical class as an angular velocity, whose axis is in the direction of the magnetic force" (Maxwell 1873b, 459). However, we have also seen that at an intermediate stage of development specific, features of the model seemed to figure so strongly in his thinking that he introduced a formal inconsistency in the set of equations that was only eliminated several years later—again through generic abstraction from a mechanism of the idle wheel–vortex model.

With hindsight, we know, as Maxwell did not, that Maxwellian electrodynamics cannot be given a Newtonian formulation. Thus, the Maxwell case presents an interesting problem for those wanting to understand the nature of the creative reasoning employed by scientists in the processes of representational change. If Maxwell really did derive the mathematical laws of the electromagnetic field in the way he presents the analysis, how is it possible that employing analogies drawn from Newtonian mechanical domains, he constructed the laws of a non-Newtonian system, electrodynamics? My answer is that employing the process of generic abstraction in model-based reasoning enabled Maxwell to abstract a system of dynamical relations of greater generality than Newtonian mechanics. In this case, generic abstraction from mechanical models enabled Maxwell to construct a system of abstract laws that when applied to the class of electromagnetic systems yield the laws of a dynamical system that is nonmechanical, that is, one that is inconsistent with the laws of the mechanical domains from which its mathematical structure was abstracted. For Maxwell, Newtonian mechanics and general (Lagrangian) dynamics were thought to be coextensive. In arriving at the general dynamical form of the field equations starting from considerations of the specific mechanisms of the models, he could think that he had demonstrated the possibility of a "mechanical explanation" for the electromagnetic phenomena. Bu the Lagrangian formalism provides no information about the underlying system. Maxwell did not, however, "turn away from mechanical models" (Siegel 1991, 54). A complete explanation would need to provide an account of the actual mechanisms in the aether that create the phenomena, and Maxwell fully expected such an account to be forthcoming. The laws he had formulated would act as a constraint on any acceptable mechanism. Maxwell saw himself as following the same strategy Newton had in formulating the universal law of gravitation: "investigat[ing] the forces with which bodies act on each other in the first place, before attempting to explain *how* that force is transmitted" [italics in original] (Maxwell 1890a). Just what kind of mechanical system remained a problem for Maxwell and he saw his theory as incomplete without that specification. With hindsight, we can see Maxwell as having abstracted away the notion of mechanism, creating a representation of a nonmechanical, dynamical system.

I do not know if Howard will feel that I have provided a satisfactory assessment of the role Maxwell's analogies played in the development of his electromagnetic theory but I thank him for inspiring me to try.

NOTES

1. See also Larmor 1937; Maxwell 1855–56; Maxwell 1856.

2. See also Chalmers 1986 which argues in response to my1984a and 1984b analyses to the contrary that the analogy was an "unproductive digression."

3. Berkson also criticizes the symmetry argument as presented by Jackson (Berkson 1974, 338–39).

4. A few commentators at that time did take seriously at least parts of the analogy, most significantly Bromberg (1968) whose analysis of the displacement current I found useful; and Berkson (1974), whose work I did not come across until much later. Later, Seigel (1986 and 1991) presented arguments in favor of the centrality of the analogical model in the 1861–62 paper. Hesse's position is more difficult to characterize because she does see analogy as playing a significant role in hypothesis formation and theory interpretation in general. However, her analysis of Maxwell's theory passes over the analysis in the 1861–62 paper—which I am arguing contains the main generative work—entirely (Hesse 1973). Although I cannot elaborate here, I believe the problem is that her theory of analogy does not allow for the kind of abstraction I will be discussing in sections 3 and 4.

5. "If you want to find out anything from the theoretical physicists about the methods they use, I advise you to stick closely to one principle: don't listen to their words, fix your attention on their deeds. To him who is a discoverer in this field, the products of his imagination appear so necessary and natural that he regards

them, and would like to have them regarded by others, not as creations of thought but as given realities" (Einstein 1973, 264).

6. See especially letters of November 13, 1854; May 18, 1855, pp. 8 and 11; and September 13, 1855, pp. 1718.

7. I have written this equation, which Maxwell wrote in component form, in modern vector notation and will do so throughout. The vector calculus was just being developed around the time of Maxwell's analysis. Note the actual physical meaning of the vector operators here: the gradient is a slope along a vortex, the curl is the rotation of the fluid vortex, and the divergence is the flowing of fluid in the medium. Maxwell reformulated his electromagnetic theory in quaternions, a form of vector calculus developed by Hamilton, in the *Treatise* (1873b).

8. I disagree, however, with his claim (Siegel 1991, 75) that Maxwell was prevented initially from making the vortices elastic since "Thomson saw elasticity as resulting from motion in a mechanical substratum." The vortices *are* a form of motion in a mechanical substratum and, as noted above, elasticity is consistent with the analysis in part 1. A more likely reason is that Maxwell had considered and rejected this alternative in the calculations for part 2 and was reluctant to modify this specific feature of the idle wheel–vortex version of the model.

9. See Nersessian 1984, 82 and Siegel 1991, 112.

10. Add MS 7655, V, c/8. University Library, Cambridge University.

11. See Siegel 1991, 174–75 for a similar point.

12. Harman mistakenly says this equation appears without the minus sign in the 1864 paper (Harman [Heimann] 1995, 161, n. 6).

REFERENCES

Berkson, W. 1974. *Fields of Force: The Development of a World View From Faraday to Einstein.* New York: John Wiley and Sons.

Bromberg, J. 1968. Maxwell's Displacement Current and His Theory of Light. *Archive for History of Exact Science* 4:218–34.

Chalmers, A. F. 1973. Maxwell's Methodology and his Application of it to Electromagnetism. *Studies in the History and Philosophy of Science* 4 (2): 107–64.

———. 1986. The Heuristic Role of Maxwell's Mechanical Model of Electromagnetic Phenomena. *Studies in the History and Philosophy of Science* 17:415–27.

Chi, M. T. H., P. J. Feltovich, and R. Glaser. 1981. Categorization and Representation of Physics Problems by Experts and Novices. *Cognitive Science* 5:121–52.

Clement, J. 1989. Learning Via Model Construction and Criticism. In *Handbook of Creativity: Assessment, Theory, and Research,* edited by G. Glover, R. Ronning, and C. Reynolds. N.Y.: Plenum

Craik, K. 1943. *The Nature of Explanation.* Cambridge: Cambridge University Press.

Darden, L. 1980. "Theory Construction in Genetics." In *Scientific Discovery: Case Studies,* edited by T. Nickles. Dordrecht: Reidel.

———. 1991. *Theory Change in Science: Strategies from Mendelian Genetics.* New York: Oxford University.

DeKleer, J., and J. S. Brown. 1983. Assumptions and Ambiguities in Mechanistic Mental Models. In *Mental models,* edited by D. G. a. A. Stevens, N.J.: Lawrence Erlbaum.

Duhem, P. 1902. *Les théories électriques de J. Clerk Maxwell: Etude historique et critique.* Paris: A. Hermann & Cie.

———. 1914. *The Aim and Structure of Physical Theory.* Translated by P. P. Weiner. New York: Atheneum.

Dunbar, K. 1995. "How Scientists Really Reason: Scientific Reasoning in Real-world Laboratories." In *The nature of Insight,* edited by R. J. Sternberg and J. E. Davidson. Cambridge, Mass.: MIT Press.

———. 1999. "How Scientists Build Models in Vivo Science." In *Model-based Reasoning in Scientific Discovery,* edited by L. Magnani, N. J. Nersessian, and P. Thagard. New York: Kluwer Academic/Plenum Publishers.

Einstein, A. 1973. *Ideas and Opinions.* New York: Dell.

Gentner, D. 1983. "Structure-mapping: A Theoretical Framework for Analogy." *Cognitive Science* 7:155–70.

———. 1989. "The Mechanisms of Analogical Learning." In *Similarity and Analogical Reasoning,* edited by S. Vosniadou and A. Ortony. New York: Cambridge University Press.

Gentner, D., S. Brem, R. W. Feruson, A. B. Markman, B. B. Levidow, P. Wolff, and K. D. Forbus. 1997. "Analogical Reasoning and Conceptual Change: A Case Study of Johannes Kepler." *The Journal of the Learning Sciences* 6 (1): 3–40.

Gentner, D. and D. R. Gentner. 1983. "Flowing Waters and Teeming Crowds: Mental Models of Electricity." In *Mental Models,* edited by D. Gentner and A. Stevens. Hillsdale, N.J.: Lawrence Erlbaum.

Gick, M. L., and K. J. Holyoak. 1980. "Analogical Problem Solving." *Cognitive Psychology* 12: 306–355.

———. 1983. Schema Induction and Analogical Transfer." *Cognitive Psychology* 15: 1–38.

Giere, R. 1994. "The Cognitive Structure of Scientific Theories." *Philosophy of Science* 61:276–96.

Giere, R. N. 1988. *Explaining Science: A Cognitive Approach.* Chicago: University of Chicago Press.

———. 1992. "Cognitive Models of Science." In vol. 15 of *Minnesota Studies in the Philosophy of Science.*

Gooding, D. 1981. "Final Steps to the Field Theory: Faraday's Study of Electromagnetic Phenomena, 1845–1850." *Historical Studies in the Physical Sciences* 11:231–75.

———. 1990. *Experiment and the Making of Meaning: Human Agency in Scientific Observation and Experiment.* Dordrecht: Kluwer.

———. 1992. "The Procedural Turn: Or Why Did Faraday's Thought Experiments Work?" In *Cognitive Models of Science,* edited by R. Giere. Minneapolis: University of Minnesota Press.

Griesemer, J. R. 1991a. "Material Models in Biology." *PSA 1990.*

————. 1991b. "Must Scientific Diagrams be Eliminable? The Case of Path Analysis." *Biology and Philosophy* 6:177–202.

Griesemer, J. R., and W. Wimsatt. 1989. "Picturing Weismannism: A Case Study of Conceptual Evolution." In *What the Philosophy of Biology Is, Essays for David Hull,* edited by M. Ruse. Dordrecht: Kluwer.

Griffith, T. W. 1999. "A Computational Theory of Generative Modeling in Scientific Reasoning." Ph.D., College of Computing, Georgia Institute of Technology, Atlanta.

Griffith, T. W., N. J. Nersessian, and A. Goel. 1996. "The Role of Generic Models in Conceptual Change." Vol. 18 of *Proceedings of the Cognitive Science Society,* 312–17. Hillsdale, N.J.: Lawrence Erlbaum.

————. 2000. "Function-follows-form Transformations in Scientific Problem Solving." Vol. 22 of *Proceedings of the Cognitive Science Society,* 196–201. Hillsdale, N.J.: Lawrence Erlbaum.

Harman (Heimann), P. M. 1995. *The Scientific Letters and Papers of James Clerk Maxwell.* 3 vols. Vol. 2. Cambridge: Cambridge University Press.

Hegarty, M. 1992. "Mental Animation: Inferring Motion from Static Diagrams of Mechanical Systems." *Journal of Experimental Psychology: Learning, Memory, and Cognition* 18 (5):1084–1102.

Hegarty, M., and M. A. Just. 1994. "Constructing Mental Models of Machines from Text and Diagrams." *Journal of Memory and Language* 32:717–42.

Hegarty, M., and V. K. Sims. 1994. "Individual Differences in Mental Animation from Text and Diagrams." *Journal of Memory and Language* 32:411–30.

Heimann, P. M. 1970. "Maxwell and the Modes of Consistent Representation." *Archive for the History of Exact Sciences* 6:171–213.

Hesse, M. 1973. "Logic of Discovery in Maxwell's Electromagnetic Theory." In *Foundations of Scientific Method: The Nineteenth Century,* edited by R. N. Giere and R. S. Westfall. Bloomington: University of Wisconsin Press.

Holland, J. H., K. J. Holyoak, R. E. Nisbett, and P. R. Thagard. 1986. *Induction: Processes of Inference, Learning, and Discovery.* Cambridge, Mass.: MIT Press.

Holmes, F. L. 1981. "The Fine Structure of Scientific Creativity." *History of Science* 19:60–70.

————. 1985. *Lavoisier and the Chemistry of Life: An Exploration of Scientific Creativity.* Madison: University of Wisconsin Press.

Holyoak, K., and P. Thagard. 1989. "Analogical Mapping by Constraint Satisfaction: A Computational Theory." *Cognitive Science* 13:295–356.

————. 1996. *Mental Leaps: Analogy in Creative Thought.* Cambridge, Mass.: MIT Press.

Hutchins, E. 1995. *Cognition in the Wild.* Cambridge, Mass.: MIT Press.

Jackson, J. D. 1962. *Classical Electrodynamics.* New York: John Wiley.

Johnson-Laird, P. N. 1982. "The Mental Representation of the Meaning of Words." *Cognition* 25:189–211.

————. 1983. *Mental Models.* Cambridge, Mass.: MIT Press.

————. 1989. "Mental Models." In *Foundations of Cognitive Science,* edited by M. Posner. Cambridge, Mass.: MIT Press.

Kitcher, P. 1993. *The Advancement of Science.* New York: Oxford University Press.

Kosslyn, S. M. 1980. *Image and Mind.* Cambridge, Mass.: Harvard University Press.

Larmor, J., ed. 1937. *The Origins of Clerk Maxwell's Electric Ideas.* Cambridge, Mass.: Cambridge University Press.

Latour, B. 1986. "Visualisation and Cognition: Thinking with Eyes and Hands." *Knowledge and Society* 6:1–40.

———. 1987. *Science in Action.* Cambridge, Mass.: Harvard University Press.

Lorentz, H. A. 1891. "Electriciteit en ether." In *Collected papers.*

Lynch, M. and S. Woolgar, eds. 1990. *Representation in Scientific Practice.* Cambridge, Mass.: MIT Press.

Mani, K. and P. N. Johnson-Laird. 1982. "The Mental Representation of Spatial Descriptions." *Memory and Cognition* 10:181–87.

Maxwell, J. C. 1855–56. "On Faraday's Lines of Force." In *Scientific Papers.*

———. 1856. "Are There Real Analogies in Nature?" In *The Life of James Clerk Maxwell,* edited by L. Campbell and W. Garrnett. London: Macmillan and Co.

———. 1861–62. "On Physical Lines of Force." In *Scientific Papers,* edited by W. D. Niven. Cambridge: Cambridge University.

———. 1864. "A Dynamical Theory of the Electromagnetic Field." In *Scientific Papers.*

———. 1873a. "Quaternions." *Nature* 9 (217):137–38.

———. 1873b. *A Treatise on Electricity and Magnetism.* 1st Oxford: Clarendon.

———. 1890a. "On Action at a Distance." In *Scientific Papers.*

———. 1890b. *The Scientific Papers of James Clerk Maxwell.* Edited by W. D. Niven. 2 vols. Cambridge: Cambridge University.

McNamara, T. P., and R. J. Sternberg. 1983. "Mental Models of Word Meaning.' *Journal of Verbal Learning and Verbal Behavior* 22:449–74.

Morrow, D. G., G. H. Bower, and S. L. Greenspan. 1989. "Updating Situation Models During Narrative Comprehension." *Journal of Memory and Language* 28:292–312.

Nersessian, N. J. 1984a. "Aether/Or: The Creation of Scientific Concepts." *Studies in History and Philosophy of Science* 15:175–212.

———. 1984b. *Faraday to Einstein: Constructing Meaning in Scientific Theories.* Dordrecht: Martinus Nijhoff/Kluwer Academic Publishers.

———. 1985. "Faraday's Field Concept." In *Faraday Rediscovered: Essays on the Life and Work of Michael Faraday,* edited by D. C. Gooding and F. A. J. L. James. London: Macmillan.

———. 1988. "Reasoning from Imagery and Analogy in Scientific Concept Formation." *PSA 1988.*

———. 1992a. "How Do Scientists Think? Capturing the Dynamics of Conceptual Change in Science." In *Minnesota Studies in the Philosophy of Science,* edited by R. Giere. Minneapolis: University of Minnesota Press.

———. 1992b. "In the Theoretician's Laboratory: Thought Experimenting as Mental Modeling." *PSA 1992.*

———. 1995. "Opening the Black Box: Cognitive Science and the History of Science." *Osiris* 10 (Constructing Knowledge in the History of Science, ed. A. Thackray). 194–211.

———. 1999. "Model-based Reasoning in Conceptual Change." In *Model-based Reasoning in Scientific Discovery,* edited by L. Magnani, N. J. Nersessian, and P. Thagard. New York: Kluwer Academic/Plenum Publishers.

————. 2000. "Abstraction via Generic Modeling in Concept Formation in Science." In *Correcting the Model: Abstraction and Idealization in Science,* edited by M. R. Jones and N. Cartwright. Amsterdam: Rodopi.

Oakhill, J., and A. Garnham, eds. 1996. *Mental Models in Cognitive Science: Essays in Honor of Philip Johnson-Laird.* Psychology Press.

Perrig, W. and W. Kintsch. 1985. "Propositional and Situational Representations of Text." *Journal of Memory and Language* 24:503–518.

Rudwick, M. J. S. 1976. "The Emergence of a Visual Language for Geological Science." *History of Science* 14:149–95.

Schwartz, D. L., and J. B. Black. 1996. "Analog Imagery in Mental Model Reasoning: Depictive Models." *Cognitive Psychology* 30:154–219.

Shelley, C. 1996. "Visual Abductive Reasoning in Archeology." *Philosophy of Science* 63:278–301.

Shepard, R. N., and L. A. Cooper. 1982. *Mental Images and Their Transformations.* Cambridge, Mass.: MIT Press.

Siegel, D. 1986. "The Origin of Displacement Current." *Historical Studies in the Physical Sciences* 17:99–145.

————. 1991. *Innovation in Maxwell's Electromagnetic Theory.* Cambridge: Cambridge University Press.

Stein, H. 1970. "On the Notion of Field in Newton, Maxwell, and Beyond." In *Historical and Philosophical Perspectives on Science,* edited by R. H. Stuewer. Minneapolis: University of Minnesota Press.

————. 1976. "On Action At a Distance: Metaphysics and Method in Newton and Maxwell." Unpublished talk, Yale University.

————. 1981. "'Subtler Forms of Matter' in the Period Following Maxwell." In *Conceptions of Ether,* edited by G. N. Cantor and M. J. S. Hodge. Cambridge: Cambridge University Press.

Steiner, M. 1989. "The Application of Mathematics to Natural Science." *The Journal of Philosophy* 86 (9):449–80.

Thagard, P. 1991. *Conceptual Revolutions.* Princeton: Princeton University Press.

Thagard, P., K. J. Holyoak, G. Nelson, and D. Gochfield. 1990. "Analog Retrieval by Constraint Satisfaction." *Artificial Intelligence* 46:259–310.

Trumpler, M. 1997. "Converging Images: Techniques of Intervention and Forms of Representation of Sodium-channel Proteins in Nerve Cell Membranes." *Journal of the History of Biology* 20:55–89.

Tweney, R. D. 1985. "Faraday's Discovery of Induction: A Cognitive Approach." In *Faraday Rediscovered,* edited by D. Gooding and F. A. J. L. James. New York: Stockton Press.

————. 1987. "What is Scientific Thinking?" Unpublished manuscript.

————. 1992. "Stopping Time: Faraday and the Scientific Creation of Perceptual Order." *Physis* 29:149–64.

Wason, P. C. 1960. "On the Failure to Eliminate Hypotheses in a Conceptual Rask." *Quarterly Journal of Experimental Psychology* 32:109–123.

————. 1968. "On the Failure to Eliminate Hypotheses in a Conceptual Rask—A Second Look." In *Thinking and Reasoning,* edited by P. C. Wason and P. N. Johnson-Laird. Cambridge: Cambridge University Press.

Woods, D. D. 1997. "Towards a Theoretical Base for Representation Design in the

Computer Medium: Ecological Perception and Aiding Human Cognition." In *The Ecology of Human-machine Systems,* edited by J. Flach et al. Hillsdale, N.J.: Lawrence Erlbaum.

ACKNOWLEDGMENTS

I appreciate the comments received on earlier versions of this paper from Abner Shimony, Jed Buchwald, and other participants in the colloquium presented at the Dibner Institute for the History of Science and Technology, and participants in the History of Science Colloquium at the Niels Bohr Institute. This work was supported by a grant from the National Science Foundation SES9810913.

[7]

Reconsidering Ernst Mach on Space, Time, and Motion

ROBERT DISALLE

The standing of Ernst Mach as a philosopher has depended, at least since 1915, on the prevailing interpretation of general relativity. After the successes of the atomic theory in the early years of the twentieth century, Mach's denial of the reality of atoms seemed to discredit his severe form of empiricism. But the advent of general relativity seemed to vindicate Mach after all. According to Einstein himself, general relativity incorporated two central insights of Mach: that motion is essentially relative, so that any meaningful statement about motion is necessarily a description of relative change of position; and that inertia is not an irreducible property of any given body, but arises from its interconnection with all the other masses in the universe. To philosophers, especially Reichenbach, and physicists, especially Einstein himself, these principles signified that general relativity was not merely a new theory of space, time, and gravitation, but a new philosophical understanding of space and time in general. Whereas Newtonian mechanics and special relativity had taken space and time, or spacetime, to be objective features of the world, general relativity had revealed the arbitrariness of all spatiotemporal descriptions. In doing so, general relativity seemed to confirm Mach's view of absolute space and time as "conceptual monsters."

In the second half of the twentieth century, however, many philosophers and physicists came to understand that general relativity does not satisfy Mach's philosophical demands as Einstein had originally hoped. First, it fails to establish the "general relativity of motion," since it does indeed distinguish geodesic motion (gravitational free-fall) from other states of motion. Thus, the theoretical distinction between rotating and nonrotating systems—what Einstein called the "epistemological defect" of the earlier theories—persists in general relativity, though in a much more complicated form; the difference between the Keplerian and Tychonic models of the solar system, which according to Reichenbach had been reduced to a matter of convention, has a physical basis in general relativity just as much as in Newtonian gravitation theory. Second, although general relativity holds that the inertial structure (the affine structure) of spacetime is affected by the distribution of matter, it no more accounts for the *origins* of inertia than did Newtonian mechanics or special relativity. This is obvious from the fact that the asymptotic structure of spacetime in general relativity is identical to that of special relativity, so that a body sufficiently isolated from the other masses of the universe would have the same inertial motion as in special relativity. Indeed, Einstein was provoked by this fact to introduce the cosmological hypothesis of a closed universe, so that the problem of the inertia of isolated bodies would be eliminated by fiat. These considerations suggest that general relativity is, in crucial philosophical respects, essentially similar to its two predecessors: all of the theories postulate an objective spatiotemporal structure and specify how physical processes are supposed to exhibit that structure. The move to general relativity rests not on the application of any Machian philosophical principle—whatever the heuristic value such principles may have had for Einstein—but on a deeper physical understanding of the relationship between gravity and inertia.

This way of understanding the relations among general relativity, the previous spacetime theories, and Mach's philosophy was suggested early on by Hermann Weyl, who pointed out that general relativity, like its predecessors, postulates a "world structure"—the geometrical structure of spacetime—without which even the idea of relative motion, let alone that of absolute motion, makes no sense. Weyl emphasized that the influence of matter and energy on this "world structure" is the distinctive feature of general relativity; the distinctive feature of the earlier spacetime theories is not their metaphysical excess, but their extension to the global spacetime structure of geometrical properties that general relativity reveals to be local. Weyl's general approach to spacetime geometry was sustained by some important mathematical physicists, especially J. L. Synge and J. A. Wheeler. Among the most prominent philosophers concerned with problems of space and time, meanwhile, the views of Weyl seem to have had little visible influence, and the view of Einstein and Reichenbach dominated at least until the 1960s.

Even as Einstein's and Reichenbach's view began to be widely questioned, it continued to determine the way in which the basic philosophical problem of space and time was (and is) formulated: relative motions are observable, while spacetime structures are "theoretical entities" invoked in order to explain them, and so the philosophical issue is whether spacetime can be justified as an ontological addendum to the known spatiotemporal relations. But the newer understanding of general relativity, and its essential philosophical continuities with earlier spacetime theories, made this "absolute versus relational" issue seem much less one-sided than before. And the Machian position, formerly regarded as the only one compatible with general relativity, now seemed to be not merely debatable; it came to be widely regarded as exemplifying the narrowness of Mach's philosophy in general. Mach's antimetaphysical polemics against absolute space and time now seemed to display a naive empiricist suspicion against any kind of scientific theorizing. Indeed, they seemed to be of a piece with Mach's published rejection of the atomic theory and relativity theory, displaying a narrow-minded prejudice against fruitful scientific theories just because they fail to meet some impossible empiricist standard.[1]

An important contribution to this reaction against Mach was the work of Howard Stein, especially his celebrated papers "Newtonian Space-Time" (1967) and "Some Philosophical Prehistory of General Relativity" (1977). Many of Stein's points have become commonplaces of the philosophy of space and time, but three of them are particularly relevant to the present discussion. First, "Newtonian Space-Time" made clear that Newton's theory of absolute space, time, and motion was not mere metaphysical nonsense, as Mach seemed to suggest, and as many twentieth-century commentators had claimed in even more extreme terms.[2] Rather, it is a straightforwardly empirical theory of spacetime structure, and the arguments of Newton's "Scholium to the Definitions" are not, therefore, naive inferences from physics to metaphysics, but, in Stein's words, "a classic case of an analysis of the empirical content of a set of theoretical notions" (1967). Second, contrary to the belief of Reichenbach and others, Mach did not provide an account of inertia that is more empirical than Newton's—even though Mach proposed to replace an "unobservable entity," absolute space, with the observable fixed stars. Instead, Mach's version of the principle of inertia arises from the conjunction of Newton's principle with some completely unempirical assumptions about the cosmic distribution of mass. Third, Mach never provided a theory of the origins of inertia in some long-range interaction between masses, nor any sort of hint about how such a theory might be possible. Instead, he merely suggested that since the "fixed stars" are *practically* indispensable for distinguishing inertial from accelerated celestial motions, they might in some unknown way be *physically* indispensable as well. Therefore we should not pretend to know whether, in the absence of the fixed stars, a particle would continue to move uniformly in a straight line or a rotating sphere would bulge at the equator from centrifugal force. In particular we should not pretend to know whether the earth would exhibit its actual centrifugal bulge even if it were at rest and the fixed stars were rotating around it—or more precisely, whether the bulge is the effect of anything other than the *relative* rotation of the earth and the stars. And, again, the claims that general relativity had provided such a theory of the origins of inertia turned out to be quite exaggerated. In light of these considerations, Stein characterized Mach's philosophical opposition to Newton as "abusive empiricism"—meaning by this not merely the prejudice against theories involving unobservable entities or far-reaching counterfactual implications, but, more important, the absurd willingness to accept empirically unmotivated hypotheses about cosmic geography, boundary conditions, and so on, just to avoid theories of that sort.

Stein's criticisms of Mach provided an urgently necessary corrective to some widespread misconceptions in the philosophy of space and time, especially the notion that Newton's theory of space and time had been a naive philosophical blunder that Mach had overcome by providing a superior analysis of rotation. And the influence of these criticisms on subsequent literature is quite obvious. As I will explain later, I believe that, in much of this literature, Stein's seminal discussions of Newton have been seriously misunderstood. Nonetheless, we rarely see now the facile dismissals of Newton's theory, or the dogmatic acceptance of the Machian critique, that were so common before Stein wrote. If anything, Machian approaches to physics are now held up to a Newtonian standard: theories that reduce absolute rotation to relative rotation have the burden of showing that they can give a convincing account of the dynamical effects that Newton described in his "water-bucket" experiment.[3] Although such theories may be thought to satisfy certain philosophical preferences better than Newton's theory does, it is clear that they have to be regarded, and judged, as alternative physical hypotheses rather than as philosophical critiques of Newton's view.

It would be unfortunate, however, to replace the former too-high estimate of Mach with an overly unsympathetic one. I believe that this was not the intention of Stein's severe critique in (1977); rather, I take it that Stein was trying to point out precisely the discrepancy between what Mach actually wrote in *Die Mechanik* and the prevailing view of his ideas, and that the severity of Stein's judgments is really directed at Mach's twentieth-century acolytes rather than at Mach himself. My purpose is to offer a more sympathetic analysis of Mach's ideas—not one that vindicates the estimate of the twentieth-century Machians, nor one on which Mach's thinking appears to be perfectly clear and perspicacious, but one that reveals Mach's serious philosophical engagement with, and surprisingly clear understanding of, the conceptual foundations of Newtonian mechanics. In particular I will pursue two themes:

I. I certainly agree with Stein that Mach's so-called "solution" to the "problem of rotation" is no more a real solution than the problem is a real problem. Instead it is a somewhat confusing jumble of three mutually inconsistent proposals. But I suggest that some of what makes it seem confusing is a reflection of the changes in Mach's views over time—changes which were gradually and not always neatly inserted into successive editions of *Die Mechanik*,[4]and which reflect positively on Mach's gradually improving understanding of the structure of Newtonian mechanics.

II. The most prominent philosophical errors that have been committed in Mach's name by twentieth-century philosophers are generally not errors of which Mach himself was guilty. I suggest that by identifying them precisely, we can better appreciate not only Mach's own views, but the philosophical context of general relativity and its relation to earlier theories of space and time.

I. It is easy to get the impression that Mach was quite alone in the nineteenth century in criticizing Newton's theory of inertia and his conceptions of space and time. In fact, Mach was one of many physicists who were questioning the empirical meaning of Newton's first law of motion. This small critical movement apparently began in earnest around 1870, when Mach and Carl Neumann, independently of one another, raised the question, "*relative to what* is the motion of a free particle supposed to be uniform and rectilinear?" And it culminated around 1885 with the articulation of the concept of inertial frame by James Thomson and Ludwig Lange (also independently of one another). This literature has been discussed in detail elsewhere,[5] but a few important points are worth mentioning here. First, various philosophical points of view animated this discussion, but one predominant view held that there was something essentially correct about Newton's view that was poorly expressed or even confused by the notion of absolute space. Implicitly, at least, this was the view of Neumann, Thomson, and Lange. They agreed with Newton that no actual body or bodies defines the reference-frame required by the law of inertia, since the law clearly states how a body will behave even in the absence of any relative motion; Neumann in particular emphasized that a rotating sphere will bulge at the equator even if there is no other body relative to which it rotates, and he regarded this as the outstanding objection to the reduction of absolute motion to relative motion. Neumann proposed, therefore, that Newton's first law becomes meaningful when conjoined with the postulate that, somewhere in the universe, there exists a rigid body or a system of coordinate axes—the "Alpha body" or "Alphasystem"— relative to which a free particle moves uniformly in a straight line. The grain of truth in Neumann's idea was fully developed in Thomson's conception of a "reference-frame," and in Lange's (more or less equivalent) conception of an "inertial system." We can express their idea thus: there exist a reference frame and a time-scale, relative to which the motion of a free particle is rectilinear and uniform. Or, to preempt the (essentially irrelevant) question about whether there really are any free particles, we can say: For any system of particles moving anyhow, there exists a frame and a time-scale with respect to which every acceleration is proportional to and in the direction of an applied force, and every such force belongs to an action-reaction pair. Moreover, any frame in uniform rectilinear motion relative to such a frame is also an inertial frame. The equivalence-class of frames thus defined is equivalent to the structure that Stein (1967) referred to as "Newtonian spacetime." The replacement of absolute space by such a structure solved the problem of absolute motion as far as Newtonian mechanics was concerned; the lingering question of a privileged rest-frame was the one posed by classical electrodynamics and answered by the special theory of relativity.

The second approach was that of Mach, who questioned the counterfactual inferences that Neumann had taken for granted. Mach pointed out that there are two possible interpretations of such inferences: either motion is absolute, as Neumann had concluded, or "our law of inertia is erroneously expressed." In the correct expression of the law, Mach wrote, "attention ought to be paid to the masses of the universe." For in the thought-experiments proposed by Newton and Neumann, who can say whether the equatorial bulge of a rotating planet must remain if the rest of the masses of the universe are eliminated? In Mach's well known words, "In thought-experiments we may modify *unimportant* circumstances in order to bring out new features in a given case; but it is not to be assumed in advance that the universe is without influence" (1901, 291).

We don't know whether we are modifying "unimportant circumstances" when we consider an alternative distribution of the stellar masses. Already in 1872, before the publication of *Die Mechanik*, Mach had considered the possibility that if the fixed stars fell into some chaotic motion among themselves, they might no longer serve as a reference system; in such a situation, the properties of inertial motion in our own system might be completely different (1872, 49).

Remarks of this sort, as Stein has pointed out, do not amount to a theory of the origins of inertia in some long-range interaction with the visible stars. At most they are hints or suggestions that such a theory might be possible. And it would be ridiculous to reject Newton's theory just because a deeper explanation of inertia could be imagined to be possible. But this is not really what Mach intended. Rather, he was attempting to explain what he took to be the "empirical content" of the law of inertia, in a peculiar sense of that phrase: not the consequences of the law for any possible empirical circumstances—as Newton or Neumann would have understood the content of the law— but the actual experiences, past and present, that constitute the practical application of the law. As far as Mach was concerned, every astronomical application of the law of inertia was, essentially, the resolution of some acceleration relative to the fixed stars into components, each of which is paired with some other acceleration relative to the fixed stars. Therefore, to interpret the laws of motion as more abstract statements about "absolute" acceleration—statements that would hold in the absence of the fixed stars—is simply to extrapolate beyond the available evidence, or more precisely, beyond the characteristic *type* of evidence on which the law is based. If, as Mach thought, a law of physics expresses a functional dependence among phenomena, then the laws of motion express the functional dependence between impressed forces and accelerations relative to the fixed stars. Expressing the laws more abstractly might enable us to apply them to imaginary situations, but it would not increase our knowledge of the functional dependence among phenomena.

On the present interpretation, Mach's famous remarks may appear perfectly reasonable, but not particularly interesting. To the Machian enthusiast of the twentieth century, they must seem obviously less interesting than genuine speculation about the origins of inertia; to anyone else, they might seem to be merely a narrow and cumbersome, even if not strictly inaccurate, way of understanding Newton's laws. As Stein remarked,

> Mach does not make it a general rule for science that in every statement based upon experience there should appear a list of all the circumstances over which we have no control (the universe being given only once), in order to avoid seeming to claim that we know that the statement would continue to be true even if these things were otherwise. Such a rule would not only grievously violate Mach's "economy of thought," it would make science impossible. (1977, 15)

One thing that can be said on behalf of Mach's remarks is that, contrary to the views of his most severe critics *and* the philosophical admirers of his epistemology, their objection against Newton is certainly not based on *phenomenalism*, or on the reduction of the principle of inertia into its phenomenal elements. Such views doubtless arise from reading Mach's critique of Newton against the background of *The Analysis of Sensations*. As we see in *Die Mechanik*, however, Mach knew that a conception of the relative positions of the planets and stars (not to mention of the masses involved) rests not on phenomenal elements, but on a complex construction involving a great deal of abstraction. His reduction of the law of inertia, then, was not to a phenomenalistic basis, but to a basis in what he took to be the historic and continuing practice of applying the law to real empirical cases. We cannot determine by experiment whether the role played by the fixed stars is fundamental or merely incidental; we only know that "hitherto they have been the sole competent means of the orientation of motions and of the descriptions of mechanical facts" (1889, 216).

In accord with this general outlook, Mach offered a revised formulation of the law of inertia: a body not subject to forces moves in such a way that its mean acceleration, with respect to "sufficiently many, sufficiently large and distant masses," vanishes (1889, 218–19). More precisely, given masses m_i at distances r_i, the motion of a "free" particle is given by

$$d^2(\Sigma_i\ m_i r_i / \Sigma_i\ m_i)/dt^2 = 0. \text{ (1889, p. 218)}$$

As Stein pointed out (1977, 17), this formulation has nothing to say about what sort of interaction might be taking place between a given particle and the surrounding masses. Moreover, on strictly "empiricist" grounds it seems to be a weaker theory than Newton's, since it combines Newton's theory with special cosmological assumptions (Stein 1977, 17–18); modern efforts to turn Mach's suggestion into a working cosmological theory have only brought the latter difficulty into sharp relief (see Earman 1989, 92–96). But these difficulties with working out Mach's proposal are not necessarily philosophical criticisms of Mach himself. Clearly such a proposal would be as hard to reconcile with phenomenalism as it is hard to realize physically—not only because of the unverified "special cosmological assumptions" involved, but more generally because it invokes cosmological distances, which presumably would not qualify as phenomena to any serious phenomenalist. But Mach did not intend to present a either a new physical theory, or a phenomenalistic interpretation of existing theory—he sought to express only the historical and practical experience that actually supported Newton's theory.

The innocuous character of Mach's views, on the present reading, might seem difficult to reconcile with the vehemence of his diatribes against Newton's conceptions of absolute space, time, and motion. But this difficulty can be overcome by a proper appreciation of Mach's historical development. When he first published *Die Mechanik* in 1883, he evidently saw three possible ways of understanding the law of inertia:

1. Newton's way, as a description of the motion of a free body relative to absolute space;

2. Neumann's way (followed also by many others), as a description that we need to supplement by postulating some body to take the place of absolute space, so that the Newtonian distinctions among states of motion will be defined relative to that body; and

3. Mach's way, as the description of the motion of a free body relative to the fixed stars.

I think that Mach can be forgiven for considering that his way was the empirical and scientific one, whereas the other two were, if not metaphysical nonsense, certainly more metaphysical than his own, precisely in Mach's pejorative sense of the term: both made the very meaning of the laws of motion seem to depend on understanding the states of motion of bodies relative to some object that could not possibly be observed. His own approach, by contrast, purports to achieve the same empirical consequences, not without theoretical assumptions altogether, but without theoretical structures that play no role in empirical reasoning.

By 1889, however, when the second edition of *Die Mechanik* appeared, Mach had become acquainted with the literature on inertial systems, and he discussed it in a lengthy appendix. By the fourth edition (1901) most of this discussion was incorporated into the body of the text. This leaves, as I suggested earlier, a somewhat jumbled impression, but with sufficient care we can trace Mach's intellectual development and get a clearer picture of his "mature" view. First, and most obviously, Mach always retained his skepticism about any extrapolation of the laws of motion beyond the known case of motion relative to the fixed stars. As a consequence, he continued to make, and elaborated upon, his two famous speculative hypotheses: that the fixed stars may be causally relevant to the apparent inertial motions of the planets; and that relative rotation, in certain circumstances, might be responsible for the effects that we attribute to absolute rotation. The existence of these possibilities argues, according to Mach, that the definitive reference-frame for the laws of motion ought to be the fixed stars. But Mach also came to see the possibility of stating the laws of motion abstractly, without expressly appealing to any reference-frame at all. This amounts to the assertion that (as Stein paraphrased it) "the laws of mechanic are to be construed as asserting that the relationships they express hold for some kinematical reference-frame" (1977, 18). Stein points out that this is precisely the correct way to regard Newton's laws, but it is hardly consistent with Mach's insistence on the general relativity of motion, and this inconsistency seems to have escaped Mach's notice.

If the inconsistency has to do with the historical evolution of Mach's views, however, it admits of a more sympathetic interpretation: in accepting the abstract formulation of the laws of motion, Mach revealed that he had come to understand something about the foundations of Newtonian mechanics that neither he, nor very many others, had understood before. And this improved understanding clearly arose from his study of the literature on inertial systems. In particular, he learned to distinguish two problems: the skeptical question about the extent of validity of Newton's laws, and the question about what Newton's laws, taken abstractly, really say about space and time. That is, Mach had learned to distinguish the "external" question about whether some other physical principles than Newton's might provide a more fundamental ground for the ordinary conception of inertia, which Newton's laws take for granted, and the "internal" question about how Newton's laws prescribe the construction of a reference-frame. In this way Mach finally understood that one could eliminate absolute space from Newtonian mechanics—and thereby advance the "antimetaphysical" aim of his book—while maintaining the essential structure of the theory.

Mach's understanding of this last point appears first in the 1889 appendix, in his comments on the work of Lange and others on the notion of inertial system. He characterized Lange's work in particular as "among the best that has been done" on the law of inertia. He even claimed to see a kinship with investigations that he himself had formerly pursued. At the same time he reiterated his methodological skepticism: "I have given up these pursuits because I have reached the conviction that through all of these modes of expression (and also through those of Streintz and Lange) one only *apparently* avoids the reference to the fixed stars and the rotation of the earth" (1889, 483). Lange (1885) had defined an "inertial system" as one in which three free particles move in straight lines, and expressed the law of inertia as the claim that, relative to an inertial system, any fourth free particle would move in a straight line. To Mach, however, it was "very questionable" whether a fourth free particle would travel in a straight line relative to an inertial system if the fixed stars did not exist, or were not sufficiently fixed (1889, 484). Here again, Mach is questioning on methodological grounds the soundness of the law of inertia as an inductive generalization from motions relative to the fixed stars. But this methodological concern did not prevent him from grasping the conceptual clarification that the concept of inertial system represented.

This is why the later editions of *Die Mechanik* contain various tributes to Newton's corollary 5 to the laws of motion, and even an expression of regret that earlier editions had not sufficient appreciation of Newton's insight on this point. To Mach, corollary 5 expressed Newton's grasp of the relativity principle, his implicit recognition of the empirical irrelevance of absolute space, and his understanding of the proper definition of an acceptable reference frame:

> In order to have a generally valid system of reference, Newton ventured Corollary V of the Principia. He thought of a . . . coordinate system for which the law of inertia holds, fixed in space without rotation relative to the fixed stars. He could also allow an arbitrary origin and uniform translation of this system . . . without losing its usefulness. Newton's laws of force would not be thereby altered; only the initial position and velocity, and the constants of integration, could vary. By this formulation Newton specified *precisely* the meaning of his *hypothetical* extension of the Galilean law of inertia. One can see that the reduction to absolute space was in no way necessary, since the reference system is just as relatively determined as in any other case. (227)

We can see that Mach now objects to absolute space not merely out of general skepticism about the laws of motion and the circumstances on which they might be contingent; he is arguing against an understanding of the Newtonian theory that is wrongheaded on its own terms. Galileo's account of inertia focuses on the persistence of motion parallel to the earth's surface, and the composition of such motion with accelerated motion. But Newton's extension of this principle—properly understood—does not merely invoke a more encompassing or abstract system, such as absolute space or the "body Alpha," to replace the earth as a standard relative to which the laws of motion hold. Rather, the extended principle is the assertion that there is an equivalence class of reference-systems, defined by the laws of motion themselves.

It is evident, then, that in Mach's mature view, the abstract presentation of the laws of motion is no longer objectionable in itself, even though it remains open to Mach's usual methodological skepticism.

> It is very much the same whether we refer the laws of motion to *absolute space,* or express them *abstractly,* without express indication of the system of reference. . . . But owing to the fact that the first way, when ever there was any actual issue at stake, was nearly always interpreted as having the same meaning as the latter, Newton's error was much less dangerous than it would otherwise have been. . . .
>
> Let us again emphasize that Newton in his oft-mentioned Corollary V, which alone has scientific value, does not refer to absolute space. (1933, 269–70.)

Earlier he had responded to arguments like Neumann's with the "Machian" suggestions we have already discussed. But the view that he arrived at by the seventh edition is philosophically quite different: "the irritating paradoxes of Neumann do not disappear until we give up absolute space, without going beyond Corollary V" (1933, 270).

By the end of his life, then, Mach had arrived at a comparatively subtle understanding of the relativity of motion, by coming to a better understanding of the significance of Galilean relativity within Newtonian physics. In particular, he had come to see that the concept of inertial system, as an expression of the Galilean principle, solves the primary philosophical problem with Newton's view and eliminates the metaphysical excess that he had identified in the theory. And he understood the more general idea of relativity as addressing a different sort of problem altogether, namely, the methodological problem of how far we can extrapolate from our knowledge of motions relative to the fixed stars. Instead of a narrow-minded polemic against Newton or a visionary proposal for an alternative physics, we see in the remarks about corollary 5 something that seems rather ironic:

Mach is actually invoking Newton as an ally against those of his contemporaries who still defend the idea of absolute space.

II. By comparison with Mach's discussions, as interpreted above, much of the twentieth-century discussion inspired by Mach's ideas represents a considerable decline in the level of philosophical clarity. By this I don't mean to criticize "Machian" ideas because they don't accord with Mach's actual way of thinking, or because they are based on misinterpretations of Mach's ideas, or because their proponents mistakenly identify them with Mach's. Nor are such ideas to be criticized merely because they have turned out, in spite of the initial beliefs of Einstein, Reichenbach, and others, not to be realized by the general theory of relativity. Nor are they to be criticized merely for failing to provide, so far, the deeper explanation of the origins of inertia that Mach was thought to have envisioned. Rather, I mean to show that they frequently involve confusions about the nature of the principle of inertia, and about the role of the concept of inertial frames in classical physics—concepts concerning which, as we have seen, Mach himself came to be relatively clear.

First, consider Einstein's philosophical objections to Newtonian mechanics and special relativity, focused on the peculiar status of inertial systems in those theories. Einstein thought it puzzling that "nature" should distinguish a particular sort of coordinate system, and a particular class of coordinate transformations, from all possible ones; that the laws of nature should be true only with respect to a restricted class of coordinate systems seemed to him inherently bizarre: "What does nature care for our coordinates?"

> What makes the situation appear particularly unpleasant is the fact that there should be infinitely many inertial systems, moving uniformly and without rotation with respect to one another, that are distinguished from all other rigid systems. (Einstein 1949, 26)

From this we gather that the infinite number of inertial systems, as well as their privileged status, puzzled Einstein, and encouraged his conviction that the special theory of relativity was merely a stage in a philosophically inevitable movement, whose end-point would be a "general theory of relativity" that made no such distinctions.

This conviction was an obvious and familiar inspiration for Einstein's theory of gravitation, and its heuristic value cannot be underestimated. If the conviction was a philosophical mistake, it has to be considered a fortunate one for the history of physics. Nonetheless, the nature of the mistake is worth identifying for the sake of a clearer picture of the relationship between general relativity and its predecessor theories of space and time. Within the framework of four-dimensional differential geometry, of course, this relationship has been made quite perspicuous in the standard literature.[6] But the comparative status of inertial frames in the various theories, and the import of Einstein's remarks, have been the subject of much confusion. One persistent confusion is the notion that the laws of motion are somehow restricted in their application because they hold only in a particular class of reference frames. A recent formulation of this view reads, "In order to apply the laws motion, we need to be sure that we have an inertial frame" (Cushing 1998, 104). If this were true, of course, any reasoning from the laws of motion would necessarily be circular, and the apparent applicability of the laws, at least to a certain degree of approximation, would be rather mysterious. But as the nineteenth-century discussions (including, as we have seen, Mach's) eventually managed to make clear, the laws of motion assert the existence of a class of frames, and therefore can't be said to be true only "relative to" those frames. If the laws hold at all, they hold "in the world," and they attribute to "the world" the characteristic that such a frame is possible. In contemporary terms, this means that the laws attribute to the world a flat spacetime structure.

Moreover, the infinite number of inertial frames is an odd thing to object to, as if it represented an unwarranted ontological inflation. As Mach had already recognized in remarks quoted above, the infinity of inertial frames merely expresses the relativity of position and velocity in the Newtonian conception of force, and it has an analogous meaning in special relativity. In other words, the equivalence class of frames exists because the theory distinguishes invariant from relative quantities, and thereby allows an infinity of equivalent descriptions of the same physical situation; the "ontology" is expressed by the invariant quantities, and is not augmented by the relative quantities, which will depend on the mode of description of the invariants. Minkowski apparently understood this situation more clearly than did Einstein, possibly because he was more thoroughly schooled than Einstein in the group-theoretical approach to geometry: his formulation of special relativity as the "postulate of the absolute world" (1908) reveals the single geometrical structure underlying the infinitely many equivalent descriptions identified by Einstein (1905). From this perspective, the apparently philosophical question—why does special relativity single out the inertial systems?—has a straightforward answer: because the theory has invariant physical quantities, and therefore has a group of nontrivial symmetries that preserve those invariants and thereby characterize the spacetime structure associated with the theory. Therefore, saying that it is inherently mysterious that nature should single out the inertial systems, as Einstein suggested, amounts to saying that it is inherently absurd that spacetime should have nontrivial symmetries—it is inherently absurd, for example, that spacetime should be flat. And this is no more intelligible, philosophically, than thinking it absurd that nature should single out (for example) the conservative systems among all possible physical systems. Moreover, on this view, it can hardly be an empirical discovery that spacetime is curved; in fact this view is more reminiscent of the apriorism often defended in the late nineteenth century (for example, in Russell 1898), in which it was considered inherently absurd that space should be inhomogeneous, than of any sort of empiricism, including Mach's "abusive" variety.

A second philosophical objection to the inertial system, also purportedly inspired by Mach, is that it functions as a "factitious cause" of inertial effects in Newtonian mechanics and special relativity. Einstein's familiar presentation of this objection is the thought-experiment in which two bodies S_1 and S_2 are in relative rotation, and S_1 is perfectly spherical while S_2 bulges at the equator as if suffering a Newtonian centrifugal force. He asks, what is the explanation for this difference between bodies whose circumstances are otherwise equivalent? "Newtonian mechanics does not give a satisfactory answer to this question. It pronounces as follows: The laws of mechanics apply to the space R_1, in respect to which the body S_1 is at rest, but not to the space R_2, in respect to which the body S_2 is at rest" (Einstein 1916, 113). Einstein's dissatisfaction stems from the notion that the "space R1," or the inertial system, is being invoked as the cause of the asymmetry between the two bodies. And this is objectionable in two respects: first, it grants a causal role to something unobservable, whereas an epistemologically satisfactory theory must seek the cause in something observable; second, it allows this cause to act without being acted upon in return—inertial systems determine the behavior of bodies, but are not in turn determined by them—and this is contrary to the basic tendency of scientific thinking.

This argument seems plausible because, after all, Newton claimed to identify absolute rotation by its "causes and effects": centrifugal forces are the effects of a rotation in absolute space. And if we replace absolute space with the equivalence class of inertial frames, we claim that centrifugal forces are caused by rotation relative to an inertial frame. What the argument overlooks was articulated clearly by Stein (1967): centrifugal forces provide, in Newton's theory, the *criterion* that *defines* absolute rotation. That is, the theory does not invoke a "factitious cause" of centrifugal effects, but takes them as defining the empirical content of the concept of absolute rotation; absolute rotation is a theoretical quantity that, Newton points out, can be calculated from the magnitude of the centrifugal forces even if no relative rotation can be observed (see 1726, 414). "Absolute rotations cause centrifugal forces," then, is not a hypothetical causal explanation of those forces, as Einstein's argument suggests, but a *definition*. Therefore it ought to be assessed according to the standard formulated by Einstein, in defense of his definition of simultaneity: "The only demand that can be made of a definition is that it provide, for every real case, an empirical decision whether the concept to be defined applies or not" (1917, 25). Newton's definition of absolute rotation obviously satisfies this demand, since Newton's laws provide an unambiguous empirical criterion of rotation and a measure of the rate of rotation.[7]

Einstein's argument reveals an important contrast between his thinking about causality and Mach's. Mach's later writings, at least, display a comparatively clear understanding of the role that inertial systems play in the Newtonian causal theory: an inertial system is not itself a cause, but constitutes the framework within which causal efficacy is measured, through the accelerations that causal agents produce in one another. And in proposing the fixed stars as the framework against which the laws of motion are to be expressed, Mach was not granting the stars the status of cause, either: he was merely claiming that, in actual practice, the stars constitute the *empirical* framework within which causal influences, at least among the celestial bodies, are measured. In other words, he was suggesting that the unobservable background against which Newton's theory analyzes causal interactions be replaced, in the statement of the laws, by a genuine empirical background. Of course, Mach's speculation about the origins of inertia suggests that the stars may be playing a causal role as well, but Mach clearly understood this as a separate issue—a question about a possible alternative theory, rather than the identification of an internal "epistemological defect" of the theory of inertial systems. Einstein, in contrast, has mixed the two issues together, and, having done so, is persuaded that the "factitious" cause, the inertial system R_1, has to be replaced by a genuine cause such as the fixed stars. To confuse these issues is to obscure the fundamental role that spacetime theories play in classical physics, as frameworks within which the very notion of physical causality is first clearly specified.

This general understanding of spacetime structures reveals the misunderstanding in Einstein's objection that spacetime should not "act" without being "acted upon"—an idea that has been invoked in subsequent literature as the "action-reaction principle" (see Brown 1996). If spacetime, or the inertial system, really were postulated in classical mechanics as a hidden cause of inertial effects, the objection would be an appropriate one, since the theory comprehends all causation as essentially *interaction*. A theoretical entity that stands in some such relation to all physical fields, yet remains a fixed and uniform background against which the states of matter and fields are defined, would be a bizarre one indeed. But, again, classical spacetime is not postulated to be such a thing; rather, it is the structure that is *implicit* in the classical notion of causal influence as what we measure by the acceleration of masses. This is no more mysterious than the fact that Minkowski spacetime (as Minkowski himself made quite clear in 1908) is not postulated as the underlying cause of the spatio-temporal relationships defined by the special theory of relativity, but is identified as the spacetime structure *implicit* in that theory. In such theories the space-time structure expresses the fundamental invariant quantities, and to that extent it expresses certain lawlike constraints on the possible evolution of physical fields—for example, that the trajectory of the center of gravity of an isolated system is a spacetime geodesic. But this does not represent an "action" of the spacetime structure on fields, in Einstein's sense, but a postulate that underlies our theory of how fields act on one another. Or, if it is a kind of action, it is a kind that is pervasive in physics, without usually provoking philosophical objections. For example, the Hilbert space structure of quantum mechanics expresses constraints on possible states of physical systems, and therefore "acts on" matter in precisely the sense that Newtonian or Minkowski spacetime does. Yet no one, to my knowledge, has ever found it philosophically objectionable that matter should not react upon the Hilbert space. Nor, apparently, is it considered an objection to Lorentz-invariant field theory that fields do not react upon the Lorentz group. In such cases it is implicitly understood that a structure of constraints on the ways in which physical systems evolve and interact is not a mysterious entity that is "acting upon" them without reciprocal effect; in the case of spacetime this understanding has been slow in coming.

Apart from the issue of whether it rests on a mistaken characterization of classical spacetime theories, Einstein's objection concerning causality has another serious difficulty: like the objection to the privileged status of inertial frames, this objection amounts to an a priori exclusion of the possibility that spacetime is flat, except in the absence of all matter. If it makes no sense to think of spacetime as unaffected by matter, then it makes no sense to think of gravity as a field acting on bodies whose inertial trajectories would be the geodesics of a flat spacetime. Perhaps there is a reasonable philosophical standpoint from which these possibilities are indeed absurd, but, as I suggested earlier, that standpoint could scarcely be characterized as a form of empiricism. In spite of his questionable philosophical pronouncements, Einstein was always quite clear on the essential role played by the equivalence principle: the general theory of relativity depends on the contingent, empirical fact that inertial and gravitational mass are equivalent. It is certainly conceivable that this might have been otherwise, and that gravity might have behaved as the other fundamental interactions behave, so that gravitational acceleration would not be independent of the mass and composition of the affected body. It is perhaps not an interesting possibility—although some yet-undetected violation of the equivalence principle might someday show the way to a quantum theory of gravity—but it certainly shows that whether gravitational and inertial mass are equivalent is an empirical question. And so it is an empirical question whether freely falling frames are distinguishable from inertial frames, and whether gravity can be identified with spacetime curvature.

For these obvious reasons it is also an empirical question whether inertial frames have a privileged status, and whether spacetime can "act on" matter without being reacted upon. More particularly, it is an empirical question whether we can really do what Newton's laws of motion say we can—namely, identify an inertial system by analyzing the interactions within a system of bodies. And from an empiricist standpoint, Einstein's great insight was that, because of the empirically established properties of the gravitational field, it is an empirical fact that we cannot identify such a system. Thus the problem with inertial systems that made general relativity possible was not that the idea of such a system makes no sense. Rather, it was the contingent fact that the equivalence principle holds, which means that the procedure for constructing an inertial system must fail to construct one, or fail to construct one that is distinguishable from a system in gravitational free-fall—unless of course, we could include the entire universe in one interacting system, and thereby be assured that there was no other gravitational source that might accelerate its center of mass. It is in fact somewhat ironic that empirical facts about the gravitational field should have been the undoing of Newton's conception of spacetime structure, since the gravitational field was his only source of information about the celestial masses, and therefore his only hope of constructing an inertial system in the first place. In any case, the role that empirical facts played in Einstein's creation of general relativity is only obscured by the insistence that the theory was necessary on some purely philosophical grounds.

The contrast between these twentieth-century Machian themes and Mach's own philosophy is quite striking. In Mach we see—at least after a careful study of the development of his thought—a relatively clear understanding of the abstract structure of Newtonian physics, and the role played by inertial systems in the theory's treatment of the interactions among bodies. Mach's vehement antimetaphysical polemics turn out to be directed not at this theory, but at the conjunction of this theory with the unnecessary structure of absolute space; his speculations about the origins of inertia are not intended as philosophical objections to the Newtonian structure, but as a reminder that other theories are possible in which the same distinctions among states of motion would not hold—and that such a theory might be perfectly compatible, in principle, with everything we know about the accelerations of masses relative to the fixed stars, that is, with all of the evidence that supports the Newtonian conception of inertia. In the twentieth-century gloss on Mach's ideas, however, we see these simple (and perhaps simplistic) empiricist platitudes transformed into philosophical prejudices against particular types of theory—prejudices that exclude, on a priori grounds that are not particularly well motivated, theoretical possibilities that are perfectly capable of being put to empirical tests.

As was already acknowledged, prejudices of this sort can play a crucial heuristic role in the search for physical theories, and the impetus that they gave to the creation of general relativity is surely more important than the confusions that they contain. Moreover, these "Machian" preferences have motivated interesting derivations from general relativity, such as the Lense-Thirring effect, and important research into alternatives to general relativity. In the 1960s, for example, Robert Dicke considered it a "logical flaw" of general relativity that made possible consideration of the rotational state of a single body in an otherwise empty universe (1964, 7); this was a primary motivation for his work on a scalar-tensor theory of gravitation (the Brans-Dicke theory), in which a hypothetical long-range scalar field might influence local inertial behavior, in the manner vaguely suggested by Mach's speculations about the fixed stars. It should be emphasized, however, that the theory's chief observable departure from general relativity is that the scalar field makes the ratio of gravitational to inertial mass depend on the gravitational self-energy of a body (see Will 1993, 41–43). Therefore, the strength (or the very existence) of the inertia-producing scalar field is an empirical question, and general relativity's lack of such a field can hardly be called a *logical* flaw. Moreover, in the limit as the strength of the Brans-Dicke field vanishes (as the measurable coupling constant goes to infinity), inertial mass does not vanish; rather, the ratio of gravitational to inertial mass approaches unity—in other words, the theory just reduces to general relativity. This means that, even if it were successful, the Brans-Dicke version of "Machian" physics would not explain the *origin* of inertia in a long range interaction.

A more recent and perhaps more authentically Machian programme is the work of Barbour and Bertotti (see Barbour 1999), who propose, in essence, a reduction of spatio-temporal structure to constraints on the relative spatial separations of particles—that is, a reduction of spacetime to spatial relations, more or less as envisioned by Mach. It is of course possible that such a theory could prove to be empirically successful, or, what is perhaps more important, more useful than general relativity in pointing the way toward a quantum theory of gravity. This is not the place to consider the merits of such a theory in detail, or the large literature on similarly Machian proposals (see for example Barbour and Pfister 1995). But, as in the case of the Brans-Dicke theory, it must be recognized that the possible empirical interest of such a theory in no sense vindicates the general *philosophical* objection to theories that postulate some inertial structure. For one thing, the spatial structure taken for granted by such theories has the same purportedly mysterious aspects that inertial frames were supposed to have in classical mechanics, especially the function of expressing general constraints on material systems without being "reacted upon." If space is somehow assumed to be exempt from this objection in a way that spacetime is not, it could only be because spatial relations are assumed to have a more solid epistemological basis than, say, the affine structure of spacetime. And, in the time of Ernst Mach, it would have appeared obvious that spatial relations are immediately observable, while the way in which momentary spaces are connected through time by dynamical principles like the law of inertia—what we would call spacetime structure—involves theoretical assumptions that require some further justification. But if any philosophical lesson of Einstein still stands, it is that this apparently obvious epistemological fact is a delusion: we are not immediately acquainted with the spatial relations among bodies at a given moment; these relations, together with the very notion of a "present moment," turn out to be theoretical constructions that implicitly rely on dynamical principles. In other words, the implication of special and general relativity that best deserves to be called "epistemological" is that the local causal and inertial structures of spacetime are empirically better known to us—are epistemologically more immediate— than the global "relative situation" of bodies at a moment. A "relational" theory that ignores this lesson may turn out to have enough advantages in predictive power and heuristic fruitfulness to make it worthwhile, on the whole, to abandon special and general relativity. But such a theory certainly can't claim any superiority on *epistemological* grounds. In effect such a theory would be, given what we understand now about simultaneity and spatial relations, Machian in letter but not in spirit: it would be a reduction of inertia to spatial relations, but it would go against Mach's methodological ideal of extrapolating as little as possible beyond what we are capable of observing.

Once again, it is not to be regretted that physical investigations are sometimes motivated by unreasonable philosophical prejudices for or against particular types of theory. It would be regrettable, however, if the philosophical contribution of someone like Ernst Mach were measured largely by the extent to which he expressed such prejudices, and the extent to which they proved fruitful for later physics. It would be likewise regrettable for the philosophical interest of general relativity to be measured largely by the extent to which it seems to satisfy such philosophical prejudices, or to support one or another traditional philosophical position. That would imply that the role of philosophy in physics is only to provide heuristic principles—metaphysical preferences, in effect—that motivate or discourage research in particular directions. And it would thereby ignore the contribution of philosophical *analysis* to the evolution of our understanding of space and time, especially the analysis of the connections between spatiotemporal structures and the physical processes that are supposed to exhibit them to us. General relativity might then be seen merely as a hypothesis that is empirically successful, independently of—if not in spite of—the philosophical preferences that inspired it, rather than as a product of philosophical insight into the relations between spacetime and the gravitational field. Clarifying the nature of that insight, however, is important for understanding the motivations for general relativity and their true relationship with the ideas of Mach.

Mach's theme of the relativity of rotation was behind Einstein's first glimpse of the possibility of using non-Euclidean geometry in physics. This came in his thought-experiment concerning a rigidly rotating disk: because of the Lorentz contraction of rods toward the edge of the disk, Einstein argued, the measured ratio between the circumference of the disk and its diameter will not have the Euclidean value π; he concluded that a rotating disk is equivalent to a nonrotating disk with a non-Euclidean geometry, or more generally, that the extension of the relativity of motion to include the relativity of rotation required the introduction of non-Euclidean geometry.[8] But this argument is far from warranting the use of non-Euclidean geometry to represent the equivalence of gravitation and inertia. That move was motivated—at least implicitly—by a philosophical argument of another kind: not the application of a Machian philosophical principle, but a conceptual analysis of our knowledge of the gravitational field and the role that it plays in our construction of inertial frames. As was noted above, the crucial step was the recognition that, because of the equivalence principle, the construction of a center of mass frame generally fails to identify an inertial frame; this implies that the analysis of a system of bodies in gravitational interaction, like Newton's dynamical analysis of the solar system, will identify not the absolute accelerations in an inertial frame, but the tidal gravitational accelerations in a freely falling frame. If free-fall trajectories are interpreted as the geodesics of spacetime—being, as Einstein showed, empirically indistinguishable from them—then these relative accelerations provide a direct measurement of the spacetime curvature. The central insight was not that certain states of motion might be represented equivalently as indications of a non-Euclidean geometry, but that a non-Euclidean geometry expresses the intrinsic structure underlying apparently equivalent representations of the same gravitational field.

This is a philosophical aspect of Einstein's investigation that does bear directly on the ideas of Mach. The essence of Mach's analysis of inertia, at least as I have interpreted it, is an analysis of the actual practice of constructing an inertial frame, a practice in which the fixed stars have played a historic role—perhaps for unknown physical reasons as well as for pragmatic ones, as Mach always emphasized. Mach's interpretation of this practice is deliberately quite restrictive in comparison with Newton's: for the methodological reasons already noted, Mach was reluctant to consider what this constructive process has in common with one that might take place in completely different circumstances, such as in the absence of the fixed stars. For the Newtonian understanding of the latter situation is entirely based on a "hypothetical extension" of the laws of motion, whereas to Mach, "it appears . . . that the circumstances in which we live, with their almost constant angles of directions to the stars, are an exceedingly special case, and I would not dare to infer from this case to a very different one" (1933, 235). But Mach's interpretation is quite restrictive also in comparison with Einstein's: by starting from the equivalence principle, Einstein's approach considers what the system studied by Newton has in common, not with a truly isolated system, but with every *local* system that admits of being treated in the same way—that is, with every system that may be treated as freely falling. Newton noted the equivalence between the treatment of, say, Jupiter and its moons as an inertial system and the treatment of the entire solar system, and even acknowledged the possibility that the latter, like the former, is only (in modern terms) a local inertial system that is involved in some interaction as part of a larger system. But only Einstein considered that this fundamental feature of gravitating systems reveals a fundamental feature of inertial systems as well, and therefore the deeper link between gravity and inertia. Thus both Einstein's and Newton's analyses of gravitational interactions involve hypothetical extensions of, or abstractions from, the empirical characteristics of accelerations relative to the fixed stars. Mach's chief merit was to have recognized Newton's theory as merely one possible such abstraction, and to have encouraged the pursuit of alternatives.

*   *   *   *   *   *

In comparison with the enthusiasm for Mach that first accompanied general relativity, as well as the subsequent reaction against it, this study places Mach's philosophy of space and time in a rather modest position. Mach's criticisms of Newton, on this interpretation, do not reflect a crude form of empiricism that is hostile to the use of abstract theory in physics. Nor—as Mach saw more clearly than his twentieth-century followers—do they constitute visionary proposals for an alternative "relational" theory of motion. To the extent that his criticisms do appear confused, they reflect some general confusion, among nineteenth-century physicists, about the status of absolute space in relation to the principles of Newtonian mechanics. And as that confusion began to clear, with the introduction of the concept of inertial system, Mach's objections to absolute space became clearly focused on the central issue, namely the incompatibility of absolute space with the Galilei-invariant structure that is implicit in Newtonian mechanics; his ability to distinguish that issue from general methodological issues concerning the laws of mechanics was, for his time, exemplary, even if his views on the methodological issues seem somewhat narrow-minded. In particular, he eventually understood the difference between a genuine philosophical objection to Newton's theory and a philosophical preference for a different kind of theory—a difference that twentieth-century Machian literature has had a tendency to blur.

It turns out that Mach's work played a subtle clarifying role in the philosophical background to general relativity, apart from the obvious motivational and heuristic role played by epistemological slogans about the relativity of motion and the importance of the "distant masses." To articulate the distinction between these two roles for philosophy in the development of physics was, I think, a central purpose of Stein's "Some philosophical pre-history of general relativity" (1977). For Stein's was far from the first account of that prehistory; the philosophical anticipations of general relativity have always been presented as part of the motivation and even the justification for the theory itself. Before Stein, however, the relevant factors in the philosophical prehistory were thought to be the epistemological slogans—that is, the philosophical demands for a theory that satisfies certain epistemological requirements, and the rejection of existing theories that are empirically adequate just because the other sort of theory can be imagined. On Stein's account, in contrast, the most significant factor is the analysis of the interconnections between geometry and physics, and of how these interconnections evolve as geometry and physics evolve—a kind of analysis that certainly does deserve to be called philosophical, but that can't be simply glossed as the application of any particular philosophical rule. And the result is not merely a more sophisticated epistemological perspective on spacetime theories, but a deeper understanding of the "world structure" and the ways in which this structure can be revealed by physical phenomena. This is the genuine philosophical prehistory of general relativity. I hope I have shown that it includes, after all, a modest contribution from Ernst Mach.

NOTES

1. See, for example, Holton 1988 and Earman 1989. Wolters (1988) makes a powerful case that Mach's famous rejection of relativity, published posthumously in the preface to his book *On the Theory of Heat*, was not written by Mach himself, and does not reflect Mach's true views. Wolters argues that the preface was a forgery by Mach's son Ludwig, a drug addict in need of money and patronage, who came under the influence of Hugo Dingler and came to support Dingler's anti-Einstein project. (See also DiSalle 1990.) One can only wonder how many other famous philosophers are unfairly blamed for bad arguments that are, in real-

ity, forgeries by drug addicts. Descartes's "cogito" and Quine's attack on the analytic-synthetic distinction come to mind.

2. For the most prominent and influential example, see Reichenbach 1957. For an account of the most extreme interpretations of Newton's views, see Stein 1967.

3. A prominent recent example is the work of Barbour and Bertotti (see Barbour 1999). For a critical general discussion of Machian literature on rotation see Earman 1989, chapter 5.

4. Citations from *Die Mechanik* will generally be from the second edition (1889), which differs from the first (1883) only in the appendix; from the fourth edition (1901), in which most of the appendix to the second edition has been incorporated into the body of the text; and from the ninth edition (1933), which is unchanged from the seventh, the last published in Mach's lifetime.

5. See Torretti 1983 and DiSalle 1988, 1991.

6. See, for example, Misner, Thorne, and Wheeler, 1973; Wald 1983; Friedman 1983; Earman 1989.

7. This is the aspect of Stein (1967) that has gone unnoticed in most of the vast literature in which that paper is cited, even where it is cited approvingly: Stein's defense of Newton is read as a heavy blow for the "absolutist" side of the absolute-relational debate, when in fact it is a destructive criticism of that debate itself; Stein showed that Newton was not arguing for the "absoluteness" of rotation in some previously-understood sense, but was presenting a definition of absolute rotation and showing how to apply it. Therefore it would not make sense to continue asking whether Newton's bucket argument settles the question whether rotation is absolute or relative, for the question is not antecedently well posed. See also DiSalle 1995.

8. For discussion of Einstein's argument see Stachel 1980; Torretti 1983; and Friedman, this volume.

REFERENCES

Barbour, J. (1999). *The End of Time: The Next Revolution in Physics.* Oxford and New York: Oxford University Press.

Barbour, J., and H. Pfister. (1995). *Mach's Principle: From Newton's Bucket to Quantum Gravity.* Vol. 6 of *Einstein Studies.* Boston: Birkhauser.

Brown, H. (1996). "Bovine Metaphysics: Remarks on the Significance of the Gravitational Phase Effect in Quantum Mechanics." In *Perspectives on Quantum Reality,* ed. R. Clifton. Dordrecht: Kluwer Academic Publishers, 183–94.

Dicke, R. (1964). *The Theoretical Significance of Experimental Relativity.* New York: Gordon and Breach.

DiSalle, R. (1988). *Space, Time and Inertia in the Foundations of Newtonian Physics.* Unpublished Ph.D. dissertation, University of Chicago.

———. (1990). "Critical Notice: Gereon Wolters' *Mach I, Mach II, Einstein, und die Relativitätstheorie. Eine Fälschung und ihre Folgen. Philosophy of Science* 57: 712–23.

————. (1991). "Conventionalism and the Origins of the Inertial Frame Concept." *PSA 1990*. East Lansing: The Philosophy of Science Association.

————. (1993). "Carl Neumann." *Science in Context* 6: 345–54.

————. (1995). "Spacetime Theory as Physical Geometry." *Erkenntnis* 45: 317–37.

Earman, J. (1989). *World Enough and Spacetime: Absolute and Relational Theories of Motion*. Boston: M.I.T. Press.

Einstein, A. (1905). "On the Electrodynamics of Moving Bodies." In Einstein et al. 1952, 35–65.

————. (1916). "The Foundation of the General Theory of Relativity." In Einstein et al. 1952, pp. 109-164.

————. (1917). *Über die spezielle und die allgemeine Relativitätstheorie (Gemeinverständlich)*. 2nd edition. Braunschweig: Vieweg und Sohn.

————. (1949). "Autobiographical Notes." In *Albert Einstein, Philosopher-Scientist*, edited by P. A. Schilpp.Vol. 7 of The Library of Living Philosophers. Chicago: Open Court.

Einstein, A., H. A. Lorentz, H. Minkowski, and H. Weyl. (1952). *The Principle of Relativity*. Translated by W. Perrett and G. B. Jeffery. New York: Dover Books.

Friedman, M. (1983). *Foundations of Space-Time Theories*. Princeton: Princeton University Press.

Holton, G. (1988). *Thematic Origins of Scientific Thought*. Cambridge, Mass.: Harvard University Press. Revised edition.

Lange, L. (1885). "Ueber das Beharrungsgesetz." *Berichte der Königlichen Sachsischen Gesellschaft der Wissenschaften zu Leipzig, Mathematisch-physische Classe* 37 (1885): 333–51.

Mach, E. (1872). *Die Geschichte und die Wurzel des Satzes von der Erhaltung der Arbeit*. Prague: J.G. Calve'sche K.-K. Univ-Buchhandlung.

————. (1889). *Die Mechanik in ihrer Entwickelung, historisch-kritisch dargestellt*. 2nd edition. Leipzig: Brockhaus.

————. (1901). *Die Mechanik in ihrer Entwickelung, historisch-kritisch dargestellt*. 4th edition. Leipzig: Brockhaus.

————. (1933). *Die Mechanik in ihrer Entwickelung, historisch-kritisch dargestellt*. 9th edition. Leipzig: Brockhaus.

Minkowski, H. (1908). "Space and Time." In Einstein et al. 1952, 75–91.

Misner, C., K. Thorne, and J. A. Wheeler. (1973). *Gravitation*. San Francisco: Freeman.

Neumann, C. (1870). *Ueber die Principien der Galilei-Newton'schen Theorie*. Leipzig: B. G. Teubner, 1870.

Newton, I. (1726). *The Principia: Mathematical Principles of Natural Philosophy*. Translated by I. Bernard Cohen and Anne Whitman. Berkeley and Los Angeles: University of California Press, 1999.

————. (1729). *Sir Isaac Newton's Mathematical Principles of Natural Philosophy and his System of the World*. 2 vols. Edited by Florian Cajori. Translated by Andrew Motte. Berkeley: University of California Press, 1962.

Reichenbach, H. (1957). *The Philosophy of Space and Time*. Translated by Maria Reichenbach. New York: Dover Publications.

Russell, B. (1897). *An Essay on the Foundations of Geometry.* Cambridge: Cambridge University Press

Stachel, J. (1980). "Einstein and the Rigidly Rotating Disk." Vol. 1, pp. 1–15. In *General Relativity and Gravitation,* edited by A. Held. New York: Plenum.

Stein, H. (1967). "Newtonian Space-Time." *Texas Quarterly* 10: 174–200.

———. (1977). "Some Philosophical Prehistory of General Relativity." In *Foundations of Space-Time Theories,* edited by John Earman, Clark Glymour, and John Stachel, 3–49. Vol. 8 of *Minnesota Studies in the Philosophy of Science.* Minneapolis: University of Minnesota Press.

Thomson, J. (1884). "On the Law of Inertia; the Principle of Chronometry; and the Principle of Absolute Clinural Rest, and of Absolute Rotation." *Proceedings of the Royal Society of Edinburgh* 12: 568–78.

Torretti, R. (1983). *Relativity and Geometry.* Oxford: Pergamon Press.

Wald, R. (1984). *General Relativity.* Chicago: University of Chicago Press.

Will, C. (1993). *Theory and Experiment in Gravitational Physics.* Cambridge: Cambridge University Press.

Wolters, G. (1987). *Mach I, Mach II, Einstein, und die Relativitätstheorie. Eine Fälschung und ihre Folgen.* Berlin: De Gruyter.

[8]

Geometry as a Branch of Physics: Background and Context for Einstein's "Geometry and Experience"

MICHAEL FRIEDMAN

Albert Einstein's celebrated paper, "Geometrie und Erfahrung," first presented as a lecture to the Prussian Academy of Sciences at Berlin in January of 1921 and then published in "expanded" form later in that same year, is a landmark in the philosophy of geometry.[1] In particular, it provided a very clear and sharp version of the distinction between "pure" and "applied"—mathematical and physical—geometry that soon became canonical in twentieth-century scientific thought.[2] According to this conception, mathematical geometry derives its certainty and purity from its "formal-logical" character as a mere deductive system operating with "contentless conceptual schemata." The primitive terms of mathematical geometry, such as "point," "line," "congruence," and so on, do not refer to objects or concepts antecedently given (by some sort of direct intuition, for example), but rather have only that purely "formal-logical" meaning stipulated in the primitive axioms. These axioms serve therefore as "implicit definitions" of the primitive terms, and all the theorems of mathematical geometry then follow purely logically from the stipulated axioms:

> Geometry treats of objects that are designated with the words line, point, etc. No kind of acquaintance or intuition of these objects is presupposed, but only the validity of those axioms which are likewise to be conceived as purely formal, i.e., as separated from every content of intuition and experience. . . . These axioms are free creations of the human spirit. All other geometrical propositions are logical consequences of the (only nominalistically conceived) axioms. The axioms first define the objects of which geometry treats. Schlick therefore designated the axioms very appropriately as "implicit definitions" in his book on theory of knowledge.
>
> The conception represented by modern axiomatics purifies mathematics from all elements not belonging to it, and thus removes the mystical obscurity that previously clung to the foundations of mathematics. But such a purified

presentation makes it also evident that mathematics as such may assert nothing
about either objects of intuitive representation or objects of reality. In axiomatic
geometry we understand by "point," "line," etc. only contentless conceptual
schemata. What gives them content does not belong to mathematics.[3]

Applied or physical geometry, by contrast, arises when one gives some definite and, as it were, extra-axiomatic interpretation of the primitive terms
via real objects of experience. But now the purity and certainty of mathematical geometry (which, in the end, rests simply on the purity and certainty of logic) is irrevocably lost, and we end up with one more *empirical*
science among others:

> In so far as the propositions of mathematics refer to reality they are not certain;
> and in so far as they are certain they do not refer to reality. Full clarity about
> the situation appears to me to have been first obtained in general by that ten-
> dency in mathematics known under the name of "axiomatics." The advance
> achieved by axiomatics consists in having cleanly separated the formal-logical
> element from the material or intuitive content. According to axiomatics only
> the formal-logical element constitutes the object of mathematics, but not the
> intuitive or other content connected with the formal-logical element.[4]

Thus, these famous words from Einstein's paper, which were clearly
intended and were standardly taken as a refutation of the Kantian conception that mathematics is the paradigm of synthetic a priori truth, are a vivid
expression of the modern axiomatic conception of geometry we now associate with the work of David Hilbert.[5]

Einstein's paper also presents a particular conception of physical as
opposed to purely mathematical geometry—a particular view of what a
"physical interpretation" of purely mathematical geometry amounts to.
We obtain such an interpretation, according to Einstein, when we relate
the axiomatic structure of pure geometry to "practically rigid bodies" and
their "situational possibilities [*Lagerungsmöglichkeiten*]." In particular, by
bringing intervals marked on such bodies (in the form of rigid measuring
rods, for example) into coincidence with one another, we can empirically
determine relations of distance and thereby obtain a physical interpretation
of "congruence." In this way, we obtain what Einstein calls "practical" as
distinct from "pure axiomatic geometry":

> It is clear that the conceptual system of axiomatic geometry alone can supply
> no assertions about the behavior of those objects of reality that we wish to des-
> ignate as practically rigid bodies. In order to be able to furnish such assertions
> geometry must go beyond its solely formal-logical character, so that experi-
> enceable objects of reality (experiences) are coordinated [*zugeordnet*] to the
> empty conceptual schemata of axiomatic geometry. In order to accomplish this
> one needs only to add the proposition:

Solid bodies relate to one another with respect to their situational possibilities as do bodies in three-dimensional Euclidean geometry; then the propositions of Euclidean geometry contain assertions about the behavior of practically rigid bodies.

The thus expanded geometry is obviously a natural science; we can actually consider it as the oldest branch of physics. Its assertions rest essentially on inductions from experience, and not only on logical inferences. We wish to call the thus expanded geometry "practical geometry" and to distinguish it in what follows from "pure axiomatic geometry." The question whether the practical geometry of the world is Euclidean or not has a clear sense, and its answer can only be supplied by experience. All length measurement in physics is practical geometry in this sense—geodetic and also astronomical, if one uses the empirical proposition that light propagates in a straight line, precisely a straight line in the sense of practical geometry.[6]

Practical geometry coordinates the "empty conceptual schemata" of pure geometry with real physical bodies (practically rigid bodies) and thereby transforms geometry into an "interpreted" empirical science.

Einstein closely associates this overall conception of geometry with his own recently developed general theory of relativity. Indeed, immediately following the passage last quoted he states: "I attach particular importance to this conception of geometry, because without it I would have found it impossible to establish the theory of relativity." And the bulk of the paper, immediately following the opening, more philosophical, part, is devoted to explaining how the general theory of relativity might lead us to recognize that the universe as a whole is spatially finite. Einstein's appeal to his own recently developed theory is in once sense perfectly natural. For it is in this theory, of course, that non-Euclidean geometry has first been successfully applied to nature, so that the idea of geometry as an empirical physical science has finally received a concrete realization. Yet it is by no means obvious, from a philosophical point of view, why the sharp distinction between pure or axiomatic and applied or physical geometry should be specifically associated with the theory of relativity. Einstein is himself well-aware, in particular, that this distinction had most recently become prominent in connection with Hilbert's celebrated axiomatization of Euclidean geometry in 1899 (see note 5)—a development that occurred some fifteen years before Einstein's own application of non-Euclidean geometry in physics. So the distinction has, on the face of it, no special relevance to this later development, and it makes just as much (or as little) sense in the context of the traditional use of Euclidean geometry in classical physics. Indeed, it makes just as much (or as little) sense in the context of everyday applications of Euclidean geometry in surveying and measurement (which Einstein calls "the oldest branch of physics") as it does in the context of the very complex and sophisticated application of non-Euclidean geometry employed by Einstein's new theory.

As a matter of fact, and perhaps even more importantly, Einstein's particular explanation of applied or physical geometry actually appears to be *more* appropriate to everyday applications in surveying and measurement that it does to the general theory of relativity. For it is not as if, in the general theory, we find that space is non-Euclidean by actually performing surveying experiments with rigid measuring rods. Rather, we postulate a highly theoretical link between gravitation and spatiotemporal curvature, on the basis of which it then emerges from some very abstract mathematics that the spatial region in the neighborhood of the sun, for example, is described by a non-Euclidean geometry. This conclusion can then be tested, of course, but the whole procedure bears very little resemblance to the surveyor's conception of applied geometry Einstein articulates in "Geometrie und Erfahrung."[7] The first point I want to emphasize, then, is that Einstein's close association of the above philosophical conception of geometry with the general theory of relativity can and should appear extremely puzzling.

* * * * * *

In order to begin making philosophical sense of Einstein's argument, we need first to consider it against the background of a preceding conception of geometry—one that was dominant in the nineteenth century, but has now, largely through the influence of "modern" views like Einstein's, receded far into the background in contemporary philosophical discussion. This earlier tradition had its home in projective geometry and group theory, and it found its canonical expression in the famous Erlanger program of Felix Klein.[8] Here, as Klein himself puts it, pure or mathematical geometry is by no means conceived as an "empty conceptual schema," but is rather understood as necessarily connected to our "spatial intuition":

> In the case of such people who are only interested in the logical side of the question, and not in the intuitive or general-epistemological side, one often finds the opinion nowadays that *the axioms are only arbitrary propositions that we set up entirely freely, and the fundamental concepts, ultimately, are also only arbitrary signs for things with which we wish to operate.* What is correct in such a view, of course, is that *within pure logic* no basis for these propositions and concepts is found, and that they therefore must be furnished or suggested from another side—precisely by the influence of intuition. However, the authors [in question] often express themselves much more one-sidedly, and so we are repeatedly forced nowadays, in connection with modern axiomatics, straightaway once again into that philosophical position which has been known since ancient times as *nominalism*. Here the interest in the things themselves and their properties is entirely lost; and one speaks only of how they are to be named and in accordance with which logical schema they are to be operated.

One then says, for example, that we *call* a triple of coordinates a point "without thereby thinking of anything," and we stipulate "arbitrarily" certain propositions that are to be valid of these points; one can set up arbitrary axioms in an entirely unlimited way, so long as one satisfies the laws of logic and takes care, above all, that there is no contradiction in the ensuing structure of theorems. I myself in no way share this standpoint, but take it to be the death of all science: *the axioms of geometry*—in my opinion—*are not arbitrary but rather rational propositions, which are motivated, in general, by spatial intuition and regulated, in their particular content, by considerations of purposiveness* [*Zweckmäßigkeitsgründe*].[9]

Mathematical geometry, on this view, describes the most general and abstract features of our perception of space. These features are not precise and specific enough to yield Euclidean geometry in particular, however, but only that structure common to the three classical geometries of constant curvature (Euclidean, hyperbolic, and elliptic), which, in terms of the Erlanger program, emerge naturally within the more general framework of projective geometry and group theory. So only considerations of convenience and expediency (especially simplicity), not deliverances of our spatial intuition, can then explain our choice of specifically Euclidean geometry. This view of geometry has obvious roots in the conception articulated and defended by Kant at the end of the eighteenth century, but it aims to *generalize* the Kantian picture to take account of nineteenth-century discoveries in projective and non-Euclidean geometry.[10]

For our present purposes, the most important result of this tradition is what we now call the Helmholtz-Lie theorem, which was first articulated by Hermann von Helmholtz in connection with his psycho-physiological researches into space perception and then rigorously proved by Sophus Lie (at the instigation of his teacher Klein) within Lie's theory of continuous groups. Helmholtz was inspired by Bernhard Riemann's great *Habilitationsvortrag* of 1854, "On the Hypotheses which Lie at the Basis of Geometry," to attempt to derive Riemann's fundamental assumption, that the line-element is of Pythagorean or infinitesimally Euclidean form, from what Helmholtz took to be the fundamental "facts" generating our perceptual intuition of space.[11] Helmholtz's starting point is that our idea of space is in no way immediately given or "innate," but instead arises by a process of perceptual accommodation or learning based on our experience of bodily motion. Since our idea of space arises kinematically, as it were, from our experience of moving up to, away from, and around the objects that "occupy" space, the space thereby constructed must satisfy a condition of "free mobility" that permits arbitrary continuous motions of rigid bodies.[12] And from this latter condition one can then derive the Pythagorean form of the line-element. In Lie's formulation, given a group of transformations on a manifold such that, intuitively, for any two "observers" or "points of view" there is exactly one transformation in the group mapping one onto the other, there is a unique—up to a scale factor—Riemannian metric on the manifold whose isometries are given precisely by the group in question.[13] Since, however, the Riemannian metric thereby constructed has a group of isometries or rigid motions mapping any point onto any other, it must have constant curvature as well. So the scope of the Helmholtz-Lie theorem (and the entire Kleinian tradition) is much less general than the full Riemannian theory of metrical manifolds, which of course also includes manifolds of *variable* curvature.

The Helmholtz-Lie theorem fixes the geometry of space—and, according to Helmholtz, thereby expresses the "necessary form of our outer intuition"—as one of the three classical geometries of constant curvature: Euclidean, hyperbolic, or elliptic.[14] But how do we know which of the three classical geometries actually holds? At this point, on Helmholtz's view, we investigate the actual behavior of rigid bodies (of rigid measuring rods, for example) as we move them around in accordance with the condition of free mobility. That physical space is Euclidean (which Helmholtz of course assumes) means that actual measurements carried out in this way are empirically found to satisfy the laws of this particular geometry to a very high degree of exactness.

It is precisely this Helmholtzian conception of geometry which then sets the stage for the contrasting "conventionalist" view articulated by Henri Poincaré—a view falling squarely within the Kleinian tradition. Indeed, Poincaré developed his philosophical conception immediately against the background of the Helmholtz-Lie theorem, and in the context of his own mathematical work on group theory and models of hyperbolic geometry.[15] Like Helmholtz, Poincaré views geometry as the abstract study of the group of motions associated with our initially crude experience of bodily "displacements." So we know, according to the Helmholtz-Lie theorem, that the space thereby constructed has one and only one of the three classical geometries of constant curvature. Poincaré disagrees with Helmholtz, however, that we can empirically determine the particular geometry of space simply by observing the behavior of rigid bodies. No real physical bodies exactly satisfy the condition of geometrical rigidity, and, what is more important, knowledge of physical rigidity presupposes knowledge of the forces acting on the material constitutions of bodies. But how can one say anything about such forces without first having a geometry in place in which to describe them? We have no option, therefore, but to *stipulate* one of the three classical geometries of constant curvature, by convention, as a framework within which we can then do empirical physics:

> Geometry is not an experimental science; experience forms merely the occasion for our reflecting upon the geometrical ideas which pre-exist in us. . . .

What we call geometry is nothing but the study of formal properties of a certain continuous group; so that we may say, space is a group. The notion of this continuous group exists in our mind prior to all experience; but the assertion is no less true of the notion of many other continuous groups; for example, that which corresponds to the geometry of Lobachevsky. There are, accordingly, several geometries possible, and it remains to be seen how a choice is made between them. Among the continuous mathematical groups which our mind can construct, we choose that which deviates least from that rough group, analogous to the physical continuum, which experience has brought to our knowledge as the group of displacements.

Our choice is therefore not imposed by experience. It is simply guided by experience. But it remains free; we choose this geometry rather than that geometry, not because it is more true, but because it is more convenient.[16]

The argument here, more specifically, is that establishing mathematical force-laws (underlying the physical notion of rigidity) presupposes that we already have a geometry in place in order to make spatial measurements; so we must *first* choose a particular geometry and then subsequently investigate physical forces.[17] Moreover, since Euclidean geometry is mathematically the simplest, Poincaré has no doubt at all that this particular stipulation will always be preferred. As he says following the above-quoted passage: "We choose the geometry of Euclid because it is the simplest. . . . [I]t is simpler because certain of its displacements are interchangeable with one another, which is not true of the corresponding displacements of the group of Lobachevsky."[18]

Now, as Einstein makes clear, Poincaré's geometrical conventionalism constitutes the immediate philosophical context against which the conception of applied or physical geometry articulated in "Geometrie und Erfahrung" is developed. Indeed, Einstein presents Poincaré's conventionalism as the only live alternative to his own view of the importance of practically rigid bodies—without which, according to Einstein, he "would have found it impossible to establish the theory of [general] relativity." Poincaré's view must be rejected precisely because it makes it impossible to see how one could empirically discover that space has a non-Euclidean geometry:

If one rejects the relation between the bodies of axiomatic Euclidean geometry and the practically rigid bodies of reality, then one easily arrives at the following conception, which the perceptive and deep thinker H. Poincaré has, in particular, embraced. Euclidean geometry is picked out from all other thinkable axiomatic geometries by simplicity. And since axiomatic geometry *alone* contains no assertions about experienceable reality, but only axiomatic geometry in connection with propositions of physics, then it may be possible and rational—no matter how reality may be constituted—to hold fast to Euclidean geometry. For one will prefer to opt for an alteration of the laws of physics rather than alter the laws of geometry in case there are contradictions between theory and experience. If one rejects the relation between the practically rigid body and geometry one will in fact not easily free oneself from the convention according to which Euclidean geometry is to be held fast as the simplest.[19]

Here Poincaré's famous example of a putatively non-Euclidean world from the fourth chapter of *Science and Hypothesis* comes naturally to mind.[20] If we lived in such a world we could either say that we lived in an infinite hyperbolic space or that we lived in the interior of a finite Euclidean sphere endowed with a peculiar temperature field affecting all bodies in the same way (contracting them to infinity as they radially approach the bounding spherical surface).[21] The implication is clearly that we would always opt for the second, Euclidean alternative, which means, in Einstein's terms, that we would not treat the bodies in question as practically rigid.

Einstein explicitly acknowledges that Poincaré's conception has much to recommend it in another very well-known passage:

> Why is the naturally suggested equivalence of the practically rigid bodies of experience and the bodies of geometry rejected by Poincaré and other investigators? Simply because the actual rigid bodies of nature, on closer consideration, are not rigid, because their geometrical behavior—i.e., their relative situational possibilities [*Lagerungsmöglichkeiten*]—depend on temperature, external forces, etc. The original, immediate relation between geometry and physical actuality appears thereby to be destroyed, and we feel ourselves forced to the following more general conception, characteristic of Poincaré's standpoint. Geometry (*G*) asserts nothing about the behavior of actual things, but only geometry together with the totality (*P*) of physical laws. We can say, symbolically, that only the sum (*G*) + (*P*) is subject to the control of experience. So (*G*) can be chosen arbitrarily, and also parts of (*P*); all of these laws are conventions. In order to avoid contradictions it is only necessary to choose the remainder of (*P*) in such a way that (*G*) and the total (*P*) together do justice to experience. On this conception axiomatic geometry and the part of the laws of nature that are raised to conventions appear as epistemologically of equal status.
>
> *Sub specie aeterni* Poincaré, in my opinion, is correct in this conception. The concept of measuring rod, and also the concept of clock that is coordinated to it in relativity theory, has no exactly corresponding object in the actual world. It is also clear that the rigid body and clock do not play the role of irreducible elements in the conceptual framework of physics, but rather the role of composite structures that cannot have any independent status in the construction of theoretical physics. But it is my conviction that these concepts must still be called upon as independent elements in the present stage of theoretical physics; for we are still far from such a secure knowledge of the theoretical foundations that would enable us to give exact theoretical constructions of such structures.[22]

Here Einstein agrees with Poincaré, as against Helmholtz, that rigid bodies cannot ultimately be taken as primitive elements from which one could empirically read off, as it were, a particular physical geometry. For the notion of physical rigidity is inextricably entangled with very difficult questions about the microstructure of matter and its dynamics. Nevertheless, Einstein is also of the opinion that we must *provisionally* take rigid bodies as primitive and thus *provisionally* side with Helmholtz, against Poincaré, on this crucial question about the interpretation of physical geometry. Only so, Einstein suggests, could the general theory of relativity have been established in the first place—antecedently, that is, to the further development of microphysics.

* * * * * *

We noted above that the nineteenth-century mathematical tradition in projective geometry and group theory that culminated in the work of Klein and Poincaré did not draw the now familiar distinction between pure or axiomatic geometry and applied or interpreted geometry associated with Hilbert and his followers.[23] Mathematical geometry, in this earlier tradition, continued to be a (fully interpreted) theory of space—the very same space in which we live, move, and perceive. It was in precisely this way, in fact, that the mathematical science of geometry differed from analysis and number theory, the much more abstract sciences of "magnitude" in general. Geometry studies a particular object of thought given to us through perception or intuition, whereas number theory and analysis study all objects of thought in any way countable or measurable. So these latter sciences, unlike geometry, are not limited by the nature of our sensible intuition.[24] While geometry studies the particular nature of our spatial intuition, however, it is concerned only with the general "form" of this intuition independently of all physical or empirical content. Just as in the earlier Kantian conception, therefore, pure or mathematical geometry becomes applied when it is used to describe the behavior of physical objects or bodies located within intuitive space. For Klein and Poincaré, the crucial divergence from the Kantian conception is that the very general features of space studied in projective geometry—that is, the general features of perceptual "perspectives" and their possible interrelationships—do not include its specific metrical character: all three classical geometries of constant curvature fit equally well into the overarching general structure. It is in applied geometry, therefore, that a particular choice of metrical geometry is then somehow determined.

As we also pointed out above, the theory of metrical manifolds created by Riemann extends far beyond the Kleinian projective framework. Riemannian manifolds of variable curvature, in particular, are not encompassed within projective geometry, and therefore signal a sharp and decisive departure from the "perspectival" connection with spatial intuition characteristic of the latter tradition. The general theory of Riemannian manifolds is rather developed analytically—initially as a theory of (continuous) number manifolds. From this point of view, we therefore have a distinction between pure and applied geometry closely analogous to the now familiar distinction associated with the work of Hilbert: pure geometry is a branch of analysis having no necessary relationship to our spatial intuition, whereas applied geometry is a branch of natural science in which we formulate and investigate (analytically formulated) hypotheses about the metrical structure of the physical world.[25] Moreover, this way of viewing the distinction between pure and applied geometry depends crucially on the greater generality of the manifolds or "spaces" under considerations. Thus, whereas Helmholtz, for example, can and does distinguish between the concept of space given by an analytic Riemannian treatment and the intuition of space arising from our actual perceptual experience,[26] both the concept and the intuition of space are nonetheless limited by constant curvature. Since Helmholtz takes the condition of free mobility as a necessary presupposition of the possibility of any concept of spatial magnitude at all, the (Riemannian) concept (of a space of constant curvature) simply formulates analytically the conditions necessarily present in our spatial intuition, and applied or physical geometry then merely has the task of determining which particular geometry of constant curvature actually holds in reality.

For Poincaré, too, the more general theory of Riemannian manifolds including cases of variable curvature extends far beyond the bounds of admissible geometries. Poincaré explicitly characterizes this theory as a purely analytic rather than a synthetic science, which, for precisely this reason, cannot itself qualify as a theory of *space*—as a branch of geometry in the traditional sense:

> If therefore the possibility of motion is admitted, there can be invented only a finite (and even a rather small) number of three-dimensional geometries. Yet this result seems contradicted by Riemann, for this savant constructs an infinity of different geometries, and that to which his name is ordinarily given [namely, elliptical geometry] is only a particular case. All depends, he says, on how the length of a curve is defined. Now, there is an infinity of ways of defining this length, and each of them may be the starting point of a new geometry.
>
> That is perfectly true, but most of these definitions are incompatible with the motion of a rigid figure, which in the theorem of Lie is supposed possible. These geometries of Riemann, in many ways so interesting, could never therefore be other than purely analytic and would not lend themselves to demonstrations analogous to those of Euclid.[27]

Genuine geometry, for Poincaré, must be synthetic and, like Euclid's original axiomatization, necessarily connected with our intuition of space.[28]

Since, however, our spatial intuition turns out to be more general than the Euclidean, no particular metrical structure (as above) is thereby determined. If we then want to apply geometry in physics, according to Poincaré, we have no choice but to *stipulate* such a metrical structure (of constant curvature) in advance.

The general theory of relativity, however, uses a four-dimensional manifold of *variable* curvature, in which straightest possible paths or geodesics represent the space-time trajectories of bodies in a gravitational field. Moreover, three-dimensional spatial geometries as well, such as the Schwarzschild geometry in the neighborhood of the sun (note 7 above), are also, in general, of variable rather than constant curvature. So general relativity, in sharp contrast to both classical physics and everyday surveying techniques, is *not* compatible with the principle of free mobility. We are forced, therefore, to use the more general Riemannian theory of metrical manifolds, wherein the Helmholtz-Lie theorem—and the entire projective tradition culminating in Klein's Erlanger program—no longer have application or meaning. The generalized conception of spatial intuition lying at the heart of this tradition is thus also incompatible with Einstein's new theory, and we are rather left with a sharp distinction between pure and applied geometry in the tradition of Riemann.

We saw that Einstein appeals to Moritz Schlick's articulation of the notion of "implicit definitions" in the latter's *Allgemeine Erkenntnislehre*.[29] And Schlick, in turn, explicitly links this notion with Hilbert's *Foundations of Geometry* (note 5 above). On this basis, Schlick erects a rigid distinction between conceptual and intuitive-perceptual knowledge that is entirely foreign to the thought of both Helmholtz and Poincaré; and, what is of even more importance in the present context, Schlick then employs what he calls a "method of coincidences" directly inspired by the general theory of relativity to rebuild the bridge between conceptual thought (now conceived as an "empty schema") and physical reality.[30]

Schlick explains how we set up the crucial relation of "designation [*Bezeichnung*]" or "coordination [*Zuordnung*]" between our abstract system of concepts (given by Hilbert-style implicit definitions) and empirical reality in a section entitled "Quantitative and Qualitative Knowledge," which is principally devoted to explaining how we construct the objective or "transcendent" spatiotemporal ordering employed in physics on the basis of the intuitive, psychologically subjective spatiotemporal ordering present in the immediately given data of consciousness.[31] The former ordering is conceptually describable and knowable [*erkennen*], whereas the latter is directly presented in nonconceptual intuitive acquaintance [*kennen*]. So what turns out to be primarily knowable in the method of coincidences is the *quantitative* structure of the "transcendent" ordering thereby effected, not the purely *qualitative* structure of the immediately given subjective data of consciousness.

We construct the "transcendent" ordering, more specifically, on the basis of singularities or coincidences in our various intuitively given sensory fields. For example, I see the tip of my pencil touch my finger in my visual field and, at the same time, feel its touch on my finger in my tactile field. The intuitive spatiality of these two sensory fields is entirely different in the two cases, and they have, as such, no intuitive spatial relations to one another. I then bring them into relation by constructing a single, nonintuitive spatial ordering containing both the pencil and my finger, where a single point in objective space (the coincidence of my finger with the pencil tip) corresponds to both singular points in the two previously independent sensory fields. In this procedure I abstract completely from the qualitative peculiarities of my sensory fields (color, tactile quality, and so on) and concentrate solely on their purely topological properties—the presence or absence of a singularity. And this focus on singularities or coincidences is also crucial from a scientific point of view, for it is precisely on the basis of such coincidences that the technique of numerical measurement now proceeds. We measure objective spatial intervals by observing the coincidences of the end-points of a measuring rod with points on a measured object; we measure objective temporal intervals by observing coincidences between events in a given natural process and pointer positions on a clock; and so on: "[A]ll measurement, from the most primitive to the most sophisticated, rests on the observation of spatio-temporal coincidences such as those described above."[32] The result, in the end, is a numerical model of our initially uninterpreted system of implicit definitions generated by actual empirical measurements of real physical processes.

In *Space and Time in Contemporary Physics*, written virtually simultaneously with *Allgemeine Erkenntnislehre*, Schlick explains the significant connection he perceives between this method of coincidences, on the one hand, and the general theory of relativity, on the other.[33] The most important chapter of the latter work, in this connection, is entitled "The General Postulate of Relativity and the Metrical Determination of the Space-Time Continuum," where Schlick draws a fundamental contrast between general relativity and both Newtonian physics and special relativity. In the case of both of the latter two theories he explains how space was presupposed to have a given and independent background metrical structure (a Euclidean metrical structure), which could then be straightforwardly determined by rigid measuring rods:

> [Space] still preserved a certain objectivity, so long as it was still tacitly thought as equipped with completely determined metrical properties. In the older

physics one based every measurement procedure, without hesitation, on the idea of a rigid rod, which possessed the same length at all times, no matter at which place and in which situation and environment it may be found, and, on the basis of this thought, all measurements were determined in accordance with the precepts of Euclidean geometry. . . . In this way, space was still left with a "Euclidean structure," as a separate and independent [*selbständig*] property, as it were, for the result of these metrical determinations was thought to be entirely independent of the physical conditions prevailing in space, e.g., of the distribution of bodies and their gravitational fields.[34]

But this is emphatically not the case in Einstein's new theory:

> If we want, therefore, to maintain the general postulate of relativity in physics, we must refrain from describing measurements and situational relations [*Lagebeziehungen*] in the physical world with the help of Euclidean methods. However, it is not that, in place of Euclidean geometry, a determinate other geometry—e.g., Lobachevskian or Riemannian—would now have to be used for the whole of space, so that our space would be treated as pseudospherical or spherical, as mathematicians and philosophers are accustomed to imagine this. Rather, the most various kinds of metrical determinations are to be employed, in general, different ones at each position, and what they are now depends on the gravitational field at each place.[35]

Space and time in general relativity now have no background geometry at all—neither Euclidean nor non-Euclidean—that would be determined independently of the distribution of matter therein; and, according to the general postulate of relativity (the principle of general covariance), the only background that remains is the topological or manifold structure of number quadruples, that is, the space-time coincidences, so that "the whole of physics can be conceived as a totality of laws in accordance with which the occurrence of these space-time coincidences takes place."[36]

In the final chapter, entitled "Relations to Philosophy," Schlick explains the significance of Einstein's new view of space and time for epistemology. He points out that the objective spatial structure employed by physics is not intuitively given, but is rather a "*conceptual construction*," that is, a "non-intuitive ordering, which we then call objective space and conceptually grasp through a manifold of numbers (coordinates)."[37] Yet this objective conceptual construction proceeds, just as in *Allgemeine Erkenntnislehre*, on the basis of the subjective spatiotemporal coincidences present in the diverse sensory fields of various individuals:

> In order to fix a point in space, one must somehow, directly or indirectly, *point* to it . . . , that is, one establishes a spatio-temporal coincidence of two otherwise separate elements. And it now turns out that these coincidences always occur in agreement for all intuitive spaces of different senses and all individu-

als; precisely so is an objective "point," independent of individual experiences and valid for all, thereby defined. . . . By closer consideration one easily finds that we attain to the construction of physical space and time exclusively through this method of coincidences and in no other way. The space-time manifold is nothing else than the totality of objective elements defined through this method. That it is precisely a four-dimensional manifold results from experience by the execution of this method itself.

This is the result of the psychological-epistemological analysis of the concepts of space and time, and we see that we encounter precisely *that* meaning for space and time which Einstein has recognized as alone essential for physics, where he has shown it to best advantage. For he rejected the Newtonian concepts, which denied the origin we have described, and based physics instead on the concept of the coincidence of events. So here physical theory and epistemology extend their hands to one another in a beautiful alliance.[38]

It is therefore Schlick's close association of what he calls the method of coincidences with Einstein's general theory of relativity that provides the context, in turn, for Einstein's own close association of a purely formal conception of mathematical geometry (again linked to Hilbertian axiomatics) with the general theory.[39]

* * * * * *

But now Einstein's appeal to rigid bodies, in the spirit of Helmholtz, as an elucidation of applied or *physical* geometry, may appear even more puzzling. For, as we have emphasized, the variably curved spatial and spatiotemporal structures employed in general relativity do not satisfy the condition of free mobility. In this sense, there are in fact no rigid bodies in this theory, that is, no bodies that can be freely transported from place to place with no change in their geometrical properties. In a manifold of variable curvature there are in general no isometries—no continuous mappings of the manifold onto itself that preserve metrical relationships—and so the impossibility of rigid bodies is grounded in the very geometry of space.[40] Indeed, there are no rigid bodies in this sense even in the space-time structure of special relativity, for moving bodies necessarily experience the Lorentz contraction, as a consequence, in this case, of the underlying kinematics of space-time.[41] How, then, can the radically new spatiotemporal structure of the general theory of relativity possibly be consistent with Helmholtz's much older viewpoint, which, as we have seen, is entirely predicated upon the possibility of free mobility?

It is time to consider the argument Einstein actually provides for the importance of invoking practically rigid bodies in the foundations of the general theory. This argument is found sandwiched between passages we have already discussed (for convenience I reproduce the entire passage):

I attach particular importance to this conception of geometry, because without it I would have found it impossible to establish the theory of relativity. Without it, namely, the following consideration would have been impossible. In a reference system that is rotating relative to an inertial system the situational laws [*Lagerungsgesetze*] of rigid bodies, due to the Lorentz contraction, do not correspond to the rules of Euclidean geometry. Therefore, the admission of non-inertial systems as equally justified must lead to the abandonment of Euclidean geometry. The decisive step in the transition to generally covariant equations would certainly not have taken place, if the above interpretation had not been taken as basis. If one rejects the relation between the bodies of axiomatic Euclidean geometry and the practically rigid bodies of reality, then one easily arrives at the following conception, which the perceptive and deep thinker H. Poincaré has, in particular, embraced. Euclidean geometry is picked out from all other thinkable axiomatic geometries by simplicity. And since axiomatic geometry *alone* contains no assertions about experienceable reality, but only axiomatic geometry in connection with propositions of physics, then it may be possible and rational—no matter how reality may be constituted—to hold fast to Euclidean geometry. For one will prefer to opt for an alteration of the laws of physics rather than alter the laws of geometry in case there are contradictions between theory and experience. If one rejects the relation between the practically rigid body and geometry one will in fact not easily free oneself from the convention according to which Euclidean geometry is to be held fast as the simplest.[42]

Here Einstein is appealing to the following line of thought. According to the principle of equivalence (between gravitation and inertia), noninertial systems of reference (accelerating and rotating systems) are equally appropriate for the description of nature as inertial systems. If we consider a rotating system in the context of special relativity, however, we find that the Lorentz contraction differentially affects measuring rods laid off along concentric circles around the origin in the plane of rotation (due to the variation in tangential linear velocity at different distances along a radius). Moreover, no Lorentz contraction is experienced by rods laid off along a radius. Therefore, the geometry in a rotating system will be found to be non-Euclidean (the ratio of the circumference to the diameter of concentric circles around the origin in the plane of rotation will differ from π and depend on the circular radius). The importance of this line of thought for Einstein is evident in virtually all of his expositions of the general theory, where it is always used as the primary motivation for introducing non-Euclidean geometry—and thus general "Gaussian" coordinates—into the theory of gravitation. It is in precisely this way, in fact, that Einstein then arrives at the requirement of general covariance.[43]

This line of thought constituted an essential part of what we might call the heuristic route Einstein followed in creating the general theory of relativity. From our contemporary point of view—that of the completed and fully articulated theory—the principle of equivalence states that bodies affected only by gravitation follow geodesic trajectories in a four-dimensional (semi-)Riemannian manifold of variable curvature. And the presence of a gravitational field, in particular, is represented by nonvanishing four-dimensional curvature, viewed as a perturbation of the "empty" space-time, flat Minkowski geometry of special relativity, where, accordingly, there are no gravitational fields. So Einstein's example of a rotating reference system, from our contemporary point of view, has nothing to do with the variably curved *space-time* geometry we now use to represent the gravitational field, but rather represents a non-Euclidean *spatial* geometry arising in a particular three-dimensional "relative space" in a flat (and therefore gravitation-free) Minkowski space-time. This use of non-Euclidean geometry then appears entirely accidental, as it were, for it does not represent any genuine four-dimensional space-time curvature.

It is important to appreciate, however, that our contemporary framework of variably curved space-time geometry is a *product* of Einstein's creation of the general theory of relativity, not something that was present and available as Einstein struggled to create this theory. To be sure, Hermann Minkowski had formulated in 1908 the (flat) four-dimensional space-time geometry of special relativity we now identify with his name.[44] Nevertheless, not only did Einstein fail to acknowledge the importance of Minkowski's work at the time,[45] but no one else at the time, including Minkowski himself, had any idea at all how to exploit the long familiar relationship between gravitation and inertia (the equality of gravitational and inertial mass) so as to extend special relativity to include a theory of gravitation.[46] Einstein alone developed and exploited this crucial relationship, by articulating a generalized equivalence between gravitational and inertial phenomena, and, on precisely this basis, then first discovered (or "created") the four-dimensional space-time geometry of variable curvature we now identify with *his* name.

As John Norton has illuminatingly shown, Einstein did not initially conceive the principle of equivalence, and the very notion of a gravitational field, as we would understand them today.[47] For Einstein, as he was first articulating and exploiting this principle, the equivalence of gravitational and inertial phenomena meant that one could model gravitational fields by inertial fields—by the "pseudo-force" fields (such as centrifugal and Coriolis fields) arising in accelerating and rotating reference systems. The reference systems in question were moving three-dimensional spaces (relative spaces) within the framework of special relativity, and so a gravitational field could be taken to be present, by definition, in any such system moving noninertially. Einstein was thereby able to investigate the properties of gravitational fields in the theory still under construction by probing, as it were, the inertial structure of what we now call Minkowski space-time.

Proceeding in this way, in the years from 1907 to 1912, Einstein investigated in detail homogeneous (what we now call static) gravitational fields corresponding to uniformly accelerating reference systems, and then turned to the more general case (what we now call stationary gravitational fields) corresponding to uniformly accelerating or rotating systems.

It was precisely at this point, as John Stachel has shown, that Einstein made the decisive breakthrough to a four-dimensional geometry of variable curvature generalizing the flat geometrical structure of Minkowski space-time.[48] Einstein encountered a non-Euclidean *spatial* geometry in the case of uniformly rotating systems by the argument sketched above, and he concluded, accordingly, that the presence of a gravitational field could involve a non-Euclidean (spatial) geometry.[49] He then recalled that non-Euclidean geometries cannot be described by standard Cartesian coordinates, but rather require more general Gaussian coordinates generated by arbitrary (nonsingular) families of continuous curves. Therefore, as Einstein puts it, the coordinates no longer have direct physical or metrical meaning (in terms of distances from mutually orthogonal planes, for example), and this idea, generalized to four dimensions, then became the requirement of general covariance.[50] Proceeding along these lines, Einstein very quickly saw that a non-Euclidean generalization of the flat Minkowski metric was exactly what he needed to develop a relativistic theory of gravitation, and, turning to the mathematician Marcel Grossmann for help, he discovered the Riemannian theory of manifolds and tensor analysis. Einstein's repeated appeal to the example of the uniformly rotating system in his official expositions of the theory (note 43 above) therefore appears to reflect the actual historical process of discovery very accurately, and to explain, in particular, how the idea of a variably curved four-dimensional space-time geometry was actually discovered in the first place.[51]

Once discovered, the use of a four-dimensional space-time structure (what we currently call affine structure) to represent the phenomena of inertia (now including gravitation as well) has proved to be extraordinarily powerful and illuminating. Perhaps its most illuminating application is a definitive clarification of the relationship between the classical Newtonian theories of inertia and gravitation and the new relativistic theories created by Einstein. For, in the first place, it has now become clear that the classical, "gravitation-free" conception of inertia is best represented by a flat four-dimensional space-time structure containing a unique succession of three-dimensional, Euclidean, instantaneous spatial hypersurfaces (planes of absolute simultaneity)—four-dimensional straight lines in this structure then represent the trajectories of bodies moving inertially under the influence of no (net) forces at all.[52] And, in the second place, it has also become clear, even more strikingly, that the original Newtonian theory of gravitation can be similarly represented by a variably curved space-time structure conceived (analogously to general relativity) as a perturbation of the flat, "gravitation-free" inertial structure just described—four-dimensional geodesic lines in this structure then represent the trajectories of bodies moving solely under the influence of gravitation. Just as the long familiar equality of gravitational and inertial mass enabled Einstein, in the general theory of relativity, to represent the effects of gravitation by a perturbation of the Minkowski space-time geometry of special relativity, this same familiar equality, in the context of what we now call Newtonian space-time, enables us to represent the Newtonian theory of gravitation as well by an analogous variably curved space-time structure.[53]

It is clear, accordingly, that the principle of equivalence, as we now understand it, is not in fact unique to the general theory of relativity. What distinguishes Einstein's "geometrical" theory of gravitation from the Newtonian theory is not the use of four-dimensional space-time curvature *per se*, but rather the use of four-dimensional space-time curvature within the framework of a *relativistic* Minkowski space-time (more exactly, within the framework of an infinitesimally Minkowskian or what we now call Lorentzian space-time). In particular, all of the classical tests of general relativity, involving empirical predictions that diverge from the Newtonian theory (note 7 above), depend precisely on the circumstance that we have now applied the principle of equivalence to the inertial structure of a Minkowskian rather than a Newtonian space-time (for the variably curved version of Newtonian gravitation theory described above is of course empirically equivalent to the more standard version). The characteristic empirical content of general relativity, in other words, depends precisely on the relativity of simultaneity.

This application to the inertial structure of Minkowski space-time is also responsible for the characteristic relationship between four-dimensional *space-time* curvature and three-dimensional *spatial* curvature in Einstein's new theory. For, in the variably curved version of Newtonian gravitation theory, space-time curvature has no implications whatsoever for purely spatial curvature: the three-dimensional spatial geometry on the (absolute) planes of simultaneity remains Euclidean or flat despite the presence of four-dimensional (affine) curvature representing the influence of gravitation.[54] Moreover, and by the same token, moving reference frames in Newtonian space-time, whether accelerating or rotating, can generate no purely spatial curvature connected to an inertial field: the planes of simultaneity remain stubbornly absolute—entirely unaffected by the motion—and so no induced three-dimensional geometry in a relative space can possibly be other than Euclidean (see note 49 above). In other words, since there is no Lorentz contraction in Newtonian space-time, there is no possibility of generating a non-Euclidean spatial geometry as

Einstein did in the case of a rotating system within the framework of Minkowski space-time: if Einstein had applied the principle of equivalence to the inertial structure of Newtonian space-time, the possibility of a non-Euclidean spatial geometry connected to the phenomenon of gravitation could never have arisen in the first place. Strange as it may appear from our contemporary, post-general-relativistic point of view, Einstein's use of (three-dimensional) moving reference systems to probe the inertial structure of (flat) Minkowski space-time therefore constituted a quite indispensable application of the principle of equivalence—without which, as we have seen, neither the characteristic empirical content of the general theory nor the idea of four-dimensional space-time curvature as such would have ever been discovered.

But what does all this have to do with Einstein's appeal to rigid bodies, in the spirit of Helmholtz, and in explicit opposition to Poincaré, in the passage from "Geometrie und Erfahrung" with which we began (note 42 above)? This appeal is best understood, I believe, not so much as a complete endorsement of Helmholtzian rigidity (which, as we have seen, is quite incompatible with the theory of relativity), but rather as a rejection of Poincaré's reasons for disputing the direct geometrical significance of such rigidity. Poincaré's stance on this issue, as we have said, is motivated by the hierarchical conception of the sciences articulated in *Science and Hypothesis*: what Poincaré calls empirical or experimental physics (involving theories of particular physical fields and forces, for example) presupposes that both the general laws of mechanics (governing all forces and motions whatsoever) and the geometry of space (in the context of which the laws of mechanics are formulated) are already antecedently established.[55] This means, in particular, that Euclidean spatial geometry must be established in advance, on the basis of its greater mathematical simplicity, before any questions concerning particular physical fields and forces have a well-defined empirical meaning. And for precisely this reason, in cases such as the notorious temperature field example, we should always blame the apparently non-Euclidean behavior of rigid bodies on interfering physical forces rather than on the geometry of space: such bodies cannot really be rigid, therefore, in the geometrical sense.

Now Poincaré's perspective on the problem of rigidity had very significant consequences for the physics of the time. For, as is well known, although Poincaré made deep contributions to the mathematical and physical articulation of the new situation in electrodynamics addressed by Einstein's formulation of the special theory of relativity in 1905, Poincaré himself clearly preferred the Lorentz-Fitzgerald version of an "aether" theory to Einstein's theory.[56] And the crucial difference between the two theories, of course, is that the Lorentz contraction, in the former theory, is viewed as a result of the (electromagnetic) forces responsible for the microstructure of matter in the context of Lorentz's theory of the electron, whereas this same contraction, in Einstein's theory, is viewed as a direct reflection—independent of all hypotheses concerning microstructure and its dynamics—of a new kinematical structure for space and time involving essentially relativized notions of duration, length, and simultaneity. In terms of Poincaré's hierarchical conception of the sciences, then, Poincaré locates the Lorentz contraction (and the Lorentz group more generally) at the level of experimental physics, while keeping the Newtonian structure of the next higher level (what Poincaré calls mechanics) completely intact. Einstein, by contrast, locates the Lorentz contraction (and the Lorentz group more generally) at precisely this next higher level, while postponing to the future all further discussion of the physical forces and material structures actually responsible for the physical phenomenon of rigidity. The Lorentz contraction, in Einstein's hands, now receives a direct *kinematical* interpretation.

The same perspective on the Lorentz contraction is at work in the example of the rotating reference system—now, however, in the context of Einstein's newly articulated principle of equivalence. Here, unlike in the original formulation of special relativity, we are confronted not only with a non-Newtonian spatiotemporal kinematics but also with a non-Euclidean spatial geometry. Just as Einstein's original interpretation of the Lorentz contraction gave it a direct kinematical significance, the present interpretation of the Lorentz contraction (in the context of the principle of equivalence) gives it a direct *geometrical* significance (again entirely independently of all further questions concerning the microstructure of matter responsible for physical rigidity). Just as Einstein's formulation of the special theory of relativity blurred the distinction between the bottom two levels in Poincaré's hierarchy by allowing dynamical problems (in electricity and magnetism) to have direct kinematical implications, Einstein's discovery of the general theory of relativity carries this same process into the next higher level by allowing dynamical problems (in gravitation) to have direct geometrical implications.

Precisely in this sense geometry, in Einstein's hands, has now become a full-fledged branch of physics. In particular, by taking the Lorentz contraction, in a rotating system, as giving us direct access to geometrical structure independently of all further questions of matter theory and microphysics, we are then quickly led to the variably curved *space-time* geometry of general relativity. The appeal to rigid bodies, in the spirit of Helmholtz and in explicit opposition to Poincaré, has in fact led, through Einstein's here quite uncanny genius, to a radically new conception of physical geometry having no precedent whatsoever in the nineteenth-century tradition from which it emerged.

*　*　*　*　*　*

The problem with which "Geometrie und Erfahrung" begins is the relationship between pure or mathematical geometry and its physical application or interpretation. Pure mathematical geometry is conceived as a formal or uninterpreted axiom system, following the tradition associated with Hilbert, whereas applied or physical geometry is conceived as a coordination of pure mathematical geometry to the empirical behavior (*Lagerungsmöglichkeiten*) of practically rigid bodies, following the tradition associated with Helmholtz. We have seen that the connections between both of these earlier traditions and the newly created general theory of relativity are extraordinarily subtle and complex. Moreover, the entire problematic of pure versus applied geometry, in its modern Hilbertian form, is itself a quite recent development. Throughout the nineteenth century, as we have seen, pure or mathematical geometry was rather understood, in the tradition of projective geometry and group theory, as the study of the most general characteristics of space—the very space in which we live, move, and perceive—arising from an abstract consideration of the "perspectival" features of spatial perception constituting the starting point of projective geometry. Mathematical geometry, in this tradition, is therefore already interpreted in the Hilbertian sense, and so the relationship between mathematical and physical geometry must also be conceived rather differently.

Here, following the example of Helmholtz, physical geometry arises from pure mathematical geometry by considering the geometrical behavior of idealized rigid bodies located within the "perspectival" space of projective geometry. And the crucial link between the two theories is the principle of free mobility, in accordance with which idealized (but still physical) rigid bodies can be freely transported within this space with no changes or distortions of their geometrical properties. The group of isometries governing geometrical space is thereby coordinated to the physical behavior of rigid bodies in such a way that (idealized) physical measurements can be taken empirically to realize the more abstract laws of mathematical geometry.[57] At this point, however, as a consequence of the Helmholtz-Lie theorem, it emerges that there are three and only three possible geometries (Euclidean, hyperbolic, and elliptic) consistent with the condition of free mobility, among which a choice must somehow be made. This choice then takes place at the level of physical geometry—the application of mathematical geometry to empirical physics—and proceeds either by attempting empirically to "read off" the particular geometry in question from actual physical measurements (as in Helmholtz) or by conventionally stipulating a particular physical geometry in advance of empirical physics on the basis of its greater mathematical simplicity (as in Poincaré).

Yet the geometry of the general theory of relativity, as we have said, lies entirely outside this projective tradition. Here we rather require the more general Riemannian point of view on the relationship between pure and applied geometry, whereby mathematical geometry is a purely analytical theory (of continuous number manifolds) having no necessary connection at all with spatial intuition or perception, and physical geometry is a branch of natural science in which we formulate and investigate hypotheses about the geometrical structure of the physical world. And this point of view, as Schlick was the first explicitly to emphasize, can then be used to incorporate the general theory of relativity, in particular, into our now canonical Hilbertian conception of the relationship between an uninterpreted formal axiom system and its physical interpretation. Moreover, as Schlick was also the first explicitly to emphasize, we now need a radically new account of how such an uninterpreted formal axiom system acquires its empirical application and meaning. Schlick finds this account, again inspired by the general theory, in his method of coincidences for establishing a coordination (*Zuordnung*) between formal geometrical concepts and concrete results of measurement; and the model for this coordination is precisely the quite general assignment of space-time coordinates allowed by the principle of general covariance.

However, although Schlick himself is not as clear on this as one might wish, especially in the first (1918) edition of *Allgemeine Erkenntnislehre*, the method of coincidences, by itself, can hardly be a sufficient account of the application of mathematical spatiotemporal concepts to empirical reality. For the very general assignment of space-time coordinates employed in the general theory of relativity is purely topological (entirely based, in Schlick's account, on spatiotemporal singularities or coincidences), and thus does not, by itself, give us any further insight into the empirical application of *metrical* concepts. But precisely these latter concepts, of course, are central to applied mathematics and, more specifically, to the use of geometrical concepts in mathematical physics.

The second (1925) edition of *Allgemeine Erkenntnislehre* adds an entirely new section, "Definitions, Conventions, and Empirical Judgements" (sec. 11), at least partly intended to rectify this situation. Where the first edition had introduced a distinction between "definitions[,] i.e., judgements that achieve a coordination by *arbitrary* stipulation," the recognition of "empirical facts," and "hypotheses," the second edition sharply distinguishes between three types of "definitions."[58] *Implicit* definitions are given within a formal axiomatic system and have nothing to do, therefore, with relating such a system to empirical reality. A *concrete* (or ostensive) definition, by contrast, involves precisely the coordination of a real (intuitively present) object to some or another given concept: "it is an entirely arbitrary stipulation that consists in introducing its own name for an object that is somehow singled out."[59] And it is clear from the context that what Schlick is here calling "concrete definitions" are very close indeed to what he had called "definitions" *simpliciter* in the corresponding passage in the first edition.[60] Now, however, Schlick carefully distinguishes such (concrete) definitions from the *conventions*—"in the narrower sense"—that are crucial for an understanding of how we achieve a coordination to empirical reality in the mathematical exact sciences.[61]

Schlick illustrates this last point with the example of time measurement. We begin, for example, by stipulating that the times during which the earth rotates once around its axis are equal (sidereal day). This, Schlick explains, is "at bottom a concrete definition, because the stipulation refers to a concrete process [taking place] in a single heavenly body given only once." We find, however, that it is simpler and more convenient to allow corrections to this stipulation (arising from tidal friction, for example) based on "the greatest possible simplicity of the laws of nature"—that is, "fundamental equations of physics" such as the laws of motion—and it is here, and only here, that we find conventions properly speaking.[62] Thus conventions in the present sense are not concrete coordinations to particular physical processes (as in what we now call operational definitions, for example), but rather fundamental principles of mathematical natural science that provide general prescriptions for establishing and correcting such concrete coordinations as physics progresses. The laws of motion, for example, supply such "coordinating principles" for the measurement of time, just as the laws of geometry do the same for the measurement of space.[63]

Now Schlick's method of coincidences, modeled on the assignment of spatiotemporal coordinates in general relativity, most closely corresponds to the use of *concrete* (or ostensive) definitions; for it proceeds, as we have seen, by coordinating number-quadruples representing points in objective space and time with intuitively present singularities in our subjective sensory fields. In a key passage added to the conclusion of the fourth (1922) edition of *Space and Time in Contemporary Physics*, however, Schlick indicates how specifically metrical concepts, by contrast, receive their coordination with reality precisely by a *convention*:

> The question has been raised whether in the *simplest* theory, which in fact only describes what is experienceable and affirmable [*das Erfahrbare, Konstatierbare*] without arbitrary addition, every *arbitrary* element is also excluded at the same time—and one could believe that this is the case in the general theory of relativity, for [this theory] actually sets forth that which is entirely independent of the choice of coordinates and indicates only the laws between space-time coincidences, and thus between what is plainly observable and holds prior to all interpretation.
>
> But even the simplest theory, which contains no single superfluous concept, is not free from arbitrariness. The designation of facts by judgements

presupposes, like every coordination, some or another arbitrary stipulations [*willkürlichen Festsetzungen*]; a *measurement*, e.g., only becomes first possible in this way. The fundamental convention [*Konvention*] for the Einsteinian world-picture is the (certainly most natural) [one] that special relativity with its Euclidean metrical determinations shall be valid in the small. Thus Poincaré's proposition remains true that we cannot arrive at laws of nature without some convention.[64]

Schlick here moves from the merely topological structure of space-time coincidences given by the principle of general covariance to the specifically metrical structure of general relativity by adding the convention or coordinating principle that the metrical structure of the general theory continuously approaches that of the special theory as the regions under consideration become arbitrarily small—or, in contemporary terminology, that the metric of general relativity is Lorentzian. As Schlick explains in the earlier chapter on "The General Postulate of Relativity and the Metrical Determination of the Space-Time Continuum," it is precisely this coordinating principle that allows us to use the principles of Euclidean geometry in the small and thereby establish a coordination with ordinary Euclidean methods of measurement, despite the circumstance that these same Euclidean methods—and even the condition of free mobility itself—must necessarily break down in general.[65]

In the following chapter, "Formulation and Significance of the Fundamental Law of the New Theory," Schlick further considers how the metrical structure of the general theory receives its "physical interpretation." In particular, he extends the continuity principle linking general relativity with the laws of special relativity in the small to include the principle of equivalence and the geodesic law of motion. The idea is that the law of inertia or natural motion, in the special theory, is represented by flat (or semi-Euclidean) geodesics in what we currently call Minkowski space-time; and, since gravitation, according to the principle of equivalence, is now represented by space-time curvature, the corresponding law of inertia or natural motion, in our new theory, must now be represented precisely by nonflat geodesics in our new, variably curved (infinitesimally Minkowskian) space-time structure. In this way, by extending the insights of Riemann and Helmholtz into the problem of giving a physical interpretation for three-dimensional, non-Euclidean *spatial* geometry, we provide an entirely novel physical interpretation for four-dimensional, non-Euclidean *space-time* geometry. The metrical structure of Einstein's theory is not only coordinated to the behavior of measuring rods (and clocks) in the small, but also to a fundamentally new type of dynamical law governing the trajectories of idealized bodies (or "test particles") subject only to gravitation.[66]

*   *   *   *   *   *

From the point of view of our contemporary, four-dimensional perspective on both theories, the principle of equivalence thus serves as fundamental coordinating principle for the variably curved inertial structure of general relativistic space-time in precisely the same sense that the Newtonian laws of motion serve as fundamental coordinating principles for the flat (four-dimensional) inertial structure of what we currently call Newtonian space-time. Just as the spatiotemporal framework articulating the basic concepts of motion employed in classical Newtonian theory (absolute acceleration and rotation) is first given physical meaning and application by the laws of motion, so the spatiotemporal framework represented by the much more abstract theory of (semi-)Riemannian four-dimensional manifolds is first given physical meaning and application by the geodesic law of motion. For it is here, and only here, that we coordinate some real physical phenomenon—freely falling trajectories in a gravitational field—to the highly abstract mathematical concept of four-dimensional space-time curvature. Without this coordinating principle, by contrast, the abstract concept of a (semi-)Riemannian four-dimensional manifold representing the inertial structure of space-time would be entirely lacking in physical interpretation, in precisely the same way that the inertial structure of what we currently call Newtonian space-time would similarly lack such an interpretation in the absence of the laws of motion.[67]

Indeed, from our contemporary perspective on four-dimensional space-time geometry we can go even further. Just as the principle of equivalence serves as fundamental coordinating principle for the variably curved inertial structure of general relativity, the principle of the invariance of the velocity of light familiar from special relativity serves as fundamental coordinating principle for the infinitesimally Minkowskian (or Lorentzian) character of the space-time metric. Just as freely falling trajectories in a gravitational field physically realize the abstract concepts of space-time geodesic and space-time curvature, trajectories representing the propagation of light rays physically realize the abstract concept of the system of null-cones on our manifold. Thus, the entire mathematical structure of the general theory of relativity is given a physical interpretation by two coordinating principles governing elementary dynamical processes—the principle of equivalence for what we now call affine or projective structure, the light principle for what we now call conformal structure.[68] So there is no need, at any point, to appeal to the behavior of such complex physical entities as rigid bodies and clocks; and Einstein's uneasy and ambiguous attitude toward such entities, in the context of qualms about the future of microphysics, turns out, at least in this respect, to have been misplaced.[69]

Nevertheless, as we have seen, Einstein's appeal to rigid bodies, in the spirit of Helmholtz, for the interpretation of applied or physical geometry, in fact played a quite indispensable role in the discovery and articulation of the general theory of relativity, without which the four-dimensional space-time structure we now use to represent this theory would never have been discovered in the first place. Precisely this circumstance, it seems to me, explains why rigid bodies—and the question of their free mobility, in particular—have continued to dominate philosophical discussion of the foundations of geometry well after the definitive formulation of the general theory.[70] We have seen that a preoccupation with rigid bodies and free mobility arises in a very special context, that of Helmholtz's attempt to link his own researches on spatial perception to recent mathematical work on non-Euclidean geometries. Here the condition of free mobility is naturally singled out from the more comprehensive system of geometrical principles as a whole, and then elevated, as it were, to the status of sole coordinating principle governing both concrete spatial measurement and the application of geometry to physics more generally. This makes excellent sense in the nineteenth-century context in which Helmholtz formulated his philosophy of geometry, but it does not extend to a perfectly general account of the geometrical coordinating principle in physics—either in early-twentieth-century relativity theory or in late-seventeenth- and eighteenth-century Newtonian physics.[71] In relativity theory, as just noted, we rather use the principle of equivalence (together with the light principle) as fundamental coordinating principle for the entirely new geometrical structure described by this theory—a structure which, strictly speaking, is quite incompatible with rigid bodies and free mobility. Einstein's own appeal to rigid bodies, in the spirit of Helmholtz, therefore provides an especially striking example of how an older perspective on geometrical constitutive principles can be subtly transformed into a radically new perspective on such principles that is actually inconsistent with the old.

I believe that Einstein's appeal to rigid bodies, made while simultaneously delicately positioning himself within the late-nineteenth-century debate on the foundations of geometry between Helmholtz and Poincaré, also explains the continuing philosophical fascination with the problematic of geometrical conventionalism inherited from this debate. Schlick, for example, introduces what we are here calling coordinating principles by explicitly invoking Poincaré's conventionalist philosophy, and by transferring Poincaré's criterion of mathematical simplicity, in particular, to the total system of mathematical-physical laws.[72] By the same token, however, whereas Poincaré's geometrical conventionalism made excellent sense in the mathematical and physical context in which it was formulated, it does not extend to a perfectly general account of the use of geometrical coordinating principles in physics. Given the principle of free mobility as sole coordinating principle, the Helmholtz-Lie theorem tells us that there are three and only three possible physical geometries still open, among which a choice must somehow be made. Since a specific geometry of constant curvature is not determined by the condition of free mobility itself, it is then quite natural to view this choice, as Poincaré did, as a free convention made on the basis of mathematical simplicity. But in other mathematical-physical contexts—such as the use of Euclidean geometry in classical Newtonian physics or the radically new use of physical geometry in general relativity—our coordinating principles do not have this particular structure. We are not presented with a choice of incompatible geometrical alternatives framed by a common or generic coordinating principle, but are rather confronted, on the contrary, by a single geometrical structure uniquely coordinated to physical reality by a single such principle (or system of principles). The idea of conventional choice, in such cases, then has no real meaning at all—and it has no real meaning, for just this reason, in the context of the novel geometrical structure of the general theory of relativity.[73]

Einstein's own engagement with the problematic of conventionalism, and, in particular, with the debate on the foundations of geometry between Helmholtz and Poincaré, was an especially timely and fruitful one. It allowed him, as we have seen, to take the critical step, via the principle of equivalence, from the interpretation of three-dimensional, non-Euclidean spatial geometry to that of four-dimensional, non-Euclidean space-time geometry. And it is in precisely this sense, we might say, that Einstein himself made the crucial transition from nineteenth- to twentieth-century philosophy of physical geometry. One unforeseen consequence, however, was that the fundamentally new perspective on the foundations of geometry actually created by Einstein in this way—the idea of geometry as fully a branch of physics—has proved much more difficult to grasp than it otherwise might.

NOTES

1. A. Einstein, "Geometrie und Erfahrung," *Preussische Akademie der Wissenschaft. Physikalisch-mathematische Klasse. Sitzungsberichte* (1921), 123–30; *Erweiterte Fassung des Festvortrages gehalten an der Preussischen Akademie der Wissenschaft zu Berlin am 27. Januar 1921* (Berlin: Springer, 1921); translated in G. Jeffrey and W. Perrett, eds., *Sidelights on Relativity* (London: Methuen, 1923), 27–55.

2. A well-known later presentation of this canonical twentieth-century view is C. Hempel, "Geometry and Empirical Science," *American Mathematical Monthly* 52 (1945); reprinted in H. Feigl and W. Sellars, eds., *Readings in Philosophical Analysis* (New York: Appleton-Century-Crofts, 1949), 238–49. Hempel's paper is

basically an elementary exposition of the first part of Einstein's; it closes by quoting Einstein's famous characterization of the relationship between mathematical certainty and empirical reality.

3. *Geometrie und Erfahrung. Erweiterte Fassung* (note 1), 4–5 (Jeffrey and Perrett, 30–31). The reference is to M. Schlick, *Allgemeine Erkenntnislehre* (Berlin: Springer, 1918). The relation between Einstein's views and Schlick's will be discussed in detail below. (All translations from German originals are my own.)

4. *Geometrie und Erfahrung*, 3–4 (Jeffrey and Perrett, 28–29).

5. D. Hilbert, *Grundlagen der Geometrie* (Leipzig: Teubner, 1899); translated from the tenth (1968) edition by L. Unger (La Salle, Ill.: Open Court, 1971). This work is prominently cited by Schlick in *Allgemeine Erkenntnislehre*, sec. 7, as the basis for the notion of "implicit definition."

6. *Geometrie und Erfahrung*, 5–6 (Jeffrey and Perrett, 31–33).

7. Here I of course have in mind the well-known classical tests of general relativity: the bending of light rays near the sun, the advance in the perihelion of Mercury, and so on. These all test the Schwarzschild solution (to Einstein's field equations) in the neighborhood of the sun, the spatial part of which is described by a spherically symmetric geometry of variable negative curvature (increasing sharply with proximity to the sun).

8. See, for example, R. Torretti, *Philosophy of Geometry from Riemann to Poincaré* (Dordrecht: Reidel, 1978), chapter 2.3.

9. F. Klein, *Elementarmathematik vom höheren Standpunkt aus. Teil II: Geometrie* (Leipzig: Teubner, 1909), pp. 383–84; translated by E. R. Hedrick and C. A. Noble (New York: Dover, 1939), 187. The context makes clear that Klein has in mind those excessively influenced by Hilbertian axiomatics. Hilbert's own work in the foundations of geometry grew out of this same projective tradition, however, and Hilbert himself famously says in his brief introduction that the problem of axiomatizing geometry amounts to "the logical analysis of our spatial intuition." (The above passage from Klein reads rather eerily like a direct criticism of Einstein's argument at the beginning of "Geometrie und Erfahrung.")

10. See Torretti, *Philosophy of Geometry*, chapter 2.3.10 for a discussion of Klein's view of spatial intuition. For a somewhat more sympathetic discussion of the relationship between (closely related) nineteenth-century views of spatial intuition and the original Kantian conception see my "Geometry, Construction, and Intuition in Kant and His Successors," in G. Scher and R. Tieszen, eds., *Between Logic and Intuition: Essays in Honor of Charles Parsons* (Cambridge: Cambridge University Press, 2000).

11. B. Riemann, "Über die Hypothesen, welche der Geometrie zugrunde liegen," *Abhandlungen der Königlichen Gesellschaft der Wissenschaften zu Göttingen* 13 (1867); *Neu herausgegeben und erläutert von H. Weyl* (Berlin: Springer, 1919); translated by H. S. White in D. Smith, ed., *A Source Book in Mathematics*, vol. 2 (New York: Dover, 1959), 411–25. H. Helmholtz, "Über die Tatsachen, die der Geometrie zum Grunde liegen," *Nachrichten von der Königlichen Gesellschaft der Wissenschaften und der Georg-August-Universität aus dem Jahre 1868* 9 (1868); reprinted in H. Helmholtz, *Wissenschaftliche Abhandlungen*, vol. 2 (Leipzig: Barth, 1883); translated by M. Lowe in R. Cohen and Y. Elkana, eds., *Hermann von Helmholtz: Epistemological Writings* (Dordrecht: Reidel, 1977), 39–71.

12. For Helmholtz's view of space-perception see G. Hatfield, *The Natural and the Normative: Theories of Space Perception from Kant to Helmholtz* (Cambridge, Mass.: MIT Press, 1990), chapter 5. For a discussion of Helmholtz's mathematical results in the context of his theory of space-perception see J. Richards, "The Evolution of Empiricism: Hermann von Helmholtz and the Foundations of Geometry," *British Journal for the Philosophy of Science* 28 (1977): 235–53. See also my "Helmholtz's *Zeichentheorie* and Schlick's *Allgemeine Erkenntnislehre*: Early Logical Empiricism and Its Nineteenth-Century Background," *Philosophical Topics* 25 (1997), 19–50.

13. For the work of Helmholtz and Lie see Torretti, *Philosophy of Geometry.*, chapter 3.1. For a philosophically and mathematically sophisticated discussion of Helmholtz and Riemann see secs. 6–7 of H. Stein, "Some Philosophical Prehistory of General Relativity," in J. Earman, C. Glymour, and J. Stachel, eds., *Minnesota Studies in the Philosophy of Science*, vol. 8 (Minneapolis: University of Minnesota Press, 1977), 3–49; footnote 29, in particular, presents an up-to-date exposition of the mathematics of the Helmholtz-Lie theorem.

14. Helmholtz characterizes space as a "subjective *form* of intuition" in the sense of Kant, and as the "*necessary* form of our outer intuition," in his famous address on "Die Tatsachen in der Wahrnehmung" of 1878. See P. Hertz and M. Schlick, eds., *Hermann v. Helmholtz: Schriften zur Erkenntnistheorie* (Berlin: Springer, 1921), 117; translated in Cohen and Elkana, *op. cit.* (note 11 above), 124. Helmholtz viewed the condition of free mobility, in particular, as a necessary condition of the possibility of spatial measurement, and thus of the application of geometry. For discussion see the works cited in notes 12 and 13 above.

15. Poincaré discovered his well-known models of hyperbolic geometry in the context of his work on "Kleinian groups" in complex analysis—which he then found, surprisingly, to include the isometries of hyperbolic geometry. Poincaré describes this famous discovery in the chapter "Mathematical Invention" in *Science et Méthode* (Paris: Flammarion, 1908); translated by G. Halsted in *The Foundations of Science* (Lancaster, Pa.: The Science Press, 1913). For a discussion of the Poincaré models see Torretti, *Philosophy of Geometry*, chapter 2.3.7.

16. H. Poincaré, "On the Foundations of Geometry," *The Monist* 9 (1898): 41–42. The connection between geometry in this sense and physics is explained at greater length in *Science and Hypothesis*: "This is the object of geometry: it is the study of a particular 'group'; but the general concept of a group preexists in our mind, at least potentially. . . . However, from among all possible groups it is necessary to choose one that will be so to speak the *standard measure* [*étalon*] to which we relate the phenomena of nature. Our experience guides us in this choice but does not impose it upon us; it allows us to recognize, not which is the truest geometry, but rather which is the most *convenient*." See *La Science et l'Hypothèse* (Paris: Flammarion, 1902), 90–91; translated in *The Foundations of Science* (see my note 15), 79–80.

17. For a detailed analysis of Poincaré's argument—from *Science and Hypothesis* (see my note 16)—along these lines see "Poincaré's Conventionalism and the Logical Positivists," chapter 4 of my *Reconsidering Logical Positivism* (Cambridge: Cambridge University Press, 1999).

18. "Foundations of Geometry" (note 16 above), 43. The distinguishing feature of the Euclidean group (of translations and rotations) to which Poincaré is

referring here is that this group alone among the three classical cases admitted by the Helmholtz-Lie theorem possesses a normal subgroup (namely, the translations).

19. *Geometrie und Erfahrung*, 7 (Jeffrey and Perrett, 33–34).

20. We know that Einstein studied *Science and Hypothesis* very closely–in 1905, in particular, as he was formulating his three initial discoveries in quantum theory, Brownian motion, and special relativity. See A. Miller, *Albert Einstein's Special Theory of Relativity* (Reading, Mass.: Addison-Wesley, 1981), chapter 2.

21. Poincaré derives the law of the temperature contraction from his own models of hyperbolic geometry, which are variants of the Beltrami-Klein model (see the reference to Torretti in note 15 above).

22. *Geometrie und Erfahrung*, 7–8 (Jeffrey and Perrett, 34–36).

23. There has been considerable confusion in the philosophical literature arising from an unwarranted assimilation of Poincaré's geometrical conventionalism to Hilbert-style views concerning the "arbitrariness" of the initial assumptions in pure axiomatics. This assimilation is rightly and ably combated in Torretti, *Philosophy of Geometry*, chapter 4.4.

24. This nineteenth-century tradition forms the background for Gottlob Frege's well-known distinction between arithmetic (having its "source" in the purely logical faculty of understanding) and geometry (having its "source" in pure intuition). See W. Demopoulos, "Frege and the Rigorization of Analysis," in W. Demopoulos, ed., *Frege's Philosophy of Mathematics* (Cambridge, Mass.: Harvard University Press, 1995), 68–88; M. Wilson, "Frege: The Royal Road from Geometry," *Noûs* 26 (1992), reprinted (with a postscript) in W. Demopoulos, *op. cit.*, 108–59; and J. Tappenden, "Geometry and Generality in Frege's Philosophy of Arithmetic," *Synthese* 102 (1995), 319–61.

25. Here I am especially indebted to comments from Howard Stein. See also his own discussion of Riemann cited in note 13 above.

26. See especially "Über die Ursprung und die Bedeutung der geometrischen Axiomen," first given as a lecture in 1870; reprinted in Hertz and Schlick, *op. cit.* (see my note 14), and, in translation, in Cohen and Elkana, *op. cit.* (see note 11 above). Nevertheless, our intuition of space, for Helmholtz, still has an "a priori" and "transcendental" character. For discussion see my papers cited in notes 10 and 12.

27. *La Science et l'Hypothèse*, 63 (*The Foundations of Science*, 63). In the popular edition of *Science and Hypothesis* translated by W. J. Greenstreet (New York: Dover, 1952), 48, the last sentence is incorrectly translated as: "These geometries of Riemann, so interesting on various grounds, can never be, therefore, purely analytical, and would not lend themselves to proofs analogous to those of Euclid"— thereby entirely reversing its sense (and the preceding sentence on p. 47 incorrectly has "variable figure" instead of "invariable figure").

28. Poincaré's discussion of "The Reasoning of Euclid" in "Foundations of Geometry" (see note 16 above), 32–34, focuses on proofs that proceed by translating and rotating figures, and thus emphasizes, once again, the central importance of the group of isometries or rigid motions in his view. For discussion see my "Geometry, Construction, and Intuition," *op. cit.* (see note 10 above), sec. 3.

29. See note 3 above. The second edition (Berlin: Springer, 1925), which contains some significant changes and additions, is translated by A. Blumberg (La Salle, Ill.: Open Court, 1985).

30. I provide a detailed discussion of this aspect of Schlick's epistemological conception in the paper first cited in note 12 above, which also contains a detailed examination of the relationship between Schlick and Helmholtz. The present exposition of Schlick's "method of coincidences" and its relation to general relativity closely follows this earlier paper.

31. This is sec. 30 of the first edition of *Allgemeine Erkenntnislehre* (sec. 31 of the second).

32. *Allgemeine Erkenntnislehre* (1918), 237 (Blumberg, 275).

33. M. Schlick, *Raum und Zeit in der gegenwärtigen Physik* (Berlin: Springer, 1917); third edition 1920, fourth edition 1922; translated by H. Brose from the third edition (Oxford: Oxford University Press, 1920); expanded to include changes in the fourth edition by P. Heath, in H. Mulder and B. van de Velde-Schlick, eds., *Moritz Schlick: Philosophical Papers*, vol. 1 (Dordrecht: Reidel, 1978), 207–69.

34. *Raum und Zeit in der gegenwärtigen Physik* (1917), 32 (Brose-Heath, 238–39).

35. *Raum und Zeit*, 33 (Brose-Heath, 240).

36. *Raum und Zeit*, 35 (Brose-Heath, 241). The passage continues: "Everything in our world-picture that *cannot* be reduced to these coincidences is deprived of physical objectivity and can just as well be replaced by something else. All world-pictures that agree with respect to the laws of these point coincidences are physically absolutely equivalent." A few pages later—*Raum und Zeit*, 38 (Brose-Heath, 243)—Schlick explicitly links this idea, and therefore his understanding of the principle of general covariance, with Einstein's famous remark from sec. 3 of his fundamental 1916 paper that general covariance "takes away from space and time the last remnant of physical objectivity." See A. Einstein, "Die Grundlage der allgemeinen Relativitätstheorie," *Annalen der Physik* 49 (1916): 776; translated by W. Perrett and G. Jeffrey, in H. Lorentz, et. al., *The Principle of Relativity*, 117. In the same passage Einstein appeals to the idea, also enunciated by Schlick in *Allgemeine Erkenntnislehre* (note 32 above), that all physical measurement rests on the observation of spatiotemporal coincidences. This point is emphasized, in turn, in the passage from *Space and Time in Contemporary Physics* with which we began.

37. *Raum und Zeit*, 53–54 (Brose-Heath, 260).

38. *Raum und Zeit*, 58 (Brose-Heath, 263). The final sentence is moved to the following page in the fourth edition.

39. Einstein studied both *Allgemeine Erkenntnislehre* and *Raum und Zeit in der gegenwärtigen Physik* rather carefully; indeed, he read (and commented upon) drafts of the latter work at several stages of its composition. For details see D. Howard, "Realism and Conventionalism in Einstein's Philosophy of Science: The Einstein-Schlick Correspondence," *Philosophia Naturalis* 21 (1984): 616–29.

40. Of course there can be a limited range of isometries even in a variably curved manifold, provided it possesses certain symmetries. The Schwarzschild

geometry, for example, is spherically symmetric, so that bodies can be isometrically transported on a spherical surface at a given radial distance from the sun. But a body cannot then be isometrically transported away from or off of the given surface.

41. This kind of limitation on free mobility is fundamentally different from the first case, however, because it depends on taking the motion in question as a real process in space and time (more precisely, in space-time) rather than as a mere mathematical transformation in a given space. Thus, in the Schwarzschild geometry, for example, free mobility is impossible even within a given spacelike hypersurface, whereas the situation in special relativity necessarily involves relationships between different spacelike hypersurfaces (planes of simultaneity).

42. *Geometrie und Erfahrung*, 6–7 (Jeffrey and Perrett, 33–34).

43. See, for example: "Die Grundlage der allgemeinen Relativitätstheorie" (see note 36 above), 774–76 (*The Principle of Relativity*, 115–17); *Über die spezielle und die allgemeine Relativitätstheorie, gemeinverständlich* (Braunschweig: Vieweg, 1917), translated by R. Lawson from the fifth edition (London: Methuen, 1920), secs. 23–28; *The Meaning of Relativity* (Princeton: Princeton University Press, 1922), 59–61. Schlick appeals to this same Einsteinian argument to motivate the introduction of non-Euclidean geometry into the theory of gravitation immediately preceding the passage cited in note 35 above.

44. For details of Minkowski's work see L. Corry, "Hermann Minkowski and the Postulate of Relativity," *Archive for the History of the Exact Sciences* 51 (1997): 273–314. Minkowski's famous paper, "Raum und Zeit," was first presented in September of 1908 and published in *Physikalische Zeitschrift* 10 (1909): 104–11; translated in *The Principle of Relativity*, op. cit., 75–91.

45. See A. Pais, *'Subtle is the Lord . . . ' The Science and the Life of Albert Einstein* (Oxford: Oxford University Press, 1982), 152.

46. Minkowski himself attempted to incorporate gravitation into the space-time structure we now identify with his name by action-at-a-distance forces operating, in different reference systems, in different planes of simultaneity. See Corry, *op. cit.* (note 44 above), 286–92.

47. J. Norton, "What Was Einstein's Principle of Equivalence?" *Studies in History and Philosophy of Science* 16 (1985); reprinted in D. Howard and J. Stachel, eds., *Einstein and the History of General Relativity* (Boston: Birkhäuser, 1989),5–47.

48. J. Stachel, "Einstein and the Rigidly Rotating Disk," in A. Held, ed., *General Relativity and Gravitation* (New York, 1980); reprinted as "The Rigidly Rotating Disk as the 'Missing Link' in the History of General Relativity," in Howard and Stachel, *op. cit.* (note 47 above), 48–62.

49. Norton, *op. cit.* (see note 47 above), sec. 3, explains how rigorously to define a three-dimensional spatial geometry in such cases—essentially, by inducing a spatial metric based on the planes of simultaneity orthogonal to the motion of the reference system at each point (if the frame is rotating there is no spacelike hypersurface orthogonal to all the lines of motion together). This procedure is well defined if and only if the system in question is in "rigid motion"—intuitively, it experiences no internal expansion or contraction—which is the case, in Minkowski space-time, if and only if the system is moving *uniformly* (with constant velocity, uniform acceleration, or uniform rotation).

50. Norton, *op. cit.*, sec. 6, suggests that general covariance was rather motivated by the "breakdown" of well-defined geometries in relative spaces moving nonuniformly (see note 49 above), and he appeals to sec. 28 of Einstein's 1917 popular exposition (see note 43 above), where Einstein rejects the idea of "rigid reference bodies." However, Einstein's rejection of "rigid reference bodies" here simply amounts to a rejection of standard, Euclidean (Cartesian) reference systems ("rigid bodies with Euclidean properties"), and is entirely based on the familiar example of the rotating system of sec. 23 (which *is* moving rigidly in Norton's sense). Contrary to Norton and in agreement with Stachel, then, I believe that the example of the uniformly (and thus rigidly) rotating system provided the key motivation for general covariance.

51. As is now well known, Einstein (and Grossmann) required several more years of struggle to find the final generally covariant theory of gravitation—during which, in particular, they were sidetracked by the now notorious "hole argument" temporarily to reject general covariance. See J. Stachel, "Einstein's Search for General Covariance, 1912-1915," in Howard and Stachel, *op. cit.,* (see note 47 above) 63–100 (first presented in 1980), and J. Norton, "How Einstein Found His Field Equations, 1912–1915," *Historical Studies in the Physical Sciences* 14 (1984), reprinted in Howard and Stachel, *op. cit.*, 101–59. It seems to me that there is now a tendency to overemphasize these years, and the hole argument, in particular, in explaining the significance and motivations of general covariance. Thus Howard, for example, in sec. 2 of his very useful paper on the Einstein-Schlick correspondence (see note 39 above) cites Einstein's remarks from sec. 27 of his 1916 popular exposition to the effect that coordinates are now "arbitrary" and devoid of "direct physical significance" as motivated primarily by the problematic of the hole argument. Once again, however, the explicit context of these remarks makes very clear that Einstein simply has the example of the rotating reference system in mind (he makes no explicit reference to the hole argument). More generally, whereas Einstein's discussions of general covariance, lack of "physical objectivity," space-time coincidences, and so on in his official expositions of the theory (see note 43 above) may indeed bear traces of the hole argument, the example of the rotating reference system is much more prominent and explicit. From this point of view, the decisive step to general covariance had already been taken in 1912, and the later problematic of the hole argument rather represents a temporary bump in the road.

52. This way of viewing the classical Newtonian theory as a "four-dimensional affine space" governed by "Galilean geometry" is first described in H. Weyl, *Raum-Zeit-Materie* (Berlin: Springer, 1918); translated from the fourth (1921) edition by H. Brose (London: Methuen, 1922). It has more recently been explained in detail, together with a penetrating analysis of the historical context of Newton's original theory of motion and gravitation, in H. Stein, "Newtonian Space-Time," *Texas Quarterly* 10 (1967): 174–200.

53. This way of viewing Newtonian gravitation theory is first described in E. Cartan, "Sur les variétés à connexion affine et la théorie de la relativité généralisée," *Annales de l'Ecole Normale Supérieure* 40 (1923): 325–412; 41 (1924): 1–25. It is further developed by K. Friedrichs, "Eine invariante Formulierung des Newtonschen Gravitationsgesetzes und des Grenzüberganges vom Einsteinschen zum Newtonschen Gesetz," *Mathematische Annalen* 98 (1927): 566–75; A.

Trautmann, "Sur la Théorie Newtonienne de la Gravitation," *Comptes Rendus de l'Académie des Sciences (Paris)* 257 (1963): 617–20; and others. This version of Newtonian gravitation theory figures prominently in my *Foundations of Space-Time Theories: Relativistic Physics and the Philosophy of Science* (Princeton: Princeton University Press, 1983).

54. Recent work on the Newtonian limit of general relativity (following the work of Friedrichs cited in note 53 above) has made this point particularly vivid. For an especially interesting consequence of this work is that the limiting process itself forces the geometry on the Newtonian planes of simultaneity to be flat or Euclidean: intuitively, as the light cones in a general relativistic space-time collapse (as the velocity of light approaches infinity), all *spatial* curvature is squeezed out. See J. Ehlers, "Über den Newtonschen Grenzwert der Einsteinschen Gravitationstheorie," in J. Nitsch, J. Pfarr, and E. Stachow, eds., *Grundlagen-probleme der modernen Physik* (Mannheim: Bibliographisches Institut, 1981), 65–84; and D. Malament, "Newtonian Gravity, Limits, and the Geometry of Space," in R. Colodny, ed., *From Quarks to Quasars: Philosophical Problems of Modern Physics* (Pittsburgh, University of Pittsburgh Press, 1986), 181–201.

55. So empirical or experimental physics presupposes general mechanics, which in turn presupposes geometry. Moreover, geometry itself presupposes the mathematical theory of (continuous) magnitude, and the latter presupposes arithmetic. In each case, sciences at lower levels in the hierarchy presuppose that the sciences located at higher levels have already been put into place; and in all cases other than arithmetic this involves a significant element of conventional choice. For details see my "Poincaré's Conventionalism" (see note 17 above).

56. Poincaré was the first, in fact, to realize that the Lorentz transformations form a group, and his 1905–1906 papers on the dynamics of the electron were fundamental contributions to Lorentzian electrodynamics: see Miller, *op. cit.* (see note 20 above), chapter 1.14. In a famous lecture from 1912, "L'éspace et le temps," Poincaré states his preference for the Lorentz-Fitzgerald theory very clearly: see *Dernières Pensées* (Paris: Flammarion, 1913), chapter 2; translated by J. Bolduc (New York: Dover, 1963).

57. Here I am especially indebted to R. DiSalle, "Spacetime Theory as Physical Geometry," *Erkenntnis* 42 (1995): 317–37. DiSalle's discussion of geometry and free mobility occurs in sec. 2 of this paper.

58. Section 11 of the second edition replaces pp. 63–65 of the first edition (see notes 3 and 29 above); the first edition distinction occurs on p. 63.

59. *Allgemeine Erkenntnislehre* (1925), 64 (Blumberg, 69). Earlier, in sec. 6 (in both editions), Schlick discusses "concrete" (or "psychological") definitions as a prelude to his treatment of implicit definitions in sec. 7.

60. When Schlick introduces what he will call "definitions" in the first edition (see note 58 above), he explains the idea as follows: "When one coordinates a system of signs to a system of objects, it is clear that one must begin in all cases by arbitrarily stipulating certain symbols for certain things. The designation of numbers by numerals or sounds by letters are conventions of this kind. They are undertaken in different ways by different peoples." See *Allgemeine Erkenntnislehre* (1918), 61 (this passage is retained in the second edition). The "arbitrary [*willkürlich*]" character of such "conventions" is simply the arbitrari-

ness of linguistic designation in general (that is, what is sometimes called "trivial semantic conventionality").

61. See *Allgemeine Erkenntnislehre* (1925), 66 (Blumberg, 71): "We call a conceptual determination and coordination that is brought about in this way a convention [*Konvention*] (in the narrower sense, for in the wider sense every definition is of course a convention [*Ubereinkunft*]). The term convention in this meaning was introduced into the philosophy of nature by Henri Poincaré, and the investigation of the essence and meaning of particular conventions in natural science belongs among the most important tasks of that discipline."

62. See *Allgemeine Erkenntnislehre*, 66–67 (Blumberg, 71–72). The discussion concludes: "It is now the greatest possible simplicity of the laws of nature that determines the *final* choice of the definition of time, and it is first in this way that the definition of the temporal unit has acquired the character of a convention in our sense. For it is no longer tied to any concrete individual process, but is rather determined by the general prescription that the fundamental equations of physics assume their simplest form."

63. Schlick does not explicitly mention the laws of motion in this context (although these must at least be included among what he calls "the fundamental equations of physics"), but he does give "the axioms of the science of space" as an example a few pages later—*op. cit.*, 69 (Blumberg, 74). I adopt the terminology of "coordinating principles" from H. Reichenbach, *Relativitätstheorie und Erkenntnis Apriori* (Berlin: Springer, 1920); translated by M. Reichenbach (Berkeley and Los Angeles: University of California Press, 1965). It is clear that debate with Reichenbach concerning this notion in the years 1920–1922 was precisely what stimulated Schlick to add the new sec. 11 to the second edition of *Allgemeine Erkenntnislehre*. D. Howard, "Einstein, Kant, and the Origins of Logical Empiricism," in W. Salmon and G. Wolters, eds., *Logic, Language, and the Structure of Scientific Theories* (Pittsburgh: University of Pittsburgh Press, 1994), 45–105, provides a lengthy discussion of this debate and, in particular, of the differences between the first and second editions of *Allgemeine Erkenntnislehre* that were stimulated by it. It seems to me, however, that Howard's discussion is often quite misleading, due, among other things, to the circumstance that it entirely misses the fundamental difference, for Schlick, between concrete definitions (with their "trivial semantic conventionality") and conventions properly speaking (that is, coordinating principles).

64. *Raum und Zeit* (1922), *op. cit.* (see note 33 above), 105 (Brose-Heath, 267).

65. See *Raum und Zeit* (1922), 62–65 (Brose-Heath, 243–45). The passage begins as follows: "If one now considers that an entirely arbitrary [*beliebige*] division of the continuum by families of surfaces can in principle serve for laying down coordinates—for the physical laws are now supposed to be invariant with respect to *arbitrary* [*beliebigen*] transformations—then every fixed point and all orientation first appears to be lost. One does not see, at first sight, how measurement is still to be possible at all, how one can in any way still manage to attribute determinate numerical values to the new coordinates—especially when these are no longer the immediate results of measurement. A comparison of measuring rods, an observation of coincidences, first becomes a *measurement*, as we saw, when we take some

or another idea as basis, introduce some or another physical presupposition, or rather make a stipulation [*Festsetzung*], whose choice strictly speaking remains, in the end, always arbitrary [*willkürlich*]—even if experience so strongly indicates it as the simplest that we practically do not hesitate." This passage occurs in both editions, so it is certainly not true that the first edition had entirely ignored the importance of conventions or coordinating principles. What is new in the later edition (as in the second edition of *Allgemeine Erkenntnislehre*) is that this point is much more explicitly emphasized and is also explicitly linked to the general theory of scientific conventions initiated by Poincaré (compare note 61 above).

66. See *Raum und Zeit* (1922), 67–72 (Brose-Heath, 246–49); essentially the same passage, including the initial references to Riemann and Helmholtz, in particular, also occurs in the first edition. There are serious technical problems involved in linking what we now call the principle of equivalence with the infinitesimally Minkowskian (Lorentzian) character of the space-time metric, and these problems were in fact one of the topics of the Einstein-Schlick correspondence: for details see Norton, "Einstein's Principle of Equivalence" (see note 47 above), secs. 9–10. The important point here, however, is that *both* the Lorentzian character of the space-time metric *and* the geodesic law of motion contribute to the physical interpretation of Einstein's theory. Both, in our terminology, provide coordinating principles for the mathematical structure of that theory.

67. See DiSalle, *op. cit.* (see note 57 above), sec. 3, for an excellent discussion of this parallel. As Stein explains in sec. 5 of his "Philosophical Prehistory" (see note 13 above), 19–20, Newton's third law of motion, the equality of action and reaction, plays a crucial role in giving physical meaning to the flat inertial structure of Newtonian space-time. For corollary 6 to the laws of motion (a Newtonian counterpart to the principle of equivalence) implies that accelerations due to free fall are locally indistinguishable from uniform inertial motions. Nevertheless, by requiring that true accelerations always come in action-reaction pairs, we can, in favorable conditions (such as those prevailing in the solar system), separate the freely falling trajectories from the (flat) inertial trajectories, and thereby give empirical meaning to the flat inertial structure. If we demand a quite general, local criterion of separation, however, corollary 6 then forces us to adopt the variably curved version of Newtonian gravitation theory, just as the principle of equivalence, applied to the inertial structure of Minkowski space-time, leads to the variably curved *relativistic* theory of gravitation discovered by Einstein. See my *Foundations of Space-Time Theories* (see note 53 above), chapter 3.4.

68. Here see especially J. Ehlers, F. Pirani, and A. Schild, "The Geometry of Free Fall and Light Propagation," in L. O'Raifeartaigh, ed., *General Relativity. Papers in Honour of J. L. Synge* (Oxford: Oxford University Press, 1972), 63–84. As Ehlers et al., point out, this approach goes back to work of Weyl's.

69. Three-dimensional spatial geometry, in particular, can thus be given a physical interpretation entirely independently of all consideration of rigid bodies: just as in the case of inertial structure, elementary dynamical processes (here governed by the light-principle) suffice. In this sense, (spatial) geometry is much more intimately connected with the fundamental laws of physics in relativity theory than it is in classical physics.

70. The canonical source for this (continuing) philosophical discussion is of course H. Reichenbach, *Philosophie der Raum-Zeit-Lehre* (Braunschweig: Vieweg, 1928); translated by M. Reichenbach and J. Freund (New York: Dover, 1957).

71. In the context of the use of Euclidean geometry in Newtonian physics— well before the discovery of the non-Euclidean geometries of constant curvature— it is most natural to view the *whole* of Euclidean geometry as providing us with our system of coordinating principles. Spatial distances in Newtonian theory, for example, are generally determined by the application of (Euclidean) geometrical optics within the solar system rather than by the transport of rigid bodies. Here I have a small disagreement with DiSalle, *op. cit.*, sec. 2, where Helmholtz's perspective on rigid body motion is taken as a quite general account of (three-dimensional) physical geometry.

72. See again notes 61, 62 above, and compare the passage to which note 64 is appended, where Schlick extends Poincaré's conventionalism to the coordinating principles of general relativity.

73. This is the main point I was attempting to make in my "Poincaré's Conventionalism" (note 17). However, as DiSalle has argued in "Intuition, Convention, and Twentieth-Century Physics" (presented at a Workshop on Intuition in Mathematics and Physics, McGill University, September 1999), one should not let this particular tension between general relativity and Poincaré's philosophy of geometry blind one to the important *continuities* between Einstein's theory and classical geometry-plus-physics with respect to the need for presuppositions or coordinating principles to give abstract mathematical concepts an empirical physical interpretation. The upshot is that the general idea of coordinating principle must be separated from the more specific idea of conventional choice.

[9]

"The Relations between Things" versus "The Things between Relations": The Deeper Meaning of the Hole Argument[1]

JOHN STACHEL

1. Introduction

I first discussed the "hole argument" in 1980, when presenting my interpretation of the reason for Einstein's rejection, from mid-1913 to mid-1915, of generally-covariant gravitational field equations.[2] Yet I have only recently come to recognize the full import of the argument, thanks to stimulation provided by the work of Rynasiewicz (1994, 1996) and Liu (1997). In effect, if not intent, their work leads to an extension of the subject of the hole argument from the spatiotemporal relations between the points of a differentiable manifold to arbitrary relations between the elements of a set, and I am grateful to both for inspiring this extension.[3] This paper considers several generalized and abstracted versions[4] of the original hole argument.

The next section briefly recalls the original hole argument, which concerns a general-relativistic space-time, and its generalization to fibered manifolds. Section 3 discusses the process of abstraction by deletion from the differentiability and continuity properties of such manifolds, and the resulting possibility of a set-theoretical version of the hole argument within the context of G-spaces.[5] Section 4 discusses the special case of a set with structure, with which the rest of the paper is concerned.[6] After a digression in section 5 on Marx's use of reflexive definitions, section 6 returns to sets with structure and the problem of individuation of the elements of the set; section 7 then develops the analogue of the hole argument for the case in which the elements of the set are individuated by the relational structure.

Most treatments of the usual hole argument involve explicit reference to coordinate systems on the manifold, and this has been the source of much confusion and misunderstanding.[7] Coordinates are a way of introducing names for the points of a manifold,[8] so abstraction from differentiability and continuity properties leads to consideration of the names of elements of the set. Section 8 discusses the relation between permutations of the elements of the set and permutations of the names of these elements, and argues against the view that the permutation of names constitutes the essence of the hole argument.

Finally, section 9 summarizes some of the conclusions of the paper, and discusses the possible significance of the observation that space-time points and elementary particles, two kinds of the entities that constitute fundamental building blocks of our models of the world, are entirely individuated by the relational structures in which they are embedded.

The appendix gives a simple introduction to the concepts of differentiable manifold, diffeomorphism, and fiber bundle that is meant to be readable by anyone familiar with the idea of a function, and provides sufficient mathematical background for reading section 2 of this paper.

2. The Hole Argument and Its Generalization

As noted in section 1, the original hole argument is concerned with a general-relativistic space-time; that is, a four-dimensional manifold M with a pseudometrical tensor field $g_{\mu\nu}$.[9] This metric (as I shall call it for short) represents not only the chronogeometrical structure of space-time but also the potentials for the inertiogravitational field.[10] Any number of other tensor fields, representing matter and/or nongravitational physical fields on space-time, may also be present on the manifold, so long as there is a "hole" H: that is, an open region of M, on which only the metric tensor field is present. The hole argument is concerned with the significance of the group Diff(M) of all *diffeomorphisms* of M,[11] and the mappings each such diffeomorphism induces on the metric, and any other tensor fields on M, when these fields are carried along with it.

Inside the hole H, the metric is assumed to obey the empty-space Einstein field equations, which have the property of being covariant. This implies that, if $g_{\mu\nu}$ is one such solution, so is every metric field induced from that solution by any diffeomorphism within the hole that reduces to the identity mapping on its boundary.[12] If we assume the points of the hole to be individuated independently of the metric field, then all of the mathematically distinct solutions generated by such diffeomorphisms must be regarded as physically distinct. This in turn implies that, even if a physical solution to the field equations is completely specified outside and on the boundary of the hole—including specification of all matter and nongravitational fields as well as the gravitational field—this cannot serve to uniquely specify a physical solution within the hole.

Proof: Consider any global solution to the field equations (that is, one that holds everywhere on M). We can always perform a diffeomorphism on this solution that reduces to the identity outside and on the boundary of the hole (namely, on $M-H$), but differs from the identity inside H. Without in any way changing the physical situation outside H, this will then produce a different solution inside H—and indeed a distinct solution for every distinct such diffeomorphism! Q.E.D.

Parenthetically, one sees why the hole argument fails for pre-general-relativistic space-times, in which the symmetry group of space-time is a finite-parameter subgroup (Lie group) of the diffeomorphism group; for example, either the ten-parameter symmetry group of Galilei-Newtonian space-time (the inhomogeneous Galilei group) or that of special-relativistic space-time (the inhomogeneous Lorentz or Poincaré group). If an element of such a group reduces to the identity element on the boundary of H, then that element must equal the identity element everywhere inside.

Going back to general relativity, the conclusion of the hole argument runs counter to the physical requirement that the gravitational field be determined by the specification of all matter and gravitational and non-gravitational fields outside the hole, together with the imposition of sufficient boundary conditions on its boundary. Remember, the hole can be made as small as we like; so to deny this requirement would mean that the only way to specify a unique model satisfying the field equations would be to specify it *everywhere* on M. But this would mean that the field equations are practically useless.[13]

So, if we are not to abandon the Einstein equations—or indeed *any* covariant set of field equations[14]—there must be a way to evade the conclusion of the hole argument. That way is clear: it is to deny that diffeomorphically related mathematical solutions to the field equations represent physically distinct solutions. In turn, this implies that one must assume that, inside H, the points of the manifold are *not* individuated independently of the $g_{\mu\nu}$ field; in other words, that these points inherit *all* their chronogeometrical (and inertiogravitational) properties and relations from that field. With that assumption, an entire equivalence class of diffeomorphically related mathematical solutions represent just one physical solution.

It is possible to generalize the original argument in (at least) two ways:[15] The dimension of the differentiable manifold may be taken as arbitrary (but still finite); and the pseudometric and other tensor fields may be replaced by an arbitrary geometric object field, which is a cross section of the corresponding fibered manifold.[16] Such a generalization allows for the treatment of fiber bundles, including jet prolongations of these bundles, and connections on bundles; and thus the inclusion of the physically important case of gauge-field theories,[17] as well as first-order, Palatini-type formulations of general relativity based on an affine connection as well as a metric.[18] In all such cases, the underlying manifold is taken as the base space of the appropriate fiber bundle.

To make clear the coordinate independence of the hole argument, Stachel (1986) gives such a generalization of it based on the concept of fiber bundles.[19] However, it overlooked what now seems to me an obvious further move. Remember that a fiber bundle consists of three elements: a base-space manifold, a total-space manifold, and a projection from the total space onto the base space. The set of all elements of the total space that project onto a particular point of the base space constitute the fiber over that point, and the entire total space consists of the union of the (disjoint) fibers over all points of the base space. A particular physical theory is represented by a particular choice of fiber bundle, and a cross section of bundle, i.e., a smoothly varying choice of a point on each fiber, represents a particular mathematical model of the theory.

Because of the gauge invariance of many physical theories, there will often be a class of cross sections that represent the same physical model of the theory, one member of the class being related to another by a transformation belonging to the gauge group. These gauge transformations are diffeomorphisms of the fiber bundle whose action is along the individual fibers. Since they have no effect on the base space, which usually represents the space-time manifold of the physical theory, I shall call these fiber diffeomorphisms *internal*. In general, an equivalence class of cross sections related by internal fiber diffeomorphisms will correspond to a single physical model.

There is another important class of fiber-bundle automorphisms that preserves the fibers, but acts to mix them up (in a continuous, differentiable way, of course). Since they induce diffeomorphisms on the base space, I call these fiber diffeomorphisms *external*. Since the hole argument is concerned with external diffeomorphisms, I shall only consider them in the rest of this paper, reserving discussion of the important question of the relation between internal and external fiber automorphisms for Stachel 2001.[20]

What is the physical significance of these external automorphisms (hereafter I shall drop the adjective "external")? One must distinguish between two possibilities: Either the points of the base space are individuated independently of cross sections of the total space or they are not. In the first case, the hole argument is valid, and a base space defined independently of the total space is needed. In the second case, the hole argument is not valid, and one can *eliminate* the base space, replacing it with the quotient space of the total space by the fibers. In other words, one starts with a total space and an equivalence relation between the points of this space that determines its fibers. Then one *defines* the underlying manifold as consisting of the points of the quotient space—each fiber representing one point of this space—endowed with the manifold structure it inherits from that of the total space.[21] Then a permutation of the fibers of the total space (i.e., a diffeomorphism of the total space that preserves the fibers) is inseparable from the corresponding permutation of the points of the underlying manifold (i.e., a diffeomorphism of the quotient manifold).

Now, in fibered manifold (or fiber bundle) language, the very formulation of the hole argument is based on the possibility of permuting the fibers of the total space without permuting the points of the base space (or *vice versa*). In a fiber-bundle-covariant theory, if any cross section of the bundle is a model, so is any cross section resulting from the first by a permutation of the fibers without a permutation of the points of the base space[22] (or, conversely, by a permutation of the points of the base space without a permutation of the fibers). So, in this quotient-space version of the theory, based on individuation of points of the underlying manifold by the cross sections of the bundle, it is impossible to *even formulate* the hole argument. Using one of Hertz's criteria for choosing between models (*Bilder*),[23] this quotient-space formulation is clearly superior to the fiber bundle approach, since it contains less "superfluous or empty relations."

Continuing the parenthetical comment above, in the case of non-general-relativistic space-times the base manifold *could* be eliminated in the same way; but there is no *need* to do so since the hole argument fails here anyway. In other words, for pre-general-relativistic space-times, it is *optional* whether we take an absolute or a relational approach to the points, while for general relativistic space-times, the hole argument makes it *mandatory* to adopt a relational approach.[24]

3. From Manifolds to Sets

Up to this point, I have been talking about the original hole argument and its generalizations, all involving various differentiable manifolds—fibered manifolds, total spaces, base spaces, etc.—and diffeomorphisms of these spaces. The point of this paper is to emphasize that the essence of the hole argument is preserved by a process of *abstraction by deletion* from the differentiable properties of these concepts.[25] If we abstract from the differentiability and even the continuity properties of a differentiable manifold,[26] we are left with a bare *set* of points; diffeomorphisms of the manifold become *permutations* of the elements of the set.[27] Covariance under the group of diffeomorphisms becomes *permutability*, in other words., invariance under the symmetric group, consisting of *all* permutations of these elements.[28] A fibered manifold, covariant under the diffeomorphism group, becomes a *G*-space, with a *G*-invariant equivalence relation, where *G* is a symmetric group.[29] It proves possible to formulate a version of the hole argument for such a *G*-space that is not merely similar in a general sense to the geometrical-object version; but is precisely analogous, in the sense that it can be derived from the latter by a process of abstraction.[30]

Further, it can be shown that the set of all relations between the elements of a set provides an example of such a *G*-space with a *G*-invariant equivalence relation. However, the rest of this paper starts from the concept of a set with some structure of relations between its elements and develops the hole argument for this case.

4. Relations between Things

Consider a set **S** of n entities[31] (the "things" of my title), which I shall label $a_1, a_2, \ldots, a_n$, among which there is an ensemble[32] **R** of M n-place relations, $R_1, R_2, \ldots, R_M$. I shall call such a relational structure[33] (**S**, **R**) a *world*.

All the entities are assumed to be of the same kind,[34] i.e., distinguishable from entities of a different kind, as are cabbages from kings,[35] electrons from protons, or points of space-time from logarithms. Two main possibilities concerning the individuation of entities of the same kind arise immediately:

1) the entities are individuated (that is, distinguishable from other entities of the same kind) prior to and without reference to the relations **R** (as, in the kind consisting of fictional human beings, Romeo is individuated from Juliet quite apart from their love relationship).[36]

2) the entities are not individuated (that is, are indistinguishable among themselves) without reference to the relations **R** (as we have seen, the hole argument forces us so to consider points of space-time; and, as we shall see, quantum statistics forces us so to consider electrons). There are two possibilities here:

a) The entities are fully individuated by the relations R; in other words, it is possible to distinguish between any two of them by their position in the relational structure.

b) The entities are only partially individuated by these relations, with the limiting case, in which they are not individuated to any extent by them.

In the course of the later discussion, both possibilities will be considered (references will be to "case 1" and "case 2" respectively), and further subdivision of case 1 will be found necessary. But before turning to such questions, and in order to formulate rigorously the hole argument for sets and relations, we need definitions of the concepts of relation, sequence, and permutation.[37]

I start with the concept of sequence: An *N-item sequence s* of elements of the set **S** may be defined as a function *s*: **N**→**S**, where **N** is the set of all natural numbers from 1 to *N*. The values of the function for the values of the integers will be written $s_1, s_2, \ldots, s_N$. The duplication of elements is allowed, of course, in other words, s_i may be the same element as s_j, where *i, j* are any two integers $\leq N$.[38]

The direct power set S^N of any set **S** consists of the product set $S \times S \times \ldots \times S$, taken *N* times, where *N* is an integer, and the cross designates the Cartesian product of two sets.[39] It is possible to consider the case when *N* is greater than *n*, but I shall assume that $N \leq n$. An element of this direct power set is then an *N*-item sequence of the *n* entities.

An *N*-place *relation* R_N is defined (extensively) as a subset of S_N (a *property* is a subset of S_1, which may be identified with **S**).[40] Thus if the sequence $a_1, a_2, \ldots, a_N$ belongs to this subset, the proposition "The relation R_N holds between the sequence of entities $a_1, a_2, \ldots, a_N$," which we abbreviate "$R_N (a_1, a_2, \ldots, a_N)$ holds," is true; and if the sequence does not belong to this subset, the proposition is false.

By a simple trick, one can reduce (since $N \leq n$, perhaps raise would be a better term) all *N*-place relations to *n*-place relations. Simply adjoin to the subset of S^N defining the desired relation *all* subsets of S^{n-N}. Then, only the occupancy of the first *N* places in a sequence of *n* entities will determine whether the *N*-place relation holds.[41] So from now on, I shall only consider *n*-place relations, namely, subsets of S^n, with the understanding that this constitutes no loss of generality; so we may drop the subscript *n*: R will always stand for an *n*-place relation, and subscripts will be used to distinguish different *n*-place relations, for examples, R_i from R_j.

A *permutation* P is a bijection on the set **S**, P: **S** → **S**. The set of a permutations forms a transformation group, the *permutation group* on the set **S**, symbolized by **S**!. This group is often called the *symmetric group* on the set **S**, particularly if the set is not finite.[42] As mentioned in section 3, if we abstract from the continuity and differentiability properties of a differentiable manifold and its diffeomorphisms, we are left with a set and its symmetry group of automorphisms.[43]

There are *n*! permutations of *n* entities; so there are *n*! possible nonduplicating sequences of the members of **S**. Let $x = (x_1, x_2, \ldots, x_n)$ be one such sequence; I shall write P_x to symbolize a permutation of the sequence *x*, and **P**x to symbolize the entire set of *n*! of them. I shall write $R_i(x)$ as an abbreviation for the relation R_i with its places filled by some definite sequence *x* of the *n* entities, and **R**(*x*) for the entire ensemble of relations with their places filled by that sequence. It is meaningful to fill these places with any sequence *x* of the **S**.[44] If the sequence *x* satisfies the relation R, I shall say that holds; if it does not, that R(*x*) does not hold. If there is a possible world in which $\mathbf{R}(a)$ holds, I shall call $\mathbf{R}(a)$ a possible state of the world; if $\mathbf{R}(a)$ holds I shall call it a state of the world.

Consider $\mathbf{R}(a)$, and a permutated sequence Pa of the $\mathbf{a}$. Then $\mathbf{R}(Pa)$ in general will not hold if $\mathbf{R}(a)$ holds. However, one can define *another* ensemble of relations, $\mathbf{PR}$, which holds for a if and only if (iff) $\mathbf{R}(P^{-1}a)$ holds, where P^{-1} is the permutation inverse to P. That is, the two ensembles of relations have the same extension, which I write $\mathbf{PR}(a) = \mathbf{R}(P^{-1}a)$. It follows trivially from this definition (by substituting Pa for a in this equation, and noting that $PP^{-1} = I$, the identity permutation), that $\mathbf{PR}(Pa)$ holds if and only if $\mathbf{R}(a)$ holds—that is, they too have the same extension, which I write $\mathbf{PR}(Pa) = \mathbf{R}(a)$. This *trivial identity*—or, if you will, tautology (since it depends only on the definition of $\mathbf{PR}$)—has been taken to constitute the essence of covariance and hence to have important bearing on the hole argument.[45] Although used in one step of the hole argument, this identity really has nothing else to do with either covariance of geometric object fields on manifolds (or permutability of relations on sets) or the hole argument. This is easily seen, since there are no circumstances under which the identity could fail to hold, while there are (different) circumstances under which covariance (permutability) and the hole argument can fail to hold. But before getting to the hole argument in section 8, we must define permutability.

A world $(\mathbf{S}, \mathbf{R})$ is called *permutable* if, whenever $\mathbf{R}(a)$ is a possible state of the world, then $\mathbf{PR}(a)$ is also a possible state of the world for *every* one of the $n!$ permutations P of $\mathbf{a}$. As implied in the previous paragraph, this is the concept for relational structures that corresponds to covariance. Perhaps a simple example, adapted from one used in Putnam (1981) and Liu (1997), will help to make clearer the meaning of the definitions of PR and of a permutable world, and of the trivial identity. Suppose the world consists of just two objects, a cat and a cherry, and only one possible binary relation $R(a_1, a_2)$ that may hold between them. R holds iff: a_1 is on a mat and a_2 is on a tree; hence the permutated relation $PR(a_1, a_2) = R(a_2, a_1)$ holds iff a_2 is on the mat and a_1 is on the tree.[46] We start by assuming that there is a possible state of the world in which the cat is on the mat, and the cherry is on the tree, that is, R(cat, cherry) is a possible state. The permutation of (cat, cherry) is (cherry, cat), so if there is also a possible state of the world in which the cherry is on the mat and the cat is on the tree, in other words, R(cherry, cat) is also a possible state, then this world is permutable.[47] Now, consider PR and Pa. Then PR(Pa) is the possible state of the world in which the cherry is on the tree and the cat is on the mat—which I think everyone will admit is trivially the same possible state of the world in which the cat is on the mat and the cherry is on the tree; that is to say, trivially identical to the possible state $R(a)$.

This example depends on that fact that the definitions of a cat and a cherry have nothing to do with the relation R of being on a mat and on a tree: the cat and the cherry are defined independently of the relation $R(a_1, a_2)$ or its inverse, or to put it another way: The relations R, PR are relations between things that are individuated independently of these relations (case 1 discussed above, the kind here being living things). But now imagine a world in which being a cat or being a cherry *depends on* whether one is on a mat or on a tree. That is, a case in which the individuation of certain things depends (entirely or essentially) on the relations between them; in other words, apart from these relations, the distinction between them cannot be made. We may say, with a slight abuse of language, that this is a case of "things between relations" rather than "relations between things."

5. Enter Marx: Reflexive Definitions

As applied to the relation just discussed, this possibility seems (and is) far-fetched.[48] However, even leaving aside space-time points and their relations, such relations have been the subject of a great deal of discussion. A recent survey, Swoyer 1999, calls them extrinsic or relational properties ("Objects have them because of their relations to other things"), and contrasts them with intrinsic properties ("properties that a thing has quite independently of its relationships to other things").[49] However, they are more commonly referred to as internal properties or relations.[50] Bhaskar (1979, 54) offers a definition:

> A relation R_{AB} may be defined as *internal* if and only if A would not be what it *essentially* is unless B is related to it in the way that it is. R_{AB} is *symmetrically internal* if the same applies also to B. ('A' and 'B' may designate universals or particulars, concepts or things, including relations.)[51]

As it stands, Bhaskar's definition applies only to dyadic relations, but could easily be extended to polyadic ones.

In the history of philosophy, there is a tradition that maintains that *all* properties of objects are defined by such relations—this view has been called the doctrine of internal relations[52]—and another tradition that denies the existence of *any* such relations. Bhaskar strives to avoid the "erroneous view that all relations are internal," which he attributes to many "rationalists, absolute idealists and mistresses of the arts of Hegelian and Bergsonian dialectics," as well as "the doctrine that all relations are external," which he attributes to "virtually the whole orthodox (empiricist and neo-Kantian) tradition in the philosophy of science."[53] He regards it as "an epistemically contingent question as to whether or not some given relation is internal" (Bhaskar 1979, 53, 54).

The two cases—relations between things and things between relations, or as I shall also say, things that are independently and things that are reflexively defined (see next paragraph)—are just extreme or limiting cases. We are generally confronted with an intermingling of the two. It may even happen that, in one context, a thing may have been sufficiently defined in a way that is independent of certain relations and stand in these relations to other things; while, in another context, other aspects of the same thing may be defined reflexively in terms of certain other relations. In order to give a full account of the world, we must be prepared to use both independent and reflexive definitions.

Since Karl Marx's important contribution to the discussion of internal relations is not well-known among philosophers,[54] I shall cite his work extensively and adopt some of his terminology.[55] In connection with his analysis of the form of value in Marx 1867, he drew attention to what he called *Reflexionsbestimmungen*, which I translate as *reflexive definitions*:[56]

> Reflexive definitions of this kind are altogether very curious. For instance, one man is king only because other men stand in the relation of subjects to him. They, on the other hand, imagine that they are subjects because he is king. (Marx [1867]1974, translation modified from Marx [1867] 1976, 149.[57]

In the social world, terms that are defined reflexively are not the least significant ones: capitalist and wage worker, landlord and tenant, creditor and debtor, metropolis and periphery are just a few significant examples. And on a more personal level: husband and wife, parent and child, brother and sister. And, of course such terms also are frequent in the natural world: north and south magnetic poles,[58] predator and prey, electrons in an atom, atoms in a chemical compound. And as we have seen, the hole argument demonstrates that, in a generally covariant theory, the points of space-time are defined reflexively.

To continue Marx's example, a king is only a king insofar as he is in the relation of kingship to his subjects, and the subjects are only subjects insofar as they are in the (inverse) relation of subjection to the king. Thus, if we permute the relationship for any pair consisting of the king and a subject, we interchange the roles of king and subject.

Of course, one might object that we could talk about George and Peter as two male human beings, and if we permute their relationship, we interchange their roles, but not their male humanity. George remains George, whether he be king or subject. But the problem is that when we speak of a king, we usually have an individual human being in mind—that is, we tend to reify the poles of such a relationship and to forget that it is only the relationship that puts them at opposite poles. Once she is crowned,

Elizabeth is Queen of England—period. And each babe born into her realm is born a loyal subject of Her Majesty.[59] Thus, the important distinction between the two cases tends to be lost. We may forget that Mary, Queen of Scots, could lose her throne before she lost her head, and that a "loyal subject" may be found to have committed treason.

Marx discussed this problem in 1843, in connection with his critique of Hegel's *Philosophy of Law*, and indeed using just such an example:

> Birth only provides a man with his *individual* existence and constitutes him in the first instance only as a *natural* individual, while political determinations [*Bestimmungen*] . . . are *social* products, born of society and not of the natural individual. Hence what is striking [in Hegel] and even *miraculous* is to conceive of an immediate identity, an immediate coincidence, between *the birth of an individual* and the individual conceived as the *individual embodiment of a particular social position or function*. In this system nature *creates* kings and *peers* directly just as it creates eyes and noses. . . . I am a man simply by my birth without the agreement of society; a particular birth can become the birth of a peer or a king only by virtue of general agreement. Only this agreement can convert the birth of a man into the birth of a king. (Marx [1843] 1975c, 174).

Elsewhere in the same text, he criticizes Hegel for forgetting

> that the essence of the 'particular person' is not his beard and blood and abstract *Physis* [i.e., physical corporeality] but his social quality. . . . It is self-evident, therefore, that in so far as individuals are to be regarded as the bearers [*Träger*] of the functions and powers of the state, it is their social and not their private capacity that should be taken into account. (Marx [1843] 1975c, 77–78, translation of "*Träger*" changed)

Marx continued to use this term "bearers" in *Capital* in the same sense:

> But here individuals are dealt with only in so far as they are personifications of economic categories, bearers of particular class relations and interests. (Marx [1867] 1996, 10, translation modified)

I shall use this term for the entities, significant aspects of which are reflexively defined, that is, refer to them as bearers of the defining relations. Indeed, this discussion leads to a further distinction within case 1 (the case of individuation of the members of **S** independently of the relational structure **R**). In addition to its **R**-independent individuation, an entity may be subject to further specifications that do depend on **R**, and which in certain contexts are essential to the characterization of these individuals. As discussed above, the social relations between biologically individuated human beings provide us with numerous examples.[60] This may be contrasted with cases in which no significant (see preceding footnote) further specification of the individuals follows from the relations **R**. Here we are again verging on discussion of what constitutes the essence of an individual and what are just accidental qualities. Aside from remarking that the above discussion suggests that the answer is context dependent (the essence of a socialized human being, for example, as contrasted with the essence of a member of the species *homo sapiens*), I shall forgo further consideration of this topic, and resume the discussion of sets with structure started in section 4.

6. Reflexively Defined Entities

A permutable world (see section 4) will be called *generally permutable* if, for every permutation P of the **a** entities in **S**, **R**(a) and **R**(Pa) represent the *same* possible state of the world. Hence, a possible state of a generally permutable world may be written **R**(a).[61] Note the difference between "permutable" and "generally permutable": In a permutable world, if **R**(a) is a possible state of the world, then **R**(Pa) is *another* (generally quite distinct) possible state of the world. In a generally permutable world, if **R**(a) is a possible state of the world then **R**(Pa) is the *same* possible state of the world for every permutation P, which we may write **R**(Pa) or—since it is permutation-independent—**R**(a). The implication is that, in a generally permutable world, insofar as the ensemble of relations **R** is concerned, the individuation of an entity such as a_1 in that world depends *entirely* on the place it occupies in the ensemble of relations **R**. Adopting the terminology introduced in the previous section, I shall say that entities in such a world are r*eflexively defined by the ensemble of relations R*, or where no ambiguity is likely, are *reflexively defined*, and refer to the entities as the *bearers* of these relations.

I shall say that the ensemble **R** of such n-place relations defines some *structure*.[62] In particular (as noted earlier), it may happen that none of the relations in the ensemble really depend on more than m places, where $m<n$, yet the relations are satisfied for all permutations of the n entities. The structure, then, is independent of its bearers in the sense that, although the relations constituting the structure must have *some* bearers, the entire set of m bearers may be replaced in part or in whole by another set without changing the structure.

An obvious example would be a certain house of cards containing six cards: one generally considers it the same house regardless of which particular six cards in the fifty-two-card poker deck are used to build it; the two cards forming the roof, for example, are defined as roof-forming cards by their position in the structure. We may say all the elements entering into such reflexively defined structures are *homogeneous with respect to the ensemble of relations* **R**, even though they may be individuated in some other respects, as are the cards in a poker deck. This presents us with another distinction within case 1: although the elements are individuated independently of the relations **R**, the resulting distinction between the elements is *irrelevant* to the relational structure.[63]

A much more interesting example is provided by the electrons in an atom, say a carbon atom. Before discussing this example, I shall make some preliminary remarks on the general question of quantum statistics and its implications for individuality.[64] Like all elementary particles, electrons are defined as such, in other words, as entities of a particular kind, by properties that are independent of their relations with each other and with other entities: their mass and spin (which may be interpreted as the Casimir invariants that define them as belonging to a particular representation of the Poincaré group),[65] and their charge, for example. But, elementary particles of the same kind are entirely indistinguishable from each other, they possess no inherent individuality. In making this association between *distinguishability* and *individuality*, I am running counter to the custom in recent discussions of quantum statistics to make a careful distinction between the two, regarding the former as an epistemological and the latter as an ontological concept: It is maintained that objects that are ontologically individuals may be epistemologically indistinguishable under all circumstances.[66] Without entering into the question of whether such a fundamental split between epistemology and ontology could ever be justified, I shall adopt the viewpoint toward scientific models recommended by Hertz:[67] of two models that are equally capable of modeling some set of phenomena, that one is to be preferred that introduces the least number of superfluous or redundant elements. By this criterion, models of many-body systems of elementary particles that treat them as indistinguishable because they are not individuated are to be preferred to models that introduce a redundant concept of individuality.[68]

Some suggest that, if the elementary particles are not individuated, then any attempt to label them is misguided.[69] On the contrary, it is just an example of the usual method of coordinatization, introduced when treating any set of entities that are numerous, yet indistinguishable (see note 81 and the appendix for more details). What is important is to realize that, in all such cases, no one coordinatization (labeling in this case) is preferred over another; and that it is precisely invariance of all relations under all permutations of the labels that guarantees this. It is entirely indifferent which six electrons out of the universe make up a particular carbon atom They are individuated, as K-shell or L-shell electrons of that atom for example, entirely by the ensemble of their relations to the carbon nucleus of the atom and to each other. Indeed, the notation for the electronic structure of an atom is based on this type of individuation.[70] This is an example of case 2, in which individuation, in so far as it is possible, is entirely based upon position in a relational structure.

We can easily generalize the discussion of relations between elements that are all of the same kind to the case, in which not all of the places in the ensemble of n-place relations $\mathbf{R}$ can be filled by homogeneous elements, that is, entities of the same kind; but in which these places break down into subsets that can be filled by homogeneous subclasses. Subdivide the set of n places in each relation into subsets of places $(n_1, n_2, \ldots, n_K)$, and the set $\mathbf{a}$ into similar subsets $a_1, a_2, \ldots, a_K$, and require $\mathbf{R}(a_1, a_2, \ldots, a_K)$ and $\mathbf{R}(P_1 a_1, P_2 a_2, \ldots, P_K a_K)$ to represent the same world under all permutations Pi of the a_i in each subset. In this way one can handle, for example, nuclear structures consisting of protons and neutrons, atomic structures taking the nuclei into account, molecular structures consisting of atoms of various types: each case involves several kinds of entities with different roles depending on their position in the structure, but entirely independent of *which particular* entities of each kind occupy the appropriate positions in the structure.[71] At the biological level, the structure of any mature living organism depends on having molecules of various types as bearers of the structure, but the process of metabolism just consists of the continual replacement of the individual molecules while preserving the relations between them.[72] But this touches upon questions involving structural change over time, consideration of which I shall defer to another occasion (but see the end of section 9).

7. The Hole Argument for Sets

By now it should be clear how the original hole argument (see section 1) can be formulated and discussed at the level of abstraction of sets with structure, and permutations of the set members and their relations. A *theory* is a rule that picks out a class of worlds: in other words., a class of ensembles of n-place relations: $\mathbf{R}, \mathbf{R}', \mathbf{R}''$, etc., whose places are filled by the members of the same set $\mathbf{S}$ of n entities $\mathbf{a}$; further, let it be a permutable theory, that is to say, if $\mathbf{R}$ is in the selected class of worlds, so is $P\mathbf{R}$ for all P.[73] As noted above, $\mathbf{R}(a)$ represents a possible state of the world. Could such a permutable theory pick out a *unique* state of the world by first specifying a unique world, i.e., one $\mathbf{R}$; and then specifying how any number m of its places *less than* $(n-1)$ are filled? [74]

Our answer will depend on whether the entities $\mathbf{a}$ are independently or reflexively defined. First suppose they are defined independently of $\mathbf{R}$. Then $\mathbf{R}(a)$ and $P\mathbf{R}(a)$ represent different states of the world (see the discussion in sections 4, 6). But applying the trivial identity (see section 4) to $P\mathbf{R}(a)$ by letting P^{-1} act on both $P\mathbf{R}$ and a, we get $\mathbf{R}(P^{-1}a)$ as a possible state equivalent to $PR(a)$, and hence *distinct from* $R(a)$. So, if the theory is permutable, and the entities are independently defined, the answer to the question is no: Specification of the relations R without specification of how all n places in them are filled cannot possibly result in a unique possible state of the world, even if it has been specified how any number of places m less than $(n-1)$ have been filled. For we can always permute the remaining $(n-m)$ entities among the corresponding number of unfilled places, with any permutation P_{n-m} giving rise to a distinct possible state $R(P_{n-m}a)$. This is the generalized hole argument.

On the other hand, if the entities are generally permutable, i.e., if the a are only the bearers of the ensemble of relations R that define them reflexively, as discussed in section 6, then $R(a)$ and $R(Pa)$ are the same state for every permutation P. So, in this case, the answer to the question is yes: Given the ensemble R, specification of how all its n places are filled is *not* necessary to specify a possible state of the world in a generally permutable theory. This is the generally permutable escape from the hole argument. In a generally permutable theory, specification of the ensemble R itself is sufficient to pick out a unique solution. Of course, this is a different question from the question of what information *is* needed in a *particular* generally permutable theory to specify R.

One can immediately apply this result to the current discussion about the individuality of elementary particles.[75] The discussion concerns the philosophical implications of the fact that elementary particles obey quantum statistics. One group maintains that each elementary particle retains its individuality, and that quantum statistics are merely the result of the fact that certain states (namely those that are neither completely symmetric nor completely antisymmetric under permutations of the particle labels) that are accessible to systems of elementary particles that are not of the same kind, are for some reason inaccessible to systems of particles that are all of the same kind. The other group maintains that quantum statistics has its origin in the lack of individuality of elementary particles. As far as I know, no one has drawn attention to the close analogy between these positions and those of the so-called space-time substantivalists on the one hand and the relationalists on the other. Nor has anyone mentioned the possibility of extending the hole argument from the discussion of the individuality of space-time points to the discussion of the individuality of elementary particles, as I shall now do.

If we take the points of our set to represent n elementary particles of the same kind, then quantum-mechanical statistics imposes the requirement that all physical relations between them be permutable.[76] Our set-theoretical hole argument shows that, if we ascribe an individuality to the particles that is independent of the ensemble of permutable relations, then no model can be uniquely specified by giving all the n-place relations R

between them unless we further specify just *which* particle occupies *each* place in these relations. For example, the rules for filling atomic shells in the ground state of an atom with electrons would have to be regarded as radically incomplete, since they do not tell us *which* electron has the different quantum numbers that characterize that state.

On the opposite assumption, that the electrons have no individuality apart from their relations to each other and to the atomic nucleus, in other words that the relations are generally permutable, specification of the quantum numbers of an ensemble of electrons is all there is to it.

8. Names in the Metalanguage: "A Rose by Any Other Name"

In order to discuss the question of the relations between things and relations and the names of these things and relations, I introduce a metalanguage for the names. Let w represent the set of names in the metalanguage of the entities a, i.e, w_1 is the name of a_1, etc., and let w be the name of the sequence a; further, let $\Re$ be the ensemble of names in the metalanguage of the ensemble of relations R, i.e., $\Re_1$ is the name of R_1, etc. So $\Re(w)$ is the assertion in the metalanguage that the entities a, in the sequence a, satisfy the relation R. Now instead of—or rather, in addition to—permuting the entities and relations, it is possible to permute their names. Suppose we only permute the names. Just as in section 4, we can introduce the permutation group P, an element P of which acts on the names w in some initial sequence (I consider the permutations to be abstract operations, and therefore do not distinguish notationally between them when acting on things and relations, and acting on their names.) Then $\Re(Pw)$ represents a new name for the same relation $R(a)$, and hence the relation now named by $\Re(w)$ is a different one from $R(a)$. Indeed, it is the name of the relation $R(P^{-1}a)$, where P^{-1} is the permutation inverse to P. We shall give the name $P\Re(w)$ to the relation $PR(a)$ that (it will be remembered) has the same extension as $R(P^{-1}a)$. It follows that $P\Re(Pw)$ names a relation with the same extension as the relation named by $\Re(w)$. This is an example of the conventionality of language: if we permute the names of all objects, and at the same time permute the names of the relations between them, we have changed nothing. To revert to our familiar example, suppose we rename cat "cherry," and cherry "cat," keeping the other names. Then "The cherry is on the mat and the cat is on the tree" means that the cat is on the mat and the cherry is on the tree. On the other hand, if we also rename mat "tree," and tree "mat," then the assertion: "the cherry is on the tree and the cat is on the mat" means that the cat is on the mat and the cherry is on the tree—in other words, just what is meant before we renamed everything. All of this renaming has nothing to do with any change in the positions of the cat or the cherry. Like Shakespeare's rose, they remain unperturbed by all this renaming.

Liu seems to think that "the inscrutability of reference" (which I prefer to call linguistic conventionalism) that follows from the permutability of names constitutes the general form of the hole argument.[77] But as we have just seen, permutability of names has nothing to do with what happens to the objects and their relations; so *per se* it can have no bearing on the hole argument, which as we have seen in sections 6 and 7 is concerned with the permutations of objects and relations, and not of their names.

One can *impose* such a bearing by demanding that, whenever names of objects and/or relations are permuted, the objects and/or relations be similarly permuted. Then, for example, $\Re(Pw)$ is the name, not of R(a), but of R(Pa). If we permute "cat" and "cherry" when we permute cat and cherry, and "on the mat" and "on the tree" when we permute on the mat and on the tree, then "the cat is on the mat and the cherry is on the tree" means the cat is on the mat and the cherry is on the tree. Thus, all arguments about permutations of things and/or relations can be rephrased as arguments about the renaming of these things and relations.

The analogy to names of set members in the original hole argument is a coordinate system, and the analogous move is to reformulate arguments about diffeomeorphisms and dragged-along geometrical objects in terms of coordinate transformations and corresponding transformations of the coordinate components of these geometrical objects. As I have argued elsewhere (see section 1 and references therein), while not wrong, this move has created more confusion about than clarification of the true significance of the hole argument.

9. Conclusion

In the course of our discussion, the following possibilities have emerged concerning the individuation of the objects in a set with structure (an example of each, given earlier, follows in parenthesis): 1) They are individuated independently of their position in relational structure under consideration, and either a) this individuation is relevant to the relations constituting that structure (cat and cherry); or b) this individuation is basically irrelevant to these relations (cards in a card house); c) in addition to their relation-independent individuation, a further specification of these individuals depends on their position in the relational structure that is crucial to some aspect of their nature (distinct human beings become king and subject). 2) They are only individuated, a) fully or b) partially, by their position in the relational structure (points of space-time, electrons in an atom).

The hole argument for generally permutable theories is valid in cases 1a and 1c, but is invalid (or can be avoided by proper formulation of the theory) in the remaining cases, in which the relational structure confers all relevant individuality to the things between relations.

I shall conclude by discussing some further implications of these distinctions. In mathematics, perhaps the most fundamental example of the distinction between cases 1 and 2 is the distinction between the real numbers and the points of a line (one-dimensional continuum). As Shafarevich (1992, 7) puts it,

> whereas all the points of a line have identical properties (the line is homogenous), and a point can only be fixed by putting a finger on it, numbers are all individuals: $3, 7/2, \sqrt{2}, \pi$ and so on.

One can see in this distinction the origin of the distinction between geometry and algebra: Geometry deals with sets, the elements of which are homogeneous; while algebra deals with sets, the elements of which are distinguishable.[78]

In physics, the most interesting case is number 2, examples of which are provided (as we have seen) by the points of space-time (in general-relativistic theories) and by the elementary particles. If we can generalize from these cases, it seems that invariance under the full permutation group for discrete symmetries, or under the homeomorphism (diffeomorphism) group for continuous (differentiable) symmetries, should be required of any theory that demands the complete indistinguishability of its fundamental objects.[79]

This is the fundamental reason why general-relativistic space-time theories are such a profound advance over previous space-time theories, based as they are upon a ten-parameter Lie group of permutation symmetries (see section 2). In these theories, the metric and affine structures introduced *a priori* impart such a "rigid" structure of relations between the points of space-time that, if a few points are individuated, the individuation of all the rest follows as a consequence. For example, if one point is distinguished (to be interpreted as the origin of an inertial frame of reference), and four vectors at that point (to be interpreted as an orthonormal triad of unit spatial vectors and a unit temporal vector in Galilei-Newtonian space-time; and as an orthonormal tetrad in the case of special-relativistic space-time) are distinguished in some way, then all other points of the space-time are thereby distinguished.[80] In this way, a mathematical *coordinatization*[81] of the points of such space-times by Galilean or Minkowskian coordinates, respectively, can be realized physically by the introduction of rods and clocks, which also serve to define units of length and time, entities that are treated as nondynamical at least as far as concerns the dynamical entities to be treated subsequently in these background space-times.

Of course, the choice of origin and vectors is not unique. Any point can be chosen as origin, as can be any orientation of the four axes that respects the needed orthonormality conditions. Physically, this corresponds to the freedom to choose an inertial frame of reference, and within that frame to choose an origin in space and time, and an orientation of three mutually orthogonal spatial axes. But this means that the points of these space-times are not indistinguishable under *all* permutations (diffeomorphisms) of the points of the underlying manifold; but *only* under those of a ten-parameter Lie subgroup.

As we have seen in section 2, in general-relativistic theories, the points of space-time are characterized as such (i.e., as a kind) independently of the particular relations in which they stand to each other; but they are *entirely* individuated in terms of the relational structure given by some fibered manifold. We have also seen that there are other entities that are similarly individuated entirely in terms of some relational structure. Within nonrelativistic quantum mechanics at least, the elementary particles are similarly individuated.[82] As noted in section 6, electrons may be characterized as a kind in a way that is independent of the relational structure in which they are imbricated, but a particular electron (or group of electrons) in an atom, for example, can only be individuated by its role in such an atomic structure. Consequently, all relations between N of these particles must be invariant under the permutation group acting on these particles.[83,84]

Consideration of these two cases suggests the following viewpoint: The basic building blocks of any model of the universe (for example, the elementary particles and the points of space-time) are individuated entirely in terms of the relational structures in which they are embedded. Only "higher-level" entities, constructed from these building blocks, come to be individuated independently.

In the past there was a tendency to use spatiotemporal location to individuate otherwise indistinguishable material entities.[85] Today, in response to the hole argument, there is a tendency in general relativity to use material entities to individuate the points of space-time.[86] The previous considerations suggest that neither of these programs can succeed at a fundamental level. Rather, individuality would emerge from some sort of interplay between spatiotemporal and material elements; or perhaps from elements that have both a material and a spatiotemporal aspect; or perhaps space-time and matter themselves would emerge from some more fundamental entities. In the latter two cases, one would again expect that these entities would lack individuality.

Individuality would be an emergent property, and the conditions for its emergence would have to be investigated. Presumably, it would involve sufficient complexity of structure to allow for effectively irreversible processes to occur (as when we "mark" our children's socks for camp). The emergence of complexity would certainly have to have a structural aspect; but whether it should be interpreted only in a synchronic structural sense, or should also have a diachronic structural, perhaps cosmological sense is certainly a fascinating question for future study.

APPENDIX

SOME MATHEMATICAL BACKGROUND

(Notes to this appendix start on page 262.)

In this paper I use some rather abstruse mathematics. In order to make it as comprehensible as possible, I start from a well-known mathematical concept, that of a function $y = f(x)$. Here x and y are variables, the *independent* and the *dependent* variable, respectively; what makes y a function of x is that, to every value of the independent variable x there corresponds a (unique) value of the dependent variable y. The set of numbers over which a variable ranges is called its *domain*. For example, if the domain of x is the positive integers $1, 2, \ldots, n, \ldots$, the set of values of the function y is called a *sequence*, usually written y_n. I shall be concerned in this appendix with the case, in which the domains of x and y are the real numbers, so that we are dealing with real functions of one real variable.

I assume most readers are familiar with the usual method of graphing such a function. We put the real numbers in correspondence with the points of a line,[1] so that we may represent the real numbers x and y by the points of two such lines. We usually draw these lines perpendicular to each other, but it is important to realize that this representation is properly associated with another sphere of ideas, two-dimensional Euclidean geometry, that is quite independent of the idea of a function. It would be better to admit that, associated with each number x, and thus with the corresponding point of the line, there is a different line, each point of which represents a possible value of y at that value of x; and that these *fibers*, as I shall call these y-lines, apart from each being identical copies of the same line, have no relation to each other unless and until some further structure is introduced. The set of all these fibers constitutes a *fiber bundle*; with its help one can represent all possible real functions of the real variable x: each such function is a *cross section* of the bundle, i.e., a selection of a point on each of its fibers.

Now I must return to the remarkable association of the real numbers with the points of a line, a concept which I introduced so blithely. We are dealing here with a profound difference between *algebra* and *geometry*, conceiving each in the broadest sense of the term,[2] and the relation between the two.

Algebra deals with sets of elements that are *distinguished* from each other, i.e., which are individuated somehow. The numbers have this property, certainly; but constitute just one example. Other examples are the elements of a group, matrices, quaternions, etc.—algebras in the broadest sense of the word.

Geometry deals with sets of elements that are *indistinguishable* from each other, i.e., that are not individuated unless and until some additional structure is given, and which draw their individuation from such structure(s).[3] The points of a line, a plane, a space of *n*-dimensions, all have this property; but for simplicity we shall continue to confine ourselves to a line. But before turning to this example, let me emphasize that the concept of a geometry is just as applicable to a discrete set (with a finite or infinite number of elements) as to a continuum.[4] The important thing is that the elements of the set not be a priori distinguished from each other by any criterion.

The setting up of a correspondence between the real numbers and the points of a line is an example of the process called *coordinatization* by Weyl.[5] But we can imagine that the fiber bundle lies over the line, without introducing any coordinatization of the latter; in other words, we can extend the concept of a function from algebra to geometry. That is, we forget about (abstract from) the numerical values x, and stipulate that, to each point of the *line*, there corresponds a fiber that is still associated with the numbers y. So, to each point of the line there is attached a replica of the totality of real numbers; a *cross section* of this bundle associates a number with each point of the line. I shall call this a *scalar bundle* over the line, for reasons that will be clearer in a moment.

In introducing the concept of a geometry above, I omitted an important point, to which I must now turn: What distinguishes one geometry from another? Let us start with a fixed set of (indistinguishable) points. This simplest answer is that it is the group of mappings of this set of points onto itself, usually called the *automorphism group*, that characterizes a geometry. Now the largest possible group of automorphisms of a given set is the *permutation group*, consisting of all one-one (i.e., invertible) mappings of the set onto itself. Thus, the automorphism group of any geometry must be a subgroup (not necessarily proper) of the permutation group.

For a set of discrete elements, one usually chooses the full automorphism group to define a geometry on this set. For continuua, the choice is much greater. One usually demands that the action of the group be *effective*, that is that for any two elements of the set there exist an element of the group that takes one into the other. This still leaves plenty of room for various possibilities. For example, plane Euclidean geometry is characterized by the Euclidean automorphism group, consisting of all translations and rotations in the plane.[6]

Returning to our primary example, the one-dimensional real line, we see that rotations make no sense, and a one-dimensional Euclidean geometry is characterized by the automorphism group consisting of all translations of the line into itself. If we attach primary significance to one coordinatization of the real line, we are effectively adopting the translation automorphism group; i.e, we are attributing *geometric* significance to the *difference* between the coordinates of two points on the line: $x_2 - x_1$ represents the distance between these two points. The privileged coordinate systems are all then determined up to a translation of coordinates: $x_1' = x_1$ + const. This is the one-dimensional analogue of what we do in two-or-three dimensional Euclidean space and four-dimensional Minkowski space-time when we adopt, respectively, Cartesian and Minkowskian coordinate systems; and translations and, respectively. rotations and pseudorotations of these coordinate systems.

Returning to the one-dimensional real line, if we take as the automorphism group the group of *homeomorphisms*, that is all *continuous* permutations (one-one mappings) of the line onto itself, we get one-dimensional topology. The line with the resulting topology is a *topological manifold*, a concept which can be generalized from one to any—even an infinite—number of dimensions.

A topological property of a topological manifold is one that remains invariant under all homeomorphisms. In the one-dimensional case, the only such invariant is the *order* of the points on the line.[7] It is easily seen that globally, there are only two possibilities for the topology of the one-dimensional line. Either it is *open* or it is *closed*.[8] The interval AB may be regarded as open, giving us a topological "straight line";[9] or its end points A and B may be identified, giving us a topological "circle."

Now let us return to our scalar bundle over the line. Since we are requiring continuity of points on the line, it seems reasonable to require continuity as we move from fiber to fiber in the bundle. So the bundle itself now forms a continuous space, usually called the *total space*, as contrasted with the line, which is called *the base space*. Topological complications may now occur in the total space as well as the base space. If the base space is an open line, then there is no problem; but if the base space is closed, then the "initial" and "final" fibers corresponding to A and B must be identified when A and B are. But they may be identified without giving the total space a twist, or after giving it one, two, ... , *n* twists, where *n* is any positive integer. Thus, we get the concept of *twisted bundles*. Locally, there is no way to tell a twisted from an untwisted bundle—there is no place where the twist(s) reside(s); it is a purely global concept. Möbius bands provide a simple way of visualizing these topological complications.[10] If the bundle is twisted, there may be no global cross sections. We have to content ourselves with defining *local cross sections* over subintervals of the line that we patch together properly in the overlaps of their intervals of definition.

If we carry out a homeomorphism on the line, what happens to the fibers of our scalar bundle, each of which lies over some point of the line? We may either leave the fibers alone, or we may carry along the fibers with the points over which they lie, when we carry out the point transformation associated with the homeomorphism. Either way, we change nothing: since the points of the line are entirely homogeneous in themselves, the only difference between them can come from the value of some scalar field, i.e., from some cross section of the scalar bundle. Suppose the value of this field is 7 at some point. Then we may characterize (i.e., individuate) this point by describing it as "the point at which the field has the value 7." If we now carry out some point transformation, it does not matter whether we carry along the scalar field or not. We may *still* characterize *some* point of the line as "the point at which the field has the value 7."

An additional complication arises if we want to do calculus on our fiber bundle. We know how to differentiate a function $y = f(x)$. How can we differentiate a function of a point $y = f(P)$? The answer seems clear: we must *coordinatize* the line by a one-one mapping between the points and the real numbers. Such a correspondence is called a coordinate system. Now if x corresponds to P, we can differentiate $f(x)$ when we want to differentiate $f(P)$.

But this does too much: it suggests that each point of the line is individuated by the number that corresponds to it. The only way to destroy this illusion without destroying our ability to do calculus is to allow *all* coordinate systems on an equal footing. The value of $f(P)$ must be independent of the coordinate system used, which means that it behaves as a *scalar* under coordinate transformations, which now explains the names scalar bundle and scalar function.

Now we must carefully distinguish between *point transformations:* transformations of the points on the line; and *coordinate transformations:* transformations of the coordinate system that associates a real number with each point of the line. While there is a clear conceptual distinction between the two, there is a correspondence between them: to each coordinate transformation, there corresponds a point transformation such that the *old* point in the *old* coordinate system has the *same* numerical coordinate as the *new* point in the *new* coordinate system.

Since we want to do calculus, we shall be forced to restrict ourselves to differentiable coordinate transformations, and in order to make this correspondence one-one, we must confine ourselves to differentiable point transformations. Thus, we get the concept of a *differentiable manifold*, a concept which can be generalized from one to any—even an infinite— number of dimensions. What happens to the fibers of our fiber bundle when we carry out a diffeomorphism of the base line? In the case of the real number bundle, there is a natural isomorphism between the fibers: the point with the value 3, let us say, occurs on every fiber. So there is a *unique* fiber automorphism, i.e., a "natural" fiber-preserving diffeomorphism of the total space, associated with each diffeomorphism of the base line. To get the full complexity of possibilities for more complicated bundles, we must complicate our picture a little, and look at a bundle consisting of all possible vectors $\mathbf{v}(p)$ over the line. There is still a natural fiber automorphism, but we must look a little more closely at how it arises. The "local" effect of the base diffeomorphism , i.e., its effect in the first neighborhood of every point p, may be described by a vector field on the line $\mathbf{e}[p]$: The diffeomorphism takes the point p to the point $p + \in \mathbf{e}[p]$, where $\in$ is a parameter of first order. We can "lift" this vector field $\mathbf{e}[p]$ from the line into the vector bundle, i.e., identify it with the vector $\in \mathbf{e}(p)$ in the bundle.[11] Then, if $\mathbf{v}(p) = \mathrm{v}\,\mathbf{e}(p)$, where v is a scalar (number), we can identify the point $\mathbf{v}(p)$ on the fiber over p with the point $\mathrm{ve}(\mathrm{p} + \in \mathbf{e}[p])$ on the fiber over $\mathrm{p} + \in \mathbf{e}[p]$.[12]

If we start from some cross section of the bundle (local or global) and carry out the fiber automorphism associated with a (local or global) diffeomorphism of the base line, this results in a *different* cross section for each such diffeomorphism. These cross sections are said to be diffeomorphically related to each other.

Now we have introduced most of the mathematical concepts needed in the paper that are not discussed in the text itself.

NOTES

1. Dedicated with affection to Howard Stein: *Il miglior fabbro*.
2. See Stachel 1986, 1987, 1989(the last is the published version of my 1980 talk) and 1993.
3. Aside from a comment in section 8 on Liu's attempt to give a linguistic turn to the hole argument, a critical discussion of their work must await another occasion.
4. These concepts have been distinguished by MacLane (1986, 434–38): "Generalization and abstraction, though closely related, are best distinguished. A 'generalization' is intended to subsume all the prior instances under some common view which includes the major properties of these instances. An 'abstraction' is

intended to pick out certain central aspects of the prior instances, and to free them from aspects extraneous to the purpose at hand" (ibid., 435–36).

5. Stachel 2001 will give a discussion of the mathematical details of the generalizations involving fibered manifolds indicated in section 2, and of the abstractions involving G-spaces indicated in section 3.

6. If one is willing to ignore references to the earlier sections, it should be possible to read the rest of this paper beginning with section 4.

7. As noted below, a major purpose of Stachel 1986 was to demonstrate that discussion of coordinates is not needed for the hole argument.

8. Of course, for a differentiable manifold, they are also a way of enabling us to do calculus on the manifold.

9. That is to say, one with Minkowski signature, which I take as (+ - - -).

10. The pseudometrical field also functions as the potentials for the metric affine connection of the manifold, which represents the inertiogravitational field of space-time (see e.g., Stachel 1994).

11. That is, the one-one bi-continuous and (sufficiently) differentiable mappings of M onto itself.

12. Outside the hole, one may introduce other fields that, together with the g field, obey the non-empty space Einstein equations, which also have the property of being covariant. But this is easily done, and I omit the details.

13. Of course, they would enable us to *reject* certain metric fields (those that are not solutions), but not to choose one among the equivalence class of diffeomorphically related metric fields that are solutions.

14. As Einstein did between mid-1913 and mid-1915 (see papers cited in footnote 1).

15. See MacLane 1986, 434: "*Generalization from cases* refers to the way in which several specific prior results may be subsumed under a single more general theorem."

16. For fibered manifolds and geometric objects, see Kolár, Michor, and Slovák 1993. If one does not require the geometric object field to be irreducible, one can treat several irreducible fields with one fibered manifold.

17. For a recent, fairly elementary treatment of this subject, see Darling 1994.

18. See, e.g., Wald 1984, 450–69.

19. Stachel 1986 essentially uses this concept, based on the treatment of fiber spaces in Hermann 1975. Although written in terms of fiber bundles, my paper actually uses only a fibered manifold and defines a geometric object as a cross section of such a manifold. (To facilitate comparison, I use the same notation here as in my earlier paper.) Earman 1989, (158–59) follows Schouten 1954, (67–68) in giving a coordinate-dependent definition of geometric object fields, and implies that a coordinate-independent definition would be very difficult. While Earman 1989 refers to Stachel 1986 in another context, it does not note that it already gave such definition.

20. I thank Abhay Ashtekar, whose questions forced me to emphasize the distinction between these two classes of automorphisms.

21. This is precisely what one usually does in the case of a principal fiber bundle (see, for example, Kobayashi and Nomizu 1996, 50); except in that case one starts from a structure group G that acts freely on the total space, and derives the

base space and the fibers from the equivalence relation on the total space induced by G. However, Kolá?, Michor, and Slovák 1993, 75–76, define a fiber bundle without the group, only demanding that the fibers be diffeomorphic to a standard fiber. For a fibered manifold, one even drops this requirement. In either case, one can start from either the fibers or an equivalence relation on the total space that defines the fibers, and define the base space as the quotient of the total space by the fibers.

22. It is trivially true that a permutation of the points of the base space, together with the corresponding permutation of the fibers that it induces, results in the same model. This is the fibered-manifold analogue of the fundamental identity discussed in section 4.

23. "Of two images of equal distinctness the more appropriate is the one which contains, in addition to the essential characteristics, the smaller number of superfluous or empty relations,—the simpler of the two" (Hertz [1894] (1956), 2). Redhead's (1975) concept of "surplus formal structure" seems to be closely related.

24. For further discussion of this point, see section 9.

25. See MacLane 1986, 436: "*Abstraction by deletion* is a straightforward process: One carefully omits parts of the data describing the mathematical concept in question to obtain the more 'abstract' concept."

26. One might pause at the level of continuity, and consider the hole argument at the level of topological manifolds; but I shall proceed directly to the level of sets. In the language of category theory (see, e.g., Adámek, Herrlich, and Strecker 1990, 22), I am applying a forgetful functor from the category of differentiable manifolds with diffeomorphisms to the category of sets with permutations.

27. When a set contains an infinite number of elements, the term "permutation" is sometimes limited to the case when only a finite number of elements are permuted, which would correspond more to a similar abstraction from the concept of local diffeomorphism. Since I shall later confine myself to finite sets, I shall not explore this distinction any further.

28. See, for example, Smith and Romanowska 1999, 78 or Neumann, Stoy, and Thompson 1994, 9.

29. See Neumann, Stoy, and Thompson 1994, 30–37.

30. See Smith and Romanowska 1999, 72.

31. For simplicity and to facilitate comparison with the arguments of Rynasiewicz 1994, 1996 and Liu 1997, and of Putnam 1981 that inspired them, I shall assume the set of entities under discussion to be finite. However, most of what is said would hold true even for sets of transfinite cardinality.

32. Although it is usually used synonymously with "set," I shall use the word "ensemble" to refer to a set-with-structure, i.e., one that may have some additional structure(s) relating its elements. (For the concept of "set-with-structure," see Smith and Romankowska 1999, 3: "Algebra can be characterized internally as the study of sets with structure"; see also MacLane 1986, 33.)

33. A set S together with an ensemble of relations R on S is called a *relational structure* (see Smith and Romanowska 1999, 72: "a relational structure (X, R) is a set X together with a multiset R of relations on X"). I shall leave open the

question of the exact nature of this relational structure, not assuming *a priori* that the relations are either entirely independent of each other, so that their order is unimportant (as the word "set" would imply), nor that there is a unique order among them (as the word "sequence" would imply). They might, for example, have the structure of a partially ordered set, or something even more general.

34. One can easily dispense with this assumption; but as holds true for most of the cases I shall discuss, I introduce it here, and only modify it later when the need arises (see section 6). Auyang 1995, 126–27 uses "kind" in a similar way, but makes a further distinction.

35. For other purposes, cabbages and kings may be grouped together, as organic things for example as distinguished from nonorganic things.

36. See *Romeo and Juliet*, act I, scenes 1 and 2 for Romeo, scene 3 for Juliet. Note that this alternative leaves open the possibility that the distinction between these entities has been established by means of some other ensemble of relations (including properties, as one-place relations) between them, so long as this ensemble is entirely independent of R.

37. See, e.g., Smith and Romankowska 1999, chapter 1. I take the definition of a set for granted, using no more than naive set theory.

38. Characters in boldface will always stand for sets, in Roman type for elements of a set, and in italic type for sequences of elements of a set.

39. I.e., the set the elements of which consist of ordered pairs of elements from each of the two sets. When applied more than once, it is easy to prove that the elements of a multiple Cartesian product are isomorphic to ordered sequences of elements from each of the sets.

40. With this definition, of course, any two intensively defined relations (including properties) that have the same extension are equivalent.

41. For example, there may be relations that are satisfied if just *one* of the places is appropriately filled by one of the entities, regardless of what permutation of the other (n -1) entities fills the remaining places; such a relation represents what is usually called a property (conversely, the word "property" is sometimes used to denote any relation). Similarly, **R** may include two-, three-, etc. -place relations that are satisfied by appropriately filling two, three, etc., particular places of an n-place relation, regardless of what permutation of the other (n - 2), (n - 3), etc. entities fills the remaining places.

42. See references in note 28.

43. The distinction between diffeomorphisms and local diffeomorphisms disappears at this level of abstraction, as does that between cross sections and local sections.

44. This means that the assertion $R_i(a)$ is either true or false, but not meaningless. Note this stipulation leaves open the possibility that the distinction between these entities is established by means of some *other* ensemble of relations between them, which is entirely independent of **R**.

45. See Earman and Norton 1987. Of course, their discussion refers to the diffeomorphism version of this identity (see the discussion in section 1, especially footnote 19); for a critique of their paper see Stachel 1993. But the permutation version clearly captures the essence of the matter at a much higher level of abstraction.

46. Note that, for two objects, there is only one possible permutation P, and it is idempotent, i.e., $P^2 = I$, so P is its own inverse.

47. To avoid confusion with Liu's use of this example, let me emphasize here that it has nothing to do with the *names* of the entities or of their relations. If we interchange the names "cat" and " cherry" and/or "mat" and "tree," in the first world the cat will still be on the mat and the cherry on the tree, and in the second the cherry will still be on the mat and the cat on the tree. For further discussion of names and things, see section 8.

48. But perhaps not entirely. Being "The Cat" in the story "The Cat in a Hat" depends on being a cat, so the set of candidates for the job is drawn from the class of cats; but being "The Cat in a Hat" depends on being a cat in a hat.

49. Swoyer, section 7, "Kinds of Properties," "Intrinsic vs. Extrinsic Properties," (1999, 42) comments: "The main questions here are whether there are any interesting intrinsic properties and how the notions of intrinsicness and extrinsicness are to be explicated." Swoyer's definition of properties includes relations, the usual properties being "monadic (one-place, nonrelational)" relations (1999, 2) .

50. Note that Swoyer 1999, after noting that "the phrase 'internal relation' has been used in different ways," uses it in quite a different way. For an excellent summary of the traditional distinction between external and internal relations, see Rorty 1967.

51. The omission of several words has been corrected from the reprint of this chapter in Archer et al. 1998, 222. A similar definition is given in Bhaskar 1993, 10. In Bhaskar 1994, 75, the definition is slightly emended: "A may be said to be internally related to B if it is a necessary condition for the existence (weak form) or essence (strong form) of B, whether or not the converse is the case (i.e., the relation is symmetric)."

52. In this century, it has been espoused largely by those in the Hegelian tradition. See, e.g., Joachim 1939 and Harris 1987. Moore 1922 is a critique of this view, arguing that it is based on a conflation of two assertions. I am indebted to Gordon 1996 for these references.

53. For similar comments and a review of arguments for and against these views, see Rorty 1967.

54. But see Ollman 1976.

55. Especially since there is a tendency these days to treat Marx, as he noted that Moses Mendelsohn treated Spinoza, as a "dead dog."

56. The standard translations use other terms: "reflex categories" (Samuel Moore-Edward Aveling translation, Marx [1867] 1996, 67 and "determinations of reflection" (Ben Fowkes translation, Marx [1867] 1976, 149).

57. The German text reads: "*Es ist mit solchen Reflexionsbestimmungen überhaupt ein eignes Ding. Dieser Mensch ist z. B. König, weil sich andre Menschen als Untertanen zu ihm verhalten. Sie glauben umgekehrt Untertanen zu sein, weil er König ist*" (Marx [1867] 1974, 72). Both Marx [1867] 1976 and Marx [1867] 1996 add a reference to Hegel's *Science of Logic*, and indeed the term is used throughout book 2, "The Doctrine of Essence"["*Die Lehre vom Wesen*"] (Hegel [1831] 1969, 17–240). In particular, the second chapter of the first section is entitled "*Die Wesenheiten oder die Reflexionsbestimmungen*" (ibid., 35–80); but it is in

the first chapter of the second section that one finds the most relevant passages, in a subheading entitled "*Die Wechselwirkung der Dinge*" (ibid., 137–39). I am grateful to Dr. Tilman Sauer for directing my attention to this section of the *Logic*. As concerns social relations, the discussion of mastership [*Herrschaft*] and servanthood [*Knechtschaft*] in Hegel's *Phaenomenologie des Geistes* [1807] 1970, 145–55 comes to mind as an immediate example of reflexively defined entities (without such a designation) well-known to Marx.

58. Marx was well aware of this example, too, and based another form of his terminology on it in his discussion of the form of value: "The relative form and the equivalent form are two intimately connected, mutually dependent and inseparable elements of the expression of value; but, at the same time, are mutually exclusive, antagonistic extremes—i.e., poles of the same expression. They are allotted respectively to the two different commodities brought into relation by that expression" (Marx [1867] 1996, 58). Marx picks up again terminology he had first used in 1843: "north pole and south pole are both *pole*; their *essence* is identical. . . . North and south are opposed aspects of *one* essence—the differentiation of one *essence* at the *height of its development*. They are *differentiated* essence. They are what they are *only* as a distinct *attribute*, and as *this* distinct attribute of the essence. . . . The difference . . . is a difference of existence." (Marx [1843] 1975c, 88). For a discussion of Marx's essentialism, see Meikle 1985.

59. Marx gives much more important examples of the reification of reflexively defined entities, notably in his discussion of capital. Most people—even most economists—believe that an automobile plant as such is an item of capital, and forget that it becomes capital only if it functions in a capitalist process of production, i.e., one in which it is employed by workers to make cars that are sold by a capitalist for profit. They forget, that is, until an industrial crisis forcibly reminds them of the fact. Indeed, even this description oversimplifies Marx's account of the nature of capital: Capital is the entire process, or circuit as Marx calls it, in which a sum of money is converted into means of production—including labor power—in the market; these means of production are put to work in a process of production in which that labor power is set to work to produce finished products; and these finished products are in turn sold on the market for a sum of money greater than the original sum invested. If any breakdown occurs in this circuit of metamorphoses of capital—and the various types of economic crises are characterized by just such breakdowns—then neither the money nor the means of production function as capital, as the cries of the victims of such destructions of capital forcibly remind us. Indeed, one function of such crises is to reduce the total value of social capital (a process that may involve *no* change in the physical properties of the means of production) to a level, at which its profitable reproduction can be resumed. (For his theory of the circulation process of capital and crises, see Marx [1884] 1975a, [1894] 1975b).

60. In a sense this is always the case: The cards in a poker deck (I stay away from the pinochle deck to avoid duplicity) are individuated by their markings (king of spades versus five of hearts); and if we build a house of cards, we might say that each card in the house is further specified by its position in the house. But this seems a rather trivial change compared, say, to the change of fortune that came to the Prince and the Pauper when they exchanged social roles. If we do not make

some sort of distinction between essential and non-essential properties and relations, perhaps a context-dependent one (see the end of Section 5), we shall end up asserting that all relations are internal– or that none are.

61. Note that we are not assuming– or excluding–the possibility that the relations R are symmetric under some or all of the permutations P.

62. Note that the word "structure" here carries no (necessarily) spatial implications.

63. This seems to settle the age-old knife puzzle: if the blade and hilt of a knife are each replaced, is it the same knife? Insofar as a knife is just a particular relation between blade and hilt, it is indeed the same knife.

64. I found Auyang 1995, the articles in Castellani 1998a, and French 2000 especially helpful in thinking about this question.

65. For a simple discussion of the use of Casimir invariants to define elementary particles, see Castellani 1998b.

66. See, for example, Gracia 1988. I am indebted to French 1998 for this reference.

67. See note 23.

68. Note that the argument in the text differs from that discussed, for example, in French 1998, which focuses on the possible existence of inaccessible quantum states that are neither totally symmetric nor totally antisymmetric under exchange of particle labels. The situation is somewhat analogous to the question of whether an ether frame is compatible with the special theory of relativity. It has been maintained that a preferred (i.e., individuated) ether (inertial) frame exists; but that it is impossible in principle to distinguish it from all other inertial frames, which are themselves individuated by their "absolute" velocities relative to the ether frame.

69. See French 1998, 102: "According to this [nonindividuality] metaphysics . . . tagging the particles with labels was a metaphysical mistake. . . . [W]e can smile indulgently at Schrödinger's faster-than-the-eye-can-see shuffling of particle labels and assert confidently that this shuffling subverts the very classical picture in which it was introduced."

70. See any textbook on atomic structure. As a result of the Pauli exclusion principle for fermions, each electron in an atom can be fully individuated by the set of its quantum numbers. Bosons cannot be so fully individuated and are best considered as field quanta; but I shall not consider this question further. Auyang 1995 has several valuable discussions of aspects of this topic.

71. Tisza (1998) points out the difference between the identity concepts of the physical and chemical atom that had already developed by the end of the nineteenth century. With reference to the latter, he notes that: "The name 'helium' is attached to the class of identical entities. You ought to realize that it is the same class identity that is at work in Q[uantum] M[echanics]." I am grateful to Professor Tisza for drawing my attention to this point.

72. Of course I am well aware that this is a simplified and idealized version of a much more complicated process. But I am trying to develop a language for discussing such processes, and at this point am not concerned with the modifications needed for a more realistic description of complex processes.

73. In case the analogy with the differentiable manifold case is not clear, let me spell it out. The set S corresponds to the manifold M; the theory corresponds to the field equations (of general relativity, for example). The solutions to these quations correspond to the worlds R , R', R", etc. PR corresponds to a solution to the field equations that is diffeomorphically related to the solution to which R corresponds.

74. Obviously, if how (n-1) places in R are filled is specified, this fixes how the n-th place is filled, and thus specifies a unique state of the world. To continue with the analogy developed in the previous note, the unfilled n-m places correspond to the hole, and we are asking the question that is analogous to whether a unique physical solution to the field equations can be specified by giving it outside of and on the boundary of the hole.

75. For reviews of this discussion, and references to advocates of both positions, see French 1998, 2000.

76. The relations will represent values of physical properties of the system of identical particles, which must remain invariant under all permutations of the particle labels. Since these physical properties are represented by bilinear functions of the state vector of the system, they will remain invariant whether the state vector remains invariant under a permutation (bosons) or changes sign (fermions). I am here adopting the state vector formulation of quantum mechanics customary in this discussion; elsewhere, I have shown how the discussion of quantum statistics may be carried out within the Feynman formulation of quantum mechanics (see Stachel 1997).

77. "I now argue . . . that *the hole argument is really a Putnam-type argument* [about "the inscrutability of reference"], *which does not give rise to radical indeterminism*" (Liu 1997).

78. This point of view is implicit in the treatment of geometry by Weyl, in particular 1949, 71–78. I have discussed this question further in Stachel 1999.

79. Auyang (1995, 162), discussing the "permutation symmetry of the aggregate of particles," notes that: "Conceptually, it is just a kind of coordinate transformation, where the coordinates are the particle indices." Aside from the reference to coordinate transformations rather than diffeomorphisms, this brings out well the parallelism under discussion. Huggett (1999) suggests that the action of the permutation group on elementary particles is analogous to the action of finite-parameter Lie groups, such as the Galilei group, on the points of space-time.

80. More explicitly, the end points relative to the origin of the four vectors fix units of length along four (three spatial and one temporal) axes. Invariance under rotational (respectively Lorentz) transformations guarantees the necessary compatibility of these units, and all other points are distinguished by the operation of the translational elements of the symmetry group on the initially distinguished origin point.

81. I here use the term "coordinatization" in the very broad sense employed by Weyl (1946, 15–16; 1949, 75) (he also uses the term "label") and Shafarevitch (1992, 7). As they indicate, there is no implication that the coordinates need be ordinary numbers. In this broad sense, the coordinatization of a space is a way of introducing the possibility of applying algebraic methods to geometric problems (see the appendix). But, the introduction of such an artificial distinction between

the otherwise indistinguishable elements of the space requires that all meaningful geometrical results be invariant under relabeling of the coordinates. Auyang (1995, 161–62) refers to the effect of permutation invariance as a "cleansing of names"; it "wipes out the artificial distinctiveness hidden in the particles' names."

82. A discussion within the context of quantum field theory is really required, but must be deferred. See Auyang 1995 for the fullest treatment of this question that I know.

83. This requirement is often referred to as the requirement that elementary particles be either bosons (obey Bose-Einstein statistics) or fermions (obey Fermi-Dirac statistics). For a discussion of how this permutation invariance affects the choice of a configuration space, see Stachel 1997, 252–53.

84. As will be shown elsewhere, it is possible to make a move in this case analogous to that in the case of fibered manifolds, discussed in section 2: By *defining* the elements of the base set as the fibers of the *G*-space, we make it *impossible* to carry out a fiber automorphism (i.e, to permute these fibers) without carrying out the corresponding automorphism (i.e., permutation of the elements) of the base space.

85. As concerns space, this approach seems to go back to Kant (see Auyang 1995, 135). Tisza (1998) discusses physicists' use of "orbital identity" to distinguish particles by their world lines in space-time.

86. This is an important problem in quantum gravity. Rovelli (1991) has emphasized the possibility of using the properties of material reference systems to individuate the points of space-time.

APPENDIX NOTES

1. A remarkable process of symbolic representation is involved here, to which I shall return shortly.

2. I here follow the treatment of Weyl 1946, 1949 and Shafarevich 1992.

3. Or, if they are individuated, this individuation is ignored or abstracted from in considering their geometrical properties.

4. For a simple discussion of finite geometries with an application to error-correcting codes, see Judith N. Cederberg, *A Course in Modern Geometries* (New York/Berlin/Heidelberg: Springer, 1989), chapter 1, "Axiomatic Systems and Finite Geometries," 1–24.

5. See note 81 above.

6. Sometimes changes of scale are also included in the definition of Euclidean geometry.

7. We omit discuss of the question of whether we should admit improper translations and homeomorphism, which reverse the order of all points on the line.

8. One might imagine including one or both of the end points, to get a semi-closed or closed interval; but the end points would constitute boundaries of the interval, so that such a choice would correspond to a manifold-with-boundary, a concept we shall not need.

9. To say the line is "infinite in length" has no meaning unless and until a metric is introduced. Remember that the entire range of real numbers from minus to plus infinity can be mapped onto any open interval, and if the "natural metric" associated with the real numbers (for which the distance between two numbers a and b is $a - b$) is adopted for the interval, then it is infinite in length.

10. For a simple introduction, see K. Weise and H. Noack, "Aspects of Topology," in H. Behnke et al., eds., *Fundamentals of Mathematics*, vol. 2, *Geometry* (Cambridge, Mass.: MIT Press 1974), 593–670.

11. It is crucial to distinguish between $\mathbf{e}[p]$, the vector on the line, and $\mathbf{e}(p)$, the vector in the total bundle space. Locally, a vector in the tangent space of the line can be identified with a vector in the line.

12. Note that in general this differs from the original value of the vector field at this point $\mathbf{v}(p+ \in \mathbf{e}[p])$.

REFERENCES

Adámek, J., H. Herrlich, and G. Strecker. (1990). *Abstract and Concrete Categories*. New York: John Wiley.

Archer, Margaret, Roy Bhaskar, Andrew Collier, Tony Lawson, and Andrew Norrie, eds. (1998). *Critical Realism: Essential Readings*. London/New York: Routledge.

Auyang, Sunny Y. (1995). *How Is Quantum Field Theory Possible?* New York/Oxford: Oxford University Press.

Bhaskar, Roy. (1979). *The Possibility of Naturalism*. Atlantic Highlands, N.J.: Humanities Press.

———. (1993). *Dialectic: The Pulse of Freedom*. London/New York: Verso.

———. (1994). *Plato Etc*. London/New York: Verso.

Castellani, Elena. ed. (1998a). *Interpreting Bodies: Classical and Quantum Objects in Modern Physics*. Princeton, N.J.: Princeton University Press.

———. (1998b). "Galilean Particles: An Example of Constitution of Objects." In Castellani 1998a, 181–94.

Darling, R. W. R. (1994). *Differential Forms and Connections*. Cambridge: Cambridge University Press.

Earman, John. (1989). *World Enough and Space-Time*. Cambridge, Mass./London: MIT Press.

Earman, John and John Norton. (1997). "What Prince Spacetime Substantivalism? The Hole Story." *British Journal for the Philosophy of Science* 38: 515–25

French, Steven. (1998). "On the Withering Away of Physical Objects." In Castellani 1998b, 93–113.

———. (2000). "Identity and Individuality in Quantum Theory." In *Stanford Encyclopedia of Philosophy*. Electronic text available at <http://plato.stanford.edu>.

Gracia, Jorge J. E. (1988). *Individuality: An Essay on the Foundations of Metaphysics*. Albany, N.Y.: State University of New York Press.

Gordon, David. (1996). "Philosophical Origins, Bibliographical Essay." In *The Philosophical Origins of Austrian Economics*. Available on the website of the

Ludwig von Mises Institute, <http://www.mises.org/philorgig/bibessay. asp>.

Harris, Errol E. (1987). *Formal, Transcendental, and Dialectical Thinking: Logic and Reality.* Albany, N.Y.: State University of New York Press.

Hegel, Georg Wilhelm Friedrich. [1807] (1970). *Werke.* Vol. 3, *Phänomelogie des Geistes.* Frankfurt a/M: Surhkamp Verlag.

———. [1831] (1969). *Werke.* Vol. 6, *Wissenschaft der Logik II.* Frankfurt a/M: Suhrkamp.

Hermann, Robert. (1975). *Gauge Fields and Cartan-Ehresmann Connections. Part A.* Brookline: Math Sci Press.

Hertz, Heinrich. [1894] (1956). *The Principles of Mechanics Presented in a New Form.* Translated by D. E. Jones and J. T. Walley. New York: Dover.

Huggett, Nick. (1999). "On the Significance of Permutation Symmetry." *Philosophy of Science* 50: 325–47.

Joachim, Harold A. (1939). *The Nature of Truth.* Oxford: University Press.

Kobayashi, Shoshichi and Katsumi Nomizu. (1996). *Foundations of Differential Geometry.* Vol. 1. New York: Wiley Classics Library Edition.

Kolár, Ivan, Peter W. Michor, and Jan Slovák. (1993). *Natural Operations in Differential Geometry.* Berlin: Springer-Verlag.

Liu, Chuang. (1997). "Realism and Spacetime: Of Arguments Against Metaphysical Realism and Manifold Realism." *Philosophia Naturalis* 33: 243–63.

MacLane, Saunders. (1986). *Mathematics: Form and Function.* New York: Springer-Verlag.

Marx, Karl. [1867] (1974). *Das Kapital/Kritik der politischen Ökonomie. Erster Band. Buch I: Der Produktionsprozeß des Kapitals.* In Karl Marx and Friedrich Engels, *Werke*, vol. 23. Berlin: Dietz Verlag.

———. [1884] (1975a). *Das Kapital/Kritik der politischen Ökonomie. Zweiter Band. Buch II: Der Zirkulationsprozeß des Kapitals,* in Karl Marx and Friedrich Engels, Werke, vol. 24. Berlin: Dietz Verlag.

———. ([1894] 1975b), *Das Kapital/Kritik der politischen Ökonomie. Dritter Band. Buch III: Der Gesamtprozeß der kapitalistischen Produktion.* In Karl Marx and Friedrich Engels, *Werke*, vol. 25. Berlin: Dietz Verlag.

———. [1843] (1975c). "Critique of Hegel's Doctrine of the State." In Rodney Livingston and Quentin Hoare (eds.), Gregor Benton (trans.), *Early Writings.* New York: Vintage Books, 57–198.

———. [1867] (1976). *Capital: A Critique of Political Economy.* Vol. 1. Translated by Ben Fowkes. Hammondsworth: Penguin.

———. [1867](1996). *Capital. Vol. I.* In Karl Marx and Friedrich Engels, *Collected Works*, vol. 35. New York: International Publishers.

Meikle, Scott. (1985). *Essentialism in the Thought of Karl Marx.* London: Duckworth.

Moore, G. E. (1922). "External and Internal Relations." In *Philosophical Studies.* London: Routledge and Kegan Paul, 276–309.

Neumann, Peter M., Gabrielle A. Stoy, and Edward C. Thompson. (1994). *Groups and Symmetries.* Oxford/New York/Tokyo: Oxford University Press.

Ollmann, Bertell. (1976). *Alienation: Marx's Conception of Man in Capitalist Society*. 2d ed. Cambridge: Cambridge University Press.

Putnam, Hilary. (1981). "A problem about reference." In *Reason, Truth, and History*. Cambridge/New York/Melbourne: Cambridge University Press, 22–48.

Redhead, Michael. (1975). "Symmetry in Intertheory Relations." *Synthese* 32: 77–112.

Rorty, Richard M. (1967). "Relations, Internal and External." In Paul Edwards (ed.), *The Encyclopedia of Philosophy*, vol. 7. New York: The Macmillan Company and The Free Press, 125–33.

Rovelli, Carlo. (1991). "What Is Observable in Classical and Quantum Gravity." *Classical and Quantum Gravity* 8: 297–316.

Rynasiewicz, Robert. (1994). "The Lesson of the Hole Argument." *British Journal for the Philosophy of Science* 45: 407–36.

———. (1996). "Is There a Syntactic Solution to the Hole Argument?" *Philosophy of Science* 63 (*Proceedings*): S55–S62.

Schouten, J. A. (1954). *Ricci-Calculus*. 2d ed. Berlin/Göttingen/Heidelberg: Springer-Verlag.

Shafarevitch, I. R. (1992). *Basic Notions of Algebra*. Berlin: Springer-Verlag.

Smith, Jonathan D. H. and Anna B. Romanowska. (1999). *Post-Modern Algebra*. New York: John Wiley and Sons.

Stachel, John. (1986). "What a Physicist Can Learn From the Discovery of General Relativity." In Remo Ruffini (ed.), *Proceedings of the Fourth Marcel Grossmann Meeting on General Relativity*. Amsterdam: Elsevier Science Publishers, 1857–62.

———.(1987). "How Einstein Discovered General Relativity: A Historical Tale with Some Contemporary Morals." In M. A. H. MacCallum (ed.), *General Relativity and Gravitation. Proceedings of the 11th Internation Conference on General Relativity and Gravitation*. Cambridge: Cambridge University Press, 200–208.

———. (1989). "Einstein's Search for General Covariance, 1912–1915." In Don Howard and John Stachel (eds.), *The History of General Relativity: Proceedings of the 1986 Osgood Hill Conference. Einstein Studies*, vol. 1. Boston/Basel/Stuttgart: Birkhäuser, 63–100.

———. (1993). "The Meaning of General Covariance: The Hole Story." In John Earman et al. (eds.), *Philosophical Problems of the Internal and External World: Essays on the Philosophy of Adolf Grünbaum*. Konstanz: Universitätsverlag/ Pittsburgh: University of Pittsburgh Press, 129–60.

———. (1994). "Changes in the Concepts of Space and Time Brought about by Relativity." In Carol C. Gould and Robert S. Cohen (eds.), *Artifacts, Representations, and Social Practice/Essays for Marx Wartofsky*. Dordrecht/Boston/London: Kluwer Academic.

———. (1997). "Feynman Paths and Quantum Entanglement: Is There Any More to the Mystery?" In Robert S. Cohen, Michael Horne, and John Stachel (eds.), *Potentiality, Entanglement and Passion-at-a-Distance, Quantum Mechanical Studies for Abner Shimony*, vol. 2. Dordrecht/Boston/London: Kluwer Academic, 245–56.

————. (1999). "Generally Covariant Space-Time Structures: What's the Point." Talk at the Minnowbrook Symposium, May 28, 1999. Text available electronically at <www.phy.syr.edu/research/hetheory/minnowbrook/stachel.html>.

————. (2001). "Fibered Manifolds, Geometric Objects, Structured Sets, *G*-Spaces, and All That: The Hole Story From Space-Time to Elementary Particles." (Preprint).

Swoyer, Chris. (1999). "Properties." In *Stanford Encyclopedia of Philosophy*. Electronic text available at <http://plato.stanford.edu>.

Tisza, Laszlo. (1998). "End of Century Task: Resolve the Paradoxes of the Beginning." *Journal of the Association of Physics Students* (Hungary), issue 5, spring 1998: 3–7.

Wald, Robert. (1984). *General Relativity*. Chicago: University of Chicago Press.

Weyl, Hermann. (1946). *The Classical Groups/ Their Invariants and Representations*, 2d ed. Princeton: Princeton University Press.

————. (1949). *Philosophy of Mathematics and Natural Science*. Translated by Olaf Helmer. Princeton: Princeton University Press.

[10]

A No-Go Theorem about Rotation in Relativity Theory[1]

DAVID B. MALAMENT

Within the framework of general relativity, in some cases at least, it is a rather delicate and interesting question just what it means to say that a body is or is not "rotating." Moreover, the reasons for this—at least the ones I have in mind—do not have much to do with traditional controversy over "absolute vs. relative" conceptions of motion. Rather they concern particular geometric complexities that arise when one allows for the possibility of spacetime curvature. The relevant distinction for my purposes is not that between attributions of "relative" and "absolute" rotation, but between attributions of rotation than can and cannot be analyzed in terms of a motion (in the limit) at a point. It is the latter—ones that make essential reference to extended regions of spacetime—that can be problematic.

The problem has two parts. First, one can easily think of different criteria for when a body is rotating. The criteria agree if the background spacetime structure is sufficiently simple, for example, in Minkowski spacetime (the regime of "special relativity"). But they do not do so in general. Second, none of the criteria fully answers to our classical intuitions. Each one exhibits some feature or other that violates those intuitions in a significant and interesting way.

My principle goal in what follows is to make the second claim precise in the form of a modest no-go theorem. To keep things simple, I'll limit attention to a special case. I'll consider (one-dimensional) rings centered about an axis of rotational symmetry, and consider what it could mean to say that the rings are *not* rotating around the axis. (It is convenient to work with the negative formulation.) The discussion will have several parts.

First, for purposes of motivation, I'll describe two standard criteria of nonrotation that seem particularly simple and natural. (I could assemble a longer list of proposed criteria, but I am more interested in formulating a general negative claim that applies to all.)[2] One involves considerations of angular momentum ("ZAM criterion"). The other is cast in terms of the "compass of inertia" on the axis ("CIA criterion"). Next, I'll characterize a large class of "generalized criteria of nonrotation" that includes the ZAM and CIA criteria. Third, I'll abstract two (seemingly) modest conditions of adequacy that one might expect a criterion of nonrotation to satisfy (the "limit condition" and the "relative rotation condition"). Finally, I'll show that no (nonvacuous)[3] "generalized criterion of nonrotation" satisfies both conditions in all relativistic spacetime models. The proof of the theorem is entirely elementary once all the definitions are in place. But it may be of some interest to *put* them in place and formulate a result of this type. The idea is to step back from the details of particular proposed criteria of nonrotation and direct attention instead to the conditions they do and do not satisfy.

I. Informal Preview

Beginning in section II, our discussion will be cast in the precise language of relativistic spacetime geometry. But first, to explain and motivate what is coming, we give a rough, preliminary description of the no-go theorem in more direct, intuitive, quasi-operational terms. This will involve a bit of hand-waving, but not much. (This section will not presuppose familiarity with the mathematical formalism of general relativity.)

Consider a ring positioned symmetrically about a central axis as in figure 1.[4] At issue is what it means to say that the ring is *not-rotating* (about that axis). The first criterion we will be considering takes the absence of inertial or dynamical effects on the axis as the standard for nonrotation. Here is one way to set things up in terms of a telescope and a water bucket. (Water buckets, to be sure, are not particularly sensitive instruments, but

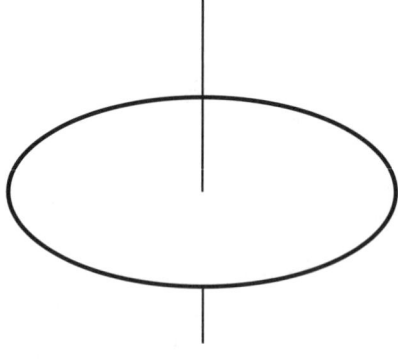

FIGURE 1

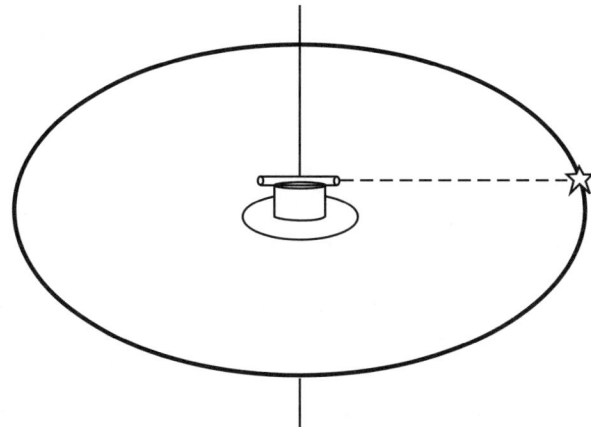

FIGURE 2

they are good enough for our purposes.) Let P be the point of intersection of the axis with the plane of the ring. Place a lazy susan at P (in the plane of the ring), bolt a half-filled water bucket to the center of the lazy susan, and bolt a tubular telescope to the water bucket. (See figure 2.) Finally, mount a light source at a point (any point) on the ring. Now consider possible rotational states of the composite apparatus on the axis (lazy susan + water bucket + telescope). There is one state in which the apparatus *tracks the ring* in the sense that an observer, standing on the lazy susan and looking through the telescope, will see the light source permanently fixed on its cross hairs. We take the ring to be *nonrotating according to the CIA criterion* if in this state (the tracking state), the water surface in the bucket is flat rather than concave.

This characterization is a bit complicated because it makes use of a telescope as well as a water bucket. The former is used to bridge the distance between the water bucket here and the ring there.

We can actually use the instruments described to ascribe an angular speed to the ring (relative to the compass of inertia on the axis). Let the composite apparatus be placed in a state of motion in which the water surface *is* flat. And (just to keep things simple), let us assume that, at some initial moment, the observer standing on the lazy susan sees the light source through the telescope. It may be the case that he continues to see it as time elapses. (This is just the case in which the ring is judged to be nonrotating according to the CIA criterion.) But, in general—assuming the ring is in *some* state of uniform rotational motion—he will see it periodically, with a characteristic interval of time Δt between sightings. (We imagine that the observer carries a stopwatch.) This interval is the time it takes for the ring to complete one rotation (relative to the CIA). So the angular speed of the ring (relative to the CIA) is just $2\pi/\Delta t$.

Now we consider a second criterion of nonrotation that is, on the face of it, very different in character from the first. There is a generic connection in mechanics, whether classical or relativistic, between (continuous) symmetries of spacetime structure and conserved quantities. Associated with the rotational symmetry of the ring system under consideration is a notion of angular momentum.[5] According to our second (ZAM) criterion, the ring is "nonrotating" precisely if the value of that angular momentum is zero at every point on the ring. The condition has an intuitive geometrical interpretation that we will review later. Here, instead, we describe an experimental test for determining whether the condition obtains. (Many other tests could be described just as well.)

Imagine that we mount a light source at some point Q on the ring, and from that point, at a given moment, emit light pulses in opposite (clockwise and counterclockwise) directions. This can be done, for example, using concave mirrors attached to the ring. Imagine further that we keep track of whether the pulses arrive back at Q simultaneously (using, for example, an interferometer). It turns out that this will be the case—they will arrive back simultaneously—if and only if the ring has zero angular momentum.

This equivalence is not difficult to verify and we will do so later. But wholly apart from the connection to angular momentum, the experimental condition described should seem like a natural criterion of nonrotation. Think about it. Suppose the ring is rotating in, say, a counterclockwise direction. (Here I am just appealing to our ordinary intuitions about "rotation.") The C pulse, the one that moves in a clockwise direction, will get back to Q before completing a full circuit of the ring because it is moving toward an approaching target. In contrast, the CC pulse is chasing a receding target. To get back to Q it will have to traverse the entire length of the ring, and then it will have to cover the distance that Q has moved in the interim time. One would expect, in this case, that the C pulse would arrive back at Q before the CC pulse. (Presumably light travels at the same speed in all directions.) Similarly, if the ring is rotating in a clockwise direction, one would expect that the CC pulse would arrive back at Q before the C pulse. Only if the ring is not rotating, should they arrive simultaneously. Thus, our experimental test for whether the ring has zero angular momentum provides what would seem to be a natural criterion of nonrotation. (Devices working on this principle, called "optical gyroscopes," are used in sensitive navigational systems. See, for example, the discussion in Ciufolini and Wheeler 1995, 365.)

We now have two criteria for whether the ring is non-rotating. It is nonrotating in the first sense if it is nonrotating with respect to the compass of inertia on the axis (as determined, say, using a water bucket and telescope). It is nonrotating in the second sense if it has zero angular momentum (as determined, say, using light pulses circumnavigating the ring in opposite directions). As we shall see later, it is a contingent matter in general relativity whether they agree or not. Whether they do so depends on the background spacetime structure in which the ring is imbedded. If it is imbedded in Minkowski spacetime, for example, it will qualify as nonrotating according to the CIA criterion iff it does so according to the ZAM criterion. But *if it is imbedded, instead, for example, in Kerr spacetime, the equivalence fails.* (We choose this example lest one imagine that the failure of agreement occurs only in pathological spacetime models that are of mathematical interest only. The Kerr solution may well describe regions of our universe, the real one, at least approximately— regions surrounding rotating black holes.)

Though the two criteria do not agree in general, it is important for our purposes that *they "agree in the limit for infinitely small rings," no matter what the background spacetime structure.* They do so in the following sense. Imagine that we have a sequence of rings R_1, R_2, R_3, . . . that share a center point P on the axis, and have radii that shrink to 0. Imagine further that each of them is non-rotating according to the ZAM criterion. Each ring R_i has a certain angular speed ω_i with respect to the compass of inertia on the axis. (We described a procedure above for measuring it.) None of the ω_i need be 0. The claim here is that (regardless of the background spacetime structure), the sequence ω_1, ω_2, ω_3, . . . must converge to 0. (We will verify the claim later.)

We have considered just two simple, natural, experimental criteria for nonrotation. We could consider others (that do not, in general, agree with either one). But the fact is, it would turn out in every case that the criterion agrees with them "in the limit for infinitely small rings" in the sense just described (no matter what the background spacetime structure).[6] This is one way to understand the claim that there *is* a robust notion of rotation (in the limit) at a point in general relativity, even if there is none that applies to extended regions of spacetime.

In any case, with these remarks as motivation, we now propose for consideration a first condition that one might expect a reasonable criterion of nonrotation to satisfy. Let us understand a "generalized criterion of nonrotation" to be, simply, a specification, for every ring, in every state of motion (or nonmotion), whether it is to qualify as "nonrotating." We don't require that it have a natural geometrical or experimental interpretation.

Limit Condition: Let R_1, R_2, R_3, . . . be a sequence of rings, each "non-rotating," that share a center point P on the axis, and have radii that converge to 0. For every i, let R_i have angular speed ω_i with respect to the compass of inertia on the axis. Then the sequence ω_1, ω_2, ω_3, . . . converges to 0.

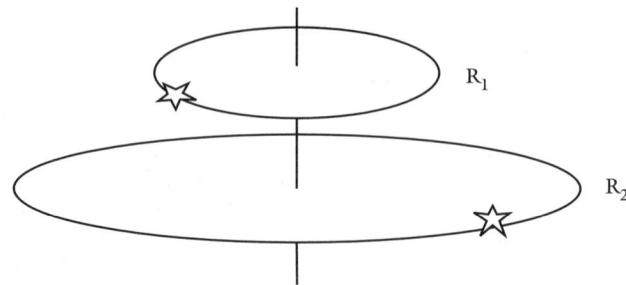

FIGURE 3

We have just asserted that the ZAM criterion satisfies this condition (regardless of the background spacetime structure). The CIA criterion does too, of course. (In the latter case, ω_i is 0, for every i. So the sequence certainly converges to 0.)

It remains to state our second condition (on a generalized criterion of nonrotation). Suppose we have two rings R_1 and R_2 centered about the axis as in figure 3. (The planes of the rings are understood to be parallel, but nor necessarily coincident.) Further suppose that "R_2 is nonrotating relative to R_1." Then, one would think, either both rings should qualify as "nonrotating," or neither should. This is precisely the requirement captured in our "relative rotation condition." It is not entirely unambiguous what it means to say that R_2 is nonrotating *relative* to R_1. But all we need is a sufficient condition for relative nonrotation of the rings. And it seems, at least, a plausible sufficient condition for this that, over time, there is no change in the distance between any point on one ring and any point on the other, i.e., the two rings move as if locked together.

> *Relative Rotation Condition:* Given two rings R_1 and R_2, if (i) R_1 is "nonrotating," and if (ii) R_2 is nonrotating relative to R_1 (in the sense that, given any point on R_2 and any point on R_1, the distance between them is constant over time), then R_2 is "nonrotating."

The relative rotation condition is really at the heart of our discussion. It seems a modest condition. But neither the CIA nor the ZAM criterion satisfies it, in general! They both do so if the rings are imbedded in Minkowski spacetime. But, as we shall see, *neither does if they are imbedded in, for example, Kerr spacetime.*

It should be clear just what is being asserted here. The situation is extremely counterintuitive. Consider the ZAM criterion. The claim is that we can have two rings, moving as if rigidly locked together, where one, but not the other, has zero angular momentum. Light pulses circumnavigating the first will arrive back at their starting point simultaneously. But pulses circumnavigating the second will not do so. (And similarly with the CIA criterion.)

We have made a number of claims involving two criteria of nonrotation, two possible conditions of adequacy on a criterion of rotation, and two spacetime models. It may help to summarize some of those claims in a table.

	In Minkowski spacetime	In Kerr spacetime
Do the CIA and ZAM criteria agree (for rings of arbitrary radius)?	Yes	No
Does the CIA criterion satisfy the limit condition?	Yes	Yes
Does the ZAM criterion satisfy the limit condition?	Yes	Yes
Does the CIA criterion satisfy the relative rotation condition?	Yes	No
Does the ZAM criterion satisfy the relative rotation condition?	Yes	No

The fact that neither the CIA nor ZAM criterion satisfies the relative rotation condition (in general), seems a significant strike against them, and it is natural to ask whether any other criterion does better. Our principal claim is that, in an interesting sense, the answer is 'no'. There *are* criteria that satisfy the relative rotation condition in Kerr spacetime.[7] But the cost of doing so is violation of the limit condition, or else the radical conclusion that no ring in *any* state of motion (or nonmotion) counts as "nonrotating."

> *Theorem* In Kerr spacetime (and other relativistic spacetime models to be discussed), there is no generalized criterion of nonrotation that satisfies the following three conditions:
> (i)   limit condition
> (ii) relative rotation condition
> (iii) nonvacuity condition: there is some ring in some state that qualifies as "nonrotating."

The result is intended to bear this interpretation. *Given any (nonvacuous) generalized criterion of nonrotation in Kerr spacetime, to the extent that it gives "correct" attributions of nonrotation in the limit for infinitely small rings—the domain where one does have a robust notion of nonrotation—it must violate the relative rotation condition.*

II. Formal Treatment

Now we start all over and cast our discussion in the language of relativistic spacetime geometry.[8] We present formal versions of the two criteria of nonrotation and the two conditions of adequacy (though not in the same order as in section I).

First we have to consider how to represent one-dimensional rings in a state of uniform rotational motion. To keep things as simple as possible, we will think of the rings as test bodies with negligible mass, imbedded in a background spacetime structure that exhibits the rotational and "time translational" symmetries of the ring system itself. More precisely, we will think of them as imbedded in a stationary, axi-symmetric spacetime model.

II.1 Stationary, Axi-Symmetric Spacetimes

We take a *(relativistic) spacetime model* to be a structure (M, g_{ab}) where M is a connected, smooth, four–dimensional manifold, and g_{ab} is a smooth, pseudo–Riemannian metric on M of Lorentz signature $(+, -, -, -)$. We say that (M, g_{ab}) is *stationary and axi-symmetric* if there exist two one-parameter isometry groups acting on M, $\{\Gamma_t: t \in \mathbf{R}\}$ and $\{\Sigma_\varphi: \varphi \in \mathbf{S}^1\}$, satisfying several conditions. (Here we identify $\mathbf{S}^1$ with the set of real numbers mod 2π.)[9]

(SAS 1) The isometries Γ_t and Σ_φ commute for all $t \in \mathbf{R}$ and $\varphi \in \mathbf{S}^1$.

(SAS 2) $\Gamma_t(p) \neq p$ for all points p in M and all $t \neq 0$. (So the orbits of all points under $\{\Gamma_t: t \in \mathbf{R}\}$ are open.)

(SAS 3) Some, but not all, points p in M have the property that $\Sigma_\varphi(p) = p$ for all φ. (Those with the property are called *axis points*. So the orbits of axis points under $\{\Sigma_\varphi: \varphi \in \mathbf{S}^1\}$ are singleton sets, and those of nonaxis points are [nondegenerate] closed curves.)

(SAS 4) The orbits of $\{\Gamma_t: t \in \mathbf{R}\}$ are timelike, and the nondegenerate orbits of $\{\Sigma_\varphi: \varphi \in \mathbf{S}^1\}$ are spacelike.[10]

The final condition is slightly more complex than the others. Let M^- be the restricted manifold that one gets by excising the (closed) set of axis points. The orbit of any point in M^- under the two-parameter isometry group $\{\Gamma_t \circ \Sigma_\varphi \colon t \in \mathbf{R}\ \&\ \varphi \in S^1\}$ is a smooth, two-dimensional, timelike[11] sub-manifold that is diffeomorphic to the cylinder $\mathbf{R} \times S^1$. Let us call it an *orbit cylinder*. So (in the tangent space) at every point p in M⁻, there is a time-like two-plane T(p) tangent to the orbit cylinder that passes through p, and a spacelike two-plane S(p) orthogonal to T(p). The final condition imposes the requirement that the set {S(p): p∈ M⁻} be integrable.

(SAS 5) (Orthogonal transitivity) Through every point in M⁻ there is a smooth, two-dimensional, spacelike submanifold Π that is tangent to S(p) at every point p in Π ∩ M⁻.

Associated with the two isometry groups $\{\Gamma_t \colon t \in \mathbf{R}\}$ and $\{\Sigma_\varphi \colon \varphi \in S^1\}$, respectively, are Killing fields τ^a and φ^a. (They arise as the "infinitesimal generators" of those groups.) It will be helpful for what follows to refor-mulate the five listed conditions in terms of these fields. (SAS 1) is equiv-alent to the assertion that the fields have vanishing Lie bracket, i.e., at all points

$$\tau^n \nabla_n \varphi^a - \varphi^n \nabla_n \tau^a = 0. \tag{1}$$

(SAS 2) comes out as the requirement that τ^a be everywhere nonzero. (SAS 3) can be understood to assert that φ^a vanishes at some (axis) points, but does not vanish everywhere. (SAS 4) is equivalent to the assertion that τ^a is everywhere timelike, and φ^a is spacelike at nonaxis points. Finally, (SAS 5) is equivalent (by Frobenius's theorem) to the assertion that the conditions

$$\varphi_{[a}\tau_b \nabla_c \tau_{d]} = 0 \tag{2a}$$

$$\tau_{[a}\varphi_b \nabla_c \tau_{d]} = 0 \tag{2b}$$

hold at every nonaxis point.[12]

The five listed conditions imply the existence of coordinate functions with respect to which the metric g_{ab} assumes a special, characteristic form.[13] The first three imply that there exist smooth maps t: M → **R** and φ: M⁻ → S¹ such that $\tau^a = (\partial/\partial t)^a$ on M, and $\varphi^a = (\partial/\partial\varphi)^a$ on M⁻. (Again, M⁻ is the restricted submanifold on which $\varphi^a \neq 0$.) The remaining con-ditions imply that, at least locally on M⁻, we can find further smooth coordinates x_2 and x_3 such that, at every point, the vectors $(\partial/\partial x_2)^a$ and $(\partial/\partial x_3)^a$ are spacelike, orthogonal to each other, and orthogonal to both $(\partial/\partial t)^a$ and $(\partial/\partial\varphi)^a$. Thus, at points in M⁻, the matrix of components of g_{ab} with respect to the coordinates (t, φ, x_2, x_3) has the characteristic form:

$\tau^a \tau_a$	$\tau^a \varphi_a$	0	0
$\tau^a \varphi_a$	$\varphi^a \varphi_a$	0	0
0	0	$(\partial/\partial x_2)^a (\partial/\partial x_2)_a$	0
0	0	0	$(\partial/\partial x_3)^a (\partial/\partial x_3)_a$

And the inverse matrix (giving the components of g^{ab}) has the form:

$(\varphi^a \varphi_a) D^{-1}$	$(-\tau^a \varphi_a) D^{-1}$	0	0
$(-\tau^a \varphi_a) D^{-1}$	$(\tau^a \tau_a)\, D^{-1}$	0	0
0	0	$[(\partial/\partial x_2)^a (\partial/\partial x_2)_a]^{-1}$	0
0	0	0	$[(\partial/\partial x_3)^a (\partial/\partial x_3)_a]^{-1}$

where $D = (\tau_a \tau^a)(\varphi_b \varphi^b) - (\tau_a \varphi^a)^2$. ($D < 0$ in M^-, since $(\varphi_b \varphi^b) < 0$ in M^- and $(\tau_a \tau^a) > 0$ everywhere.)

For future reference, we note that $\nabla^a t$ ($= g^{ab} \nabla_b t = g^{ab} (dt)_b$) can be expressed as

$$\nabla^a t = (\varphi^n \varphi_n) D^{-1} (\partial/\partial t)^a + (-\tau^n \varphi_n) D^{-1} (\partial/\partial \varphi)^a = (D^{-1}) [(\varphi^n \varphi_n) \tau^a - (\tau^n \varphi_n)\, \varphi^a]$$

in M^-. Hence

$$(\nabla_a t)(\nabla^a t) = (\varphi^n \varphi_n) D^{-1}$$

and therefore

$$(\nabla^a t) [(\nabla_n t)(\nabla^n t)]^{-1} = \tau^a - (\tau^n \varphi_n)(\varphi^m \varphi_m)^{-1} \varphi^a \tag{3}$$

in M^-.

Of special interest is the case where $[(\tau^n \varphi_n)(\varphi^m \varphi_m)^{-1}]$ is constant on M^-.[14] We will say then that the background spacetime (M, g_{ab}) is *static*. This is a slightly nonstandard way of formulating the definition.[15] But if $[(\tau^n \varphi_n)(\varphi^m \varphi_m)^{-1}]$ *is* constant, $\tau'^a = \tau^a - (\tau^n \varphi_n)(\varphi^m \varphi_m)^{-1} \varphi^a$ is a smooth time-like Killing field on M that is hypersurface orthogonal. (It is a Killing field since any linear combination of two Killing fields is one. It is hypersurface orthogonal by (3).)

An example to which we will turn repeatedly is Kerr spacetime (see, e.g., O'Neill 1995). In Boyer-Lindquist coordinates (t, φ, r, θ)—again with $\tau^a = (\partial/\partial t)^a$ and $\varphi^a = (\partial/\partial \varphi)^a$—the nonzero components are

$$\tau^a\tau_a \qquad\qquad = \quad 1 - 2\,M\,r\,\rho^{-2} \tag{5a}$$

$$\tau^a\varphi_a \qquad\qquad = \quad 2\,M\,r\,a\,(\sin^2\theta)\,\rho^{-2} \tag{5b}$$

$$\varphi^a\varphi_a \qquad\qquad = \quad -\,[r^2 + a^2 + 2\,M\,r\,a^2\,(\sin^2\theta)\,\rho^{-2}]\,(\sin^2\theta) \tag{5c}$$

$$(\partial/\partial r)^a(\partial/\partial r)_a \;=\; -\rho^2\Delta^{-1}$$

$$(\partial/\partial\theta)^a(\partial/\partial\theta)_a \;=\; -\rho^2$$

where

$$\rho^2 = r^2 + a^2\,(\cos^2\theta) \tag{5d}$$

$$\Delta = r^2 - 2\,M\,r + a^2. \tag{5e}$$

(Here a and M are positive constants.) Axis points are those at which $(\sin^2\theta) = 0$. It is not the case that τ^a is everywhere timelike and φ^a is everywhere spacelike on M^-. But these conditions do obtain in restricted regions of interest, e.g., in the open set where $r > 2\,M$. If we think of Kerr spacetime as representing the spacetime structure surrounding a rotating black hole, our interest will be in small rings that are positioned close to the axis of rotational symmetry (where $(\sin^2\theta)$ is small) and far away from the center point (where r is large). There we can sidestep complexities having to do with horizons and singularities.

II.2 Striated Orbit Cylinders and the ZAM Criterion

Assume we have fixed, once and for all, a stationary, axi-symmetric spacetime (M, g_{ab}) with isometry groups $\{\Gamma_t\colon t\in\mathbf{R}\}$ and $\{\Sigma_\varphi\colon \varphi\in\mathbf{S}^1\}$ and corresponding Killing fields τ^a and φ^b. The first of these fields defines a temporal orientation on (M, g_{ab}). We will work with that one in what follows.[16]

We want to represent (one-dimensional) rings, centered about the axis of rotational symmetry. We do so using the "orbit cylinders" introduced above. Recall that these were characterized as the orbits of points in M^- under the two-parameter isometry group $\{\Gamma_t\circ\Sigma_\varphi\ t\in\mathbf{R}\ \&\ \varphi\in\mathbf{S}^1\}$. Here is an equivalent formulation.

> *Definition* An *orbit cylinder* is a smooth, two-dimensional, timelike submanifold in M^-, diffeomorphic to the cylinder $\mathbf{R}\times\mathbf{S}^1$, that is invariant under the action of all maps Γ_t and Σ_φ.

Clearly, we are thinking about the life history of a ring, not its state at a given "time."

Let C be an orbit cylinder representing ring R. To represent the rotational state of R, we need to keep track of the motion of individual points

FIGURE 4

on it. Each such point has a worldline that can be represented as a time-like curve on C. So we are led to consider, not just C, but C together with a congruence of smooth timelike curves on C.[17]

We want to think of the ring as being in a state of rigid rotation (with the distance between points on the ring remaining constant). So we are further led to restrict attention to just those congruences of timelike curves on C that are invariant under all isometries Γ_t and Σ_φ. Equivalently (moving from the curves themselves to their tangent vectors), we are led to consider smooth, future-directed timelike vector fields ξ^a on C that are invariant under all these maps. Since each such field is determined by its value at any one point on C (and since the tangent plane to C at any point is spanned by the vectors τ^a and φ^a there), ξ^a must be of the form $(k_1\tau^a + k_2\varphi^a)$, where k_1 and k_2 are constants, and $k_1 > 0$.[18] We lose nothing if we rescale ξ^a by a positive factor and write it in the form $(\tau^a + k\varphi^a)$. So we are led to the following definition.

> *Definition* A *striated orbit cylinder* is a pair (C, k), where C is an orbit cylinder, and k is a number such that the vector field $(\tau^a + k\varphi^a)$ is timelike on C.

We call the integral curves of $(\tau^a + k\varphi^a)$ on C "striation lines," and call k their "slope factor."

Now we can formulate our fundamental question: *Under what conditions does a striated orbit cylinder count as nonrotating?* The first proposal we consider is the following.[19]

Definition A striated orbit cylinder (C, k) is nonrotating according to the *ZAM* (*zero angular momentum*) *criterion* if the vector field $(\tau^a + k\varphi^a)$ is orthogonal to φ^a, i.e., if $k = - (\tau^a\varphi_a)(\varphi^n\varphi_n)^{-1}$.

The connection with angular momentum is immediate. The stated condition is equivalent to the assertion that every point on the ring has 0 angular momentum with respect to the rotational Killing field φ^a.[20]

The criterion should seem like a reasonable one. It seems plausible to regard a striated orbit cylinder as nonrotating iff the striation lines (representing the worldlines of points on the ring) do not "wrap around the cylinder." And the latter condition is plausibly captured in the requirement that the striation lines be everywhere orthogonal to equatorial circles on the cylinder, i.e., have no component in the direction of those circles. But how does one characterize "equatorial circles" in the present context? If the background geometry were Euclidean, e.g., if we were dealing with ordinary barber shop poles, we could characterize an equatorial circle as a closed curve of shortest length on the cylinder that is not contractible to a point. That characterization does not carry over to the present context where the background metric has Lorentz signature. But an alternate, equivalent one does. In the Euclidean case, we can equally well characterize an equatorial circle as the orbit of a point under the group of rotations that leave fixed the central axis of the cylinder. Lifting that characterization to the present context, we are led to construe the orbits of points under $\{\Sigma_\varphi: \varphi \in S^1\}$ as "equatorial circles." These are just the integral curves of the field φ^a. So the requirement that striation lines not "wrap around the cylinder" is plausibly captured in the condition that they be everywhere orthogonal to the field φ^a. That is precisely the ZAM criterion of nonrotation.

Consider now the operational test described in the preceding section for whether a ring is rotating according to the ZAM criterion. We can verify that it works with a simple calculation.[21] Let (C, k) be a striated orbit cylinder. We have to keep track of three curves on C. (See figure 5). The first is a striation line γ that represents the worldline of a fixed point on the ring from which light is emitted and absorbed. The other two are null curves λ_1 and λ_2 on C that represent the worldlines of photons that start at that point, traverse the ring in opposite directions, and then arrive back at it. (Call them "photon 1" and "photon 2.") Let p_0 be the initial emission point at which the three curves intersect. Let p_1 be the intersection point of γ with λ_1 at which the first photon is reabsorbed. And let p_2 be the corresponding intersection point of γ with λ_2. We have to verify that the photons arrive back at the same instant iff (C, k) is nonrotating according to the ZAM criterion, i.e.,

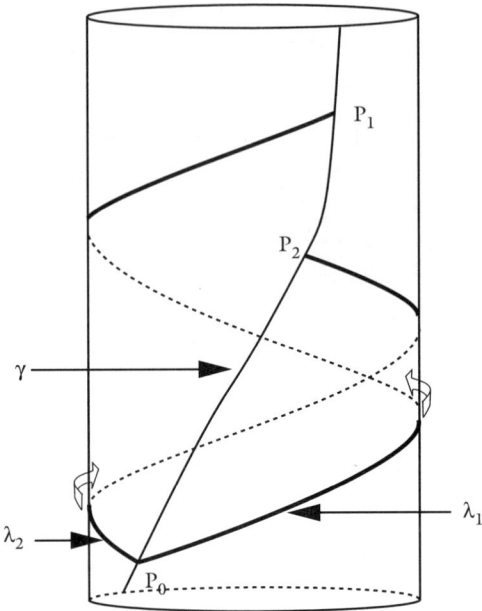

FIGURE 5

$$p_1 = p_2 \Leftrightarrow k = - (\tau^a \varphi_a)(\varphi^n \varphi_n)^{-1}. \tag{6}$$

The tangent fields to the curves γ, λ_1, and λ_2 (after rescaling by a positive constant) can be written in the form $(\tau^a + k\varphi^a)$, $(\tau^a + l_1\varphi^a)$, and $(\tau^a + l_2\varphi^a)$. Since the first is timelike, and the second two are null, we have $l_i \neq k$ and

$$(\tau^a + l_i\varphi^a)(\tau_a + l_i\varphi_a) = 0 \tag{7}$$

for $i = 1, 2$. Consider the scalar function φ': $C \rightarrow S^1$ defined by $\varphi' = (\varphi - kt)(\mathrm{mod}\ 2\pi)$. It is a circular coordinate that is adapted to (C, k) in the sense that it is constant along striation lines.[22] Let the (t, φ') coordinates of the points p_0, p_1, p_2 be (t_0, φ'_0), (t_1, φ'_0), and (t_2, φ'_0). We can verify (6) by considering the respective rates at which φ' changes along λ_1 and λ_2 as a function of t.[23] Without loss of generality, assume that it increases from φ'_0 to $(\varphi'_0 + 2\pi)$ along λ_1, and decreases from φ'_0 to $(\varphi'_0 - 2\pi)$ along λ_2. Then, the total increase (resp. decrease) along λ_1 (resp. λ_2) can be expressed as:[24]

$$2\pi = (t_1 - t_0)\, (d\varphi'/dt)_{|\mathrm{on}\ \lambda_1} = (t_1 - t_0)\, (l_1 - k)$$
$$-2\pi = (t_2 - t_0)\, (d\varphi'/dt)_{|\mathrm{on}\ \lambda_2} = (t_2 - t_0)\, (l_2 - k).$$

So

$$(t_1 - t_2) = 2\pi \, (l_1 + l_2 - 2k) \, (l_1 - k)^{-1} \, (l_2 - k)^{-1}.$$

But it follows from (7) that

$$l_1 = [-(\tau^a \varphi_a) + (-D)^{1/2}] \, (\varphi^n \varphi_n)^{-1}$$

$$l_2 = [-(\tau^a \varphi_a) - (-D)^{1/2}] \, (\varphi^n \varphi_n)^{-1}$$

where (as above) $D = (\tau_a \tau^a)(\varphi_b \varphi^b) - (\tau_n \varphi^n)^2$. So,

$$p_1 = p_2 \Leftrightarrow (t_1 - t_2) = 0 \Leftrightarrow (l_1 + l_2 - 2k) = 0 \Leftrightarrow k = -(\tau^a \varphi_a) \, (\varphi^n \varphi_n)^{-1},$$

which confirms (6).

II.3 Generalized Criteria of Nonrotation and the Relative Rotation Condition

Now we turn to "generalized criteria of nonrotation." Using our current terminology, the definition comes out this way.

> *Definition* A *generalized criterion of nonrotation* is a specification, for every striated orbit cylinder (C, k), whether it is to count as "nonrotating" or not.

We do not assume that generalized criteria of nonrotation bear a natural geometrical or experimental interpretation. Nor do we assume that given an orbit cylinder C, they render (C, k) "nonrotating" for at least one k, or at most one k. Clearly, the ZAM criterion of nonrotation qualifies as a generalized criterion of such.

Next consider the relative rotation condition. Intuitively, it asserts that if we have two rings (with the same axis of symmetry), then if the first qualifies as "nonrotating," and if the second is nonrotating relative to the first, then the second ring also qualifies as "nonrotating." As mentioned above, all we need here is a sufficient condition for relative nonrotation of the two rings; and it seems, at least, a plausible sufficient condition for this that, over time, there be no change in the distance between any point on one ring and any point on the other, i.e., the two rings move as if locked together.

Suppose we have two striated orbit cylinders (C_1, k_1) and (C_2, k_2), suppose γ_1 is a striation line of the first, and γ_2 is a striation line of the second. There are various ways we might try to measure the "distance between γ_1 and γ_2." For example, we might bounce a photon back and forth between them and keep track of how much time is required for the round trip—as measured by a clock following one of the striation lines. But no matter what procedure we use, the measured distance will be constant over time if γ_1 and γ_2 are (up to reparametrization) integral curves of a common Killing field (or, equivalently, orbits of a common one-parameter group of isometries). For any measurement procedure can be characterized in terms of some set of relations and functions that are definable in terms g_{ab} (e.g., the set of null geodesics, the length of a timelike curve) and *all* such relations and functions will be preserved under the elements of the isometry group (since these all preserve g_{ab}). So we seem to have a plausible sufficient condition for the relative nonrotation of (C_1, k_1) and (C_2, k_2)—namely, that there exist a (single) Killing field κ^a defined on M whose restriction to C_1 is proportional to $(\tau^a + k_1\varphi^a)$, and whose restriction to C_2 is proportional to $(\tau^a + k_2\varphi^a)$. But the latter condition holds immediately if $k_1 = k_2$, since, for any constant k, $(\tau^a + k\varphi^a)$ is itself a Killing field defined on M.

The upshot of this long-winded argument is the proposal that it is plausible to regard (C_2, k_2) as nonrotating relative to (C_1, k_1) if $k_1 = k_2$. So we are led to the following formulation of the relative rotation condition.

Relative Rotation Condition For all k, and all striated orbit cylinders (C_1, k) and (C_2, k) sharing k as their slope factor, if (C_1, k) qualifies as "nonrotating," so does (C_2, k).[25]

It follows easily that *the ZAM criterion of nonrotation satisfies the relative rotation condition iff the background stationary, axi-symmetric spacetime structure is static, i.e., if the function $[(\tau^a\varphi_a)(\varphi^n\varphi_n)^{-1}]$ is constant on M^-*.[26] In Kerr spacetime, by (5b) and (5c),

$$- (\tau^a\varphi_a)(\varphi^n\varphi_n)^{-1} = (2 M r a)[(r^2 + a^2) \rho^2 + 2 M r a^2 (\sin^2\theta)]^{-1}. \qquad (8)$$

The right hand side expression is not constant over any open set. So we see that *the ZAM criterion does not satisfy the relative rotation condition in Kerr spacetime, or the restriction of Kerr spacetime to any open set.* (We have been taking for granted that a and M are both strictly positive. It also follows directly from (8) that the ZAM criterion *does* satisfy the relative rotation condition in Schwarzschild spacetime (a = 0 and M > 0) and Minkowksi spacetime (a = 0 and M = 0).)

II.4 The CIA Criterion

Next we consider how to capture the CIA criterion of nonrotation in the language of spacetime geometry. Let (C, k) be a striated orbit cylinder.

The Killing field $(\tau^a + k\varphi^a)$ that determines the striation lines on C is defined on all of M. It seems a natural proposal to construe (C, k) as non-rotating if the twist (or vorticity) of $(\tau^a + k\varphi^a)$ vanishes at axis points. This is very close to being the CIA criterion. But there is a problem. It is true that given any axis point p, there is one and only k such that $(\tau^a + k\varphi^a)$ has vanishing twist at p. (We verify this in lemma 2.) But it turns out that that critical value need not be the same at all axis points. It is not, for example, in Kerr spacetime. (In the end, it is this one fact that lies at the heart of our mini no-go theorem.) So we need to direct attention to some particular axis point and take the test to be whether $(\tau^a + k\varphi^a)$ has vanishing twist *there*. The natural choice is the "centerpoint" of the ring, the point that lies at "the intersection of the axis with the plane of C." (That is where we previously placed the experimental apparatus consisting of lazy susan + waterbucket + telescope. Recall figure 2.) The question, then, is how to construe the expressions in quotation marks.

One natural way to do so is in terms of light signals traveling from the ring to the axis. There is exactly one point on the axis at which the incoming light signals arrive so as to be perpendicular to the axis. That one point is a natural candidate for the "centerpoint" of the ring, and we will treat it as such in what follows. But a bit of work is necessary to set everything up.

Let ϵ^{abcd} be a volume element,[27] and let σ^a be the smooth field defined by

$$\sigma^a = \epsilon^{abcd}\tau_b\nabla_c\varphi_d$$

We claim that at every axis point p, σ^a gives the "direction of the axis of rotation" as determined relative to τ^a. The interpretation is supported by the following lemma that collects several simple facts about σ^a for future reference. It implies that at axis points, σ^a is, up to a constant, the only nonzero vector, orthogonal to τ^a, that is kept invariant by all isometries Σ_φ.

Lemma 1 At all points:

    (i) σ^a is orthogonal to τ^a and φ^a

    (ii) $\mathbf{L}_\varphi(\sigma^a) = \mathbf{0} = \mathbf{L}_\tau(\sigma^a)$ (Here $\mathbf{L}_\varphi$ and $\mathbf{L}_\tau$ are Lie derivative operators.)

    (iii) $\tau_{[a}\nabla_b\varphi_{c]} = (1/6)\,\epsilon_{abcd}\sigma^d$.

At axis points:

    (iv) $\sigma^a \neq \mathbf{0}$

    (v) $\nabla_a\varphi_b = (1/2)(\tau^n\tau_n)^{-1}\,\epsilon_{abcd}\tau^c\sigma^d$

Given any field ψ^a, if $\mathbf{L}_\varphi(\psi^a) = \mathbf{0}$ at an axis point, then at the point it must be of form $\psi^a = k_1\tau^a + k_2\sigma^a$.

Proof (i) ϵ^{abcd} is totally antisymmetric. So $\sigma^a\tau_a = \epsilon^{abcd}\tau_a\tau_b\nabla_c\varphi_d = 0$, and
$$\sigma^a\varphi_a = \epsilon^{abcd}\varphi_a\tau_b\nabla_c\varphi_d = \epsilon^{abcd}\varphi_{[a}\tau_b\nabla_c\varphi_{d]}.$$

But $\varphi_{[a}\tau_b\nabla_c\varphi_{d]} = \mathbf{0}$, by (2b). So $\sigma^a\varphi_a = 0$. (ii) The Lie operators $\mathbf{L}_\tau$ and $\mathbf{L}_\varphi$ annihilate τ^a and φ^a, by (1), and annihilate g_{ab} and $\in^{abcd}$ because τ^a and φ^a are Killing vectors. So they annihilate all fields definable in terms of τ^a, φ^a, g_{ab}, and $\in^{abcd}$, including σ^a. (iii) follows by a simple computation:

$$\in_{abcd}\sigma^d = \in_{abcd}\in^{dmpq}\tau_m\nabla_p\varphi_q = (3!)\,\delta_a{}^{[m}\delta_b{}^p\delta_c{}^{q]}\,\tau_m\nabla_p\varphi_q = 6\tau_{[a}\nabla_b\varphi_{c]}.^{28}$$

For (iv), suppose $\sigma^a = \mathbf{0}$ at an axis point p. Then, by (iii), $\tau^a\tau_{[a}\nabla_b\varphi_{c]} = \mathbf{0}$ at p. Expanding this equation and using the fact that $\nabla_b\varphi_c = -\nabla_c\varphi_b$ (since φ^a is a Killing field), we have

$$(\tau^a\tau_a)\nabla_b\varphi_c + \tau_c\tau^a\nabla_a\varphi_b - \tau_b\tau^a\nabla_a\varphi_c = 0.$$

But the second and third terms are $\mathbf{0}$ at p, since $\tau^a\nabla_a\varphi_b = \varphi^a\nabla_a\tau_b$ (by equation (1)) and $\varphi_b = \mathbf{0}$ at p. So, since τ^a is timelike, $\nabla_b\varphi_c = \mathbf{0}$ at p. But this is impossible. For given any Killing field κ^a, if both κ_a and $\nabla_a\kappa_b$ vanish at a point, κ^a must vanish everywhere. (See Wald 1984, 443.) And we know that φ^a does not vanish everywhere. (v) follows from (iii) and a computation very close to the one just used for (iv). Finally, assume that $\mathbf{L}_\varphi(\psi^a) = \mathbf{0}$ at p. Then, at p, $\psi^a\nabla_a\varphi_b = \varphi^a\nabla_a\psi_b = \mathbf{0}$ since $\varphi^a = \mathbf{0}$ at p. Hence, by (v), $\in_{abcd}\psi^a\tau^c\sigma^d = \mathbf{0}$ at p. It follows that the three vectors ψ^a, τ^a and σ^a are linearly dependent at p and, so, ψ^a can be expressed there as a linear combination of the other two vectors. ■

Let C be an orbit cylinder. Let γ be an integral curve of τ^a on which φ^a vanishes.[29] It represents the worldline of a point[30] on the axis of rotation. We say that γ is the *centerpoint* of C if, for all future-directed null geodesics running from a point on C to a point on γ, if λ^a is the tangent field to the null geodesic, then, at the latter (arrival) point, λ^a is orthogonal to σ^a.[31]

In what follows, we take for granted that orbit cylinders *have* unique centerpoints. The assumption is harmless because it will suffice for our purposes to restrict attention to regions of spacetime near axis points (e.g., within convex sets) and there they certainly do.[32]

To complete our definition of the CIA criterion we need the following lemma.

Lemma 2 Let p be an axis point. Then there is a unique k such that the Killing field $\xi^a = (\tau^a + k\varphi^a)$ has vanishing twist at p, i.e., such that $\xi_{[a}\nabla_b\xi_{c]} = \mathbf{0}$ at p. Its value is given by:

$$k_{crit}(p) = -[(\nabla_b\tau_c)(\nabla^b\varphi^c)]\,[(\nabla_m\varphi_n)(\nabla^m\varphi^n)]^{-1}.$$

Proof Since $\varphi^a = \mathbf{0}$ at p, what we need to show is that there is a unique k such that

$$\tau_{[a}\nabla_b\tau_{c]} + k\tau_{[a}\nabla_b\varphi_{c]} = 0 \qquad (9)$$

at p. We know from clauses (iii) and (iv) of lemma 1 that $\tau_{[a}\nabla_b\varphi_{c]} \neq \mathbf{0}$ at p. So uniqueness is immediate. For existence, consider the twist vector field of τ^a defined by

$$\omega^a = \epsilon^{abcd}\tau_b\nabla_c\tau_d.$$

ω^a is orthogonal to τ^a and is Lie derived by φ^a, i.e., $\mathbf{L}_\varphi(\omega^a) = \mathbf{0}$. (The proof is almost exactly the same as for σ^a in clauses (i) and (ii) of lemma 1.) Hence, by the final assertion in lemma 1, $\omega^a = k_2\sigma^a$ at p, for some number k_2. It follows by clause (iii) of lemma 1, and the counterpart statement for ω^a and $\tau_{[a}\nabla_b\tau_{c]}$, that

$$\tau_{[a}\nabla_b\tau_{c]} = (1/6)\,\epsilon_{abcd}\omega^d = k_2\,(1/6)\,\epsilon_{abcd}\sigma^d = k_2\,\tau_{[a}\nabla_b\varphi_{c]}.$$

Thus (9) will be satisfied if we take $k = -k_2$.

Now assume that k *does* satisfy (9) at p. Contracting with $\tau^a\nabla^b\varphi^c$, and then dividing by $(\tau^c\tau_c)$, yields

$$[(\nabla_b\tau_c)(\nabla^b\varphi^c)] + k\,[(\nabla_b\varphi_c)(\nabla^b\varphi^c)] = 0.$$

(Almost all terms drop out because $\tau^a\nabla_a\varphi_b = 0$.) So to complete the proof we need only verify that $(\nabla_b\varphi_c)(\nabla^b\varphi^c) \neq \mathbf{0}$ at p. But this follows, since by clause (v) of lemma 1,

$$(\nabla_b\varphi_c)(\nabla^b\varphi^c) = (1/2)(\tau^n\tau_n)^{-1}\epsilon_{bcpq}(\nabla^b\varphi^c)\tau^p\sigma^q = -(1/2)(\tau_n\tau^n)^{-1}(\sigma_q\sigma^q)$$

at p, and by clauses (i) and (iv), $(\sigma_q\sigma^q) < 0$ at p. ∎

Lemma 2 has a simple geometric interpretation. Equation (9) is equivalent to:

$$\omega^a + k\sigma^a = \mathbf{0}.$$

So, when the dust clears, the lemma asserts that, at every axis point, the twist vector ω^a (of τ^a) is co-alligned with the axis direction vector σ^a. The critical value k is just a proportionality factor.

If p is an axis point, and γ is the integral curve of τ^a that passes through p, the function k_{crit} is constant on γ. (This follows since the condition that $(\tau^a + k\varphi^a)$ is twist free is definable in terms of τ^a, φ^a, and g_{ab}, and these are all preserved by the isometries Γ_t). So, in particular, if γ is the centerpoint of an orbit cylinder C, we can write '$k_{crit}(\gamma)$' without ambiguity. Now we have all the pieces in place for our definition.

Definition A striated orbit cylinder (C, k) is nonrotating according to the *CIA criterion* if $k = k_{crit}(\gamma)$, where γ is the centerpoint of C.

We now consider the conditions under which the CIA criterion satisfies the relative rotation condition, and the conditions under which our two criteria of nonrotation agree. We take them in order.[33] Since every axis point is the centerpoint of some orbit cylinder, the CIA criterion satisfies the relative rotation condition iff the function k_{crit} assumes the same value at all axis points. But there is a more instructive way to formulate the later condition.

Lemma 3 The function $f = [-(\tau^a \varphi_a)(\varphi^n \varphi_n)^{-1}]$ can be smoothly extended from M^- to all of M. The value of the extension at an axis point p is $k_{crit}(p)$.

Proof The proof that f can be smoothly extended to M is long, and we omit the details.[34] But the rest of the proof is easy. Consider the field $\tau'^a = \tau^a + f\varphi^a$ defined on M^-. By (3), it is hypersurface orthogonal, i.e., of form $\tau'_a = g \nabla_a h$. So, it must have vanishing twist.[35] Therefore, at all points in M^-,

$$0 = \tau'_{[a} \nabla_b \tau'_{c]} = \tau_{[a} \nabla_b \tau_{c]} + f\tau_{[a} \nabla_b \varphi_{c]} + \varphi_{[c} \tau_a \nabla_b]\, f + f\varphi_{[a} \nabla_b \tau_{c]} + f^2 \varphi_{[a} \nabla_b \varphi_{c]}.$$

Let k' be the limiting value of f at p. Then, at p we have

$$0 = \tau_{[a} \nabla_b \tau_{c]} + k' \, \tau_{[a} \nabla_b \varphi_{c]}$$

(since $\varphi_a = 0$ at p). But we saw in the proof of lemma 3 that there is a unique k that satisfies equation (9). So $k' = k_{crit}(p)$. ∎

It follows immediately from lemma 3 that *the CIA criterion satisfies the relative rotation condition iff $[-(\tau^a \varphi_a)(\varphi^n \varphi_n)^{-1}]$ has the same limit values at all axis point.* At axis points in Kerr spacetime, the limit value of $[-(\tau^a \varphi_a)(\varphi^n \varphi_n)^{-1}]$ is

$$2\, M\, r\, a\, [r^2 + a^2]^{-2}$$

(Recall (8).) Clearly, this function is not constant over any interval of values for r. So we see that *the CIA criterion does not satisfy the relative rotation condition in Kerr spacetime, or the restriction of Kerr spacetime to an open set containing an axis point.*

We also see if that *if the background stationary, axi-symmetric spacetime is static, then the CIA criterion satisfies the relative rotation condition.* (If $[-(\tau^a \varphi_a)(\varphi^n \varphi_n)^{-1}]$ is constant, then certainly the function has the same limit values at all axis point.) It turns out, however, that *the converse is false.*[36]

Next consider the conditions under which our two criteria agree (for all rings). What is required is that, for all orbit cylinders C, the value of $[-(\tau^a\varphi_a)(\varphi^n\varphi_n)^{-1}]$ on C be equal to the value of k_{crit} at the centerpoint of C. Recalling how centerpoints are defined, and making use of lemma 3, we see that *the CIA and ZAM criteria of nonrotation agree (for all rings) iff the function $[-(\tau^a\varphi_a)(\varphi^n\varphi_n)^{-1}]$ is constant on all null geodesics that terminate at axis points and have tangents there orthogonal to the axis direction σ^a.*[37] It follows that *they do not agree in Kerr spacetime, or any open set in Kerr spacetime containing an axis point.*[38]

II.4 The Limit Condition and the Theorem

Finally, we turn to the limit condition. Let (C_1, k_1), (C_2, k_2), (C_3, k_3), . . . be a sequence of striated orbit cylinders that share a common centerpoint γ, and that converge to γ. (We can take the second condition to mean that each point on γ is the accumulation point of a sequence of points p_1, p_2, p_3, . . . , with p_i on C_i.) For all i, let $\tau_i^a = \tau^a + k_i\varphi^a$, and let ω_i^a be its associated twist field

$$\omega_i^a = \epsilon^a_{\;bcd}\tau_i^b \, \nabla^c \tau_i^d.$$

We can take the limit condition to assert that, if each (C_i, k_i) qualifies as "nonrotating," then the sequence ω_1^a, ω_2^a, ω_3^a, . . . converges to 0 on γ. This captures the requirement that the measured angular velocity of (C_i, k_i) relative to the compass of inertia on γ goes to 0. An equivalent formulation is the following.[39]

> *Limit Condition* Let (C_1, k_1), (C_2, k_2), (C_3, k_3), . . . be a sequence of striated orbit cylinders that share a common centerpoint γ, and that converge to γ. If each of the (C_i, k_i) qualifies as "nonrotating," then $\lim_{i\to\omega} k_i = k_{crit}(\gamma)$.

It follows immediately, of course, that *the CIA criterion satisfies the limit condition* (in all stationary, axi-symmetric spacetime models). For if each (C_i, k_i) qualifies as nonrotating according to that criterion, $k_i = k_{crit}(\gamma)$ for all i. (One does not need to take a limit to reach $k_{crit}(\gamma)$.) It also follows immediately from lemma 3 that *the ZAM criterion satisfies the limit condition* (in all stationary, axi-symmetric spacetime models). For if each (C_i, k_i) qualifies as nonrotating according to that criterion, k_i is equal to the value of the function $[-(\tau^a\varphi_a)(\varphi^n\varphi_n)^{-1}]$ on C_i, for all i. And the sequence of *those* values converges to $k_{crit}(\gamma)$ by lemma 3 (and the fact that the C_i converge to γ).

We can, now, finally, state our principal result.

Theorem Assume the background stationary, axi-symmetric spacetime model is one (like Kerr spacetime) in which there exist axis points p and p' such that $k_{crit}(p) \neq k_{crit}(p')$. Then there is no generalized criterion of nonrotation that (in the model) satisfies the following three conditions:

    (i)   limit condition

    (ii) relative rotation condition

    (iii) nonvacuity condition: there is at least one striated orbit cylinder that qualifies as "nonrotating."

Proof: Let γ and γ' be the integral curves of τ^a containing p and p', and let $C_1, C_2, C_3, \ldots$ and $C'_1, C'_2, C'_3, \ldots$ be sequences of orbit cylinders that converge to γ and γ' respectively. Now assume there is a generalized criterion of nonrotation G that satisfies all three conditions in the model. Let (C, k) be a striated orbit cylinder that qualifies as "nonrotating" according to G. For all i sufficiently large, (C_i, k) and (C'_i, k) are striated orbit cylinders, i.e., $(\tau^a + k\varphi^a)$ is timelike on C_i and C'_i. So, by the relative rotation condition, (C_i, k) and (C'_i, k) qualify as nonrotating according to G for all i sufficiently large. Therefore, by the limit condition applied to the sequences $(C_1, k), (C_2, k), (C_3, k), \ldots$ and $(C'_1, k), (C'_2, k), (C'_3, k), \ldots$, it must be the case that $k = k_{crit}(p)$ and $k = k_{crit}(p')$. But this contradicts our hypothesis that $k_{crit}(p) \neq k_{crit}(p')$. So our nonexistence claim follows. ∎

The implication in the theorem is reversible. For if the value of k_{crit} *is* the same at all axis points, then the CIA criterion satisfies all three of the stated conditions. (Even then, of course, it need not be the case that the CIA criterion agrees with ZAM criterion, or that the latter satisfies the relative rotation condition.)

I have argued that, in the context of general relativity, the concept of rotation is a delicate and interesting one. Perhaps it is worth saying, in conclusion, that I intend no stronger claim. There is no suggestion here that the no-go theorem poses a deep interpretive problem (or any problem at all) for the foundations of general relativity, nor that we have to give up talk about rotation in general relativity. The point is just that, depending on the circumstances, we may have to disambiguate different criteria of rotation, and may have to remember that they all leave our classical intuitions far behind.

NOTES

1. I wish to thank David Garfinkle, Robert Wald, and especially, Robert Geroch for helpful discussion of the issues raised in the paper. I am also grateful to John Norton for assistance in preparing figures 4 and 5.

2. For an extended discussion of other criteria, see Page 1998 and the references cited there.

3. It turns out that a generalized criterion of rotation can satisfy both the limit and relative rotation conditions vacuously if, according to the criterion, no ring, in *any* state of motion (or nonmotion), qualifies as "nonrotating."

4. Here and throughout section I, we make free appeal to our commonsense (Euclidean) intuitions about the geometry of space. We take for granted that we understand, for example, what it means to say that the plane of the ring is orthogonal to the axis, that the axis is at the center of the ring, etc. Later, in section II, we will have to consider how to capture these conditions within the framework of four-dimensional spacetime geometry.

5. We will later restrict attention to spacetimes that are stationary and axisymmetric. It is the presence of the latter axial (or rotational) symmetry that gives rise to a notion of angular momentum. (See note 20.)

6. I am only thinking here of experimental procedures that can be performed locally, on or near the ring and axis. Procedures performed, for example, at "spatial infinity" are excluded.

7. For example, we can take an arbitrary ring in an arbitrary state of uniform rotational motion and dub it "nonrotating." Then we can take other rings to be "nonrotating" precisely if they are nonrotating relative to *that* one (in the sense described).

8. In what follows, we presuppose familiarity with the basic mathematical formalism of general relativity, and make use of the so-called "abstract index notation" (see Wald 1984).

9. Thus, Γ_0 and Σ_0 are the identity map on M, and

$$\Gamma_t \circ \Gamma_{t'} = \Gamma_{(t + t')} \text{ and } \Sigma_\varphi \circ \Sigma_{\varphi'} = \Sigma_{(\varphi + \varphi') \, (\text{mod } 2\pi)}$$

for all t, t' in $\mathbf{R}$, and all φ, φ' in $\mathbf{S}^1$.

10. Strictly speaking, this condition rules out standard examples of interest, including Kerr spacetime. We are, in effect, limiting attention to restricted regions of interest in those spacetimes where the condition holds. (See the final paragraph of this section.)

11. That is, at every point there is a timelike vector tangent to the submanifold. Equivalently, the restriction of g_{ab} to the submanifold has signature $(1, -1)$.

12. For a proof of the equivalence, see Wald 1984, 163.

13. See the discussion in Wald 1984, 162–65.

14. Note that the definition does not depend on the initial choice of timelike Killing field τ^a in this sense: given any other choice $\tau^{*a} = (k_1 \tau^a + k_2 \varphi^a), \tau^{*a}$ satisfies the constancy condition iff τ^a does.

15. Usually one says that a spacetime is "static" if there exists a timelike Killing field κ^a (defined everywhere or, at least in some domain of interest) that is hypersurface orthogonal, i.e., such that $\kappa_a = f (\nabla_a g)$ for some functions f and g. (In this case κ^a is orthogonal to the g = constant hypersurfaces.)

16. That is, a timelike vector α^a at a point will qualify as future directed if $\alpha^a \tau_a$ > 0.

17. In what follows, we will not always bother to distinguish between (parametrized) curves and the images of such curves. Strictly speaking, it is the latter in which we are usually interested.

18. Since ξ^a is timelike and future directed, it must be the case that

$$(k_1\tau^a + k_2\varphi^a)(k_1\tau_a + k_2\varphi_a) > 0 \text{ and } \tau^a(k_1\tau_a + k_2\varphi_a) > 0.$$

These two conditions imply that $k_1 > 0$.

19. Rings nonrotating according to this criterion might also be called "locally nonrotating." That terminology is often used in the literature. (See, for example, Bardeen 1970, 79, and Wald 1984, 187.)

20. Given any Killing field κ^a in any relativistic spacetime model (not necessarily stationary and axi-symmetric), and any timelike curve with (normalized) four-velocity ξ^a, we associate with the two a scalar field $(\kappa^a\xi_a)$ on the curve. If the curve represents a point particle, then we call $(\kappa^a\xi_a)$ the "energy" of the particle (relative to κ^a) if κ^a is timelike, and call it the "angular momentum" of the particle (relative to κ^a) if κ^a corresponds to a rotational symmetry. In the special case of a free particle with geodesic worldline, the canonically associated magnitude $(\kappa^a\xi_a)$ is constant on the curve (i.e., is conserved) since

$$\xi^n\nabla_n(\kappa^a\xi_a) = \kappa^a\xi^n\nabla_n\xi_a + \xi^a\xi^n\nabla_n\kappa_a = \mathbf{0}.$$

(The first term in the sum vanishes because the curve is a geodesic $(\xi^n\nabla_n\xi^a = \mathbf{0})$; the second does so because κ^a is a Killing field $\nabla_{(n}\kappa_{a)} = \mathbf{0}$.)

In the case at hand, we are considering a rotational Killing field φ^a and points on the ring with four-velocity $f(\tau^a + k\varphi^a)$, where $f = [(\tau^a + k\varphi^a)(\tau_a + k\varphi_a)]^{-1/2}$. The angular momentum of the points (with respect to φ^a) is $f(\tau^a + k\varphi^a)\varphi_a$. Clearly, this magnitude vanishes precisely if $(\tau^a + k\varphi^a)$ is orthogonal to φ_a.

21. See also Bardeen 1970 and Ashtekar and Magnon 1975. Ours is a simple, low-brow calculation. The discussion in the second reference is much more general and insightful. (Readers may want to skip the calculation. It is not needed for anything that follows.)

22. To confirm that it is constant along them, note that

$$(\tau^a + k\varphi^a)\nabla_a(\varphi - kt) = \tau^a\nabla_a(-kt) + (k\varphi^a)\nabla_a\varphi = 0.$$

23. t increases along all three curves since

$$(\tau^n + k\varphi^n)\nabla_n t = (\tau^n + l_i\varphi^n)\nabla_n t = 1$$

for i = 1, 2.

24. Note that if λ_i is parametrized by s, then

$$(d\varphi'/dt) = (d\varphi'/ds)/(dt/ds) = [(\tau^n + l_i\varphi^n)\nabla_n\varphi'][(\tau^n + l_i\varphi^n)\nabla_n t]^{-1} = (l_i - k).$$

25. It might seem preferable to state the condition this way. For all striated orbit cylinders (C, k) and all orbit cylinders C', if (C, k) qualifies as "nonrotating," then so does (C', k). But there is a problem with this formulation. It takes for granted that (C', k) is a striated orbit cylinder in the first place, i.e., that the field $(\tau^a + k\varphi^a)$ is timelike on C'.

26. (Recall our slightly nonstandard definition of "static" in section II.1.) The "if" half of the proof is straightforward. The proof of the converse involves one small complication. Here is the argument in detail. Assume that the ZAM criterion satisfies the relative rotation condition. Let p_1 and p_2 be any points in M^-, let C_1 and C_2 be the orbit cylinders that contain them, and let k_1 and k_2 be the values of the function $[-(\tau^a\varphi_a)(\varphi^n\varphi_n)^{-1}]$ at p_1 and p_2. We must show that $k_1 = k_2$. We don't know (initially) that either (C_1, k_2) or (C_2, k_1) qualifies as a striated orbit cylinder. But, by moving sufficiently close to the axis, we can find a point p_3 such that, if C_3 is the orbit cylinder that contains p_3, (C_3, k_1) and (C_3, k_2) both qualify as striated orbit cylinders. (For *any* value of k, the vector field $(\tau^a + k\varphi^a)$ is timelike at points sufficiently close to axis points.) (C_1, k_1) and (C_2, k_2) are both nonrotating according to the ZAM criterion. So, by the relative rotation condition, (C_3, k_1) and (C_3, k_2) are both nonrotating according to that criterion. So k_1 and k_2 must both be equal to the value of $[-(\tau^a\varphi_a)(\varphi^n\varphi_n)^{-1}]$ at p_3. Therefore, $k_1 = k_2$.

27. Volume elements always exist locally, and that is sufficient for our purposes.

28. For facts such as $\in_{abcd} \in^{dmpq} = (3!) \, \delta_a{}^{[m}\delta_b{}^p\delta_c{}^{q]}$, see Wald 1984, 433.

29. If φ^a vanishes at one point on an integral curve of τ^a, it necessarily vanishes at all points. This follows from equation (1).

30. There is a certain ambiguity in terminology here. We have taken an "axis point" to be a point in M at which $\varphi^a = 0$. But here we have in mind an "axis point" in the sense of figure 2 (i.e., a point in a three-dimensional space). It is represented by a timelike curve in M. In what follows, when referring to "axis points," it should be clear from context (and notation) which is intended.

31. Note that if the stated condition holds for one future-directed null geodesic running from a point on C to a point on γ, it holds for all. For the entire class of such null geodesics is generated from any one under the action of the isometry group $\{\Gamma_t \circ \Sigma_\varphi : t \in \mathbf{R} \,\&\, \varphi \in S^1\}$. Note too that the requirement that λ^a be orthogonal to σ^a at the arrival point is equivalent to the (slightly more intuitive) requirement that, at that point, the component of λ^a orthogonal to τ^a (representing the "spatial direction" of the incoming light signal relative to τ^a) be orthogonal to σ^a.

32. Here is a rough sketch of the proof. Suppose p is an axis point and suppose λ^a is a past-directed null vector at p that is orthogonal to the axis direction σ^a. We can extend λ^a to a past directed null geodesic. Let q be any point on that geodesic and let C(q) be the orbit cylinder that contains q, i.e., the orbit of q under the isometry group $\{\Gamma_t \circ \Sigma_\varphi : t \in \mathbf{R} \,\&\, \varphi \in S^1\}$. Then, "by construction," C has a centerpoint γ (with p on it).

There is a smooth two-dimensional timelike submanifold S through p that consists entirely of axis points. (At every point of S, the tangent plane to S is spanned by τ^a and σ^a, where σ^a is as in the preceding paragraph.) If we let λ^a range over all past-directed null vectors at points of S that are orthogonal to σ^a, and consider all points q on the past-directed null geodesics they determine (or at least all such points sufficiently close to p), we sweep out an open set O. The argument in the preceding paragraph shows that every orbit cylinder through every point in O has a centerpoint. Uniqueness follows from that fact that, at least locally, given any point q, there is a unique point p on S such that there is a future directed null geodesic that runs from q to p and whose tangent vector at q is orthogonal to σ^a.

33. To support the operational interpretation of the CIA criterion presented in section I, one can proceed as follows. Let (C, k) be a striated orbit cylinder, let γ be the centerpoint of C, and let γ' be a striation line on the cylinder, representing the point on the ring (say R) at which a light source is mounted. (So, both γ and γ' are integral curves of the field $\tau'^a = \tau^a + k\varphi^a$.) Finally, let λ^a be a future directed null geodesic field, the integral curves of which run from γ' to γ. (The latter represent light signals emitted at R and received at the center point.) The entire field of integral curves is generated from any one of them under the action of the isometry group associated with τ'^a, i.e., the field λ^a is Lie derived by τ'^a. Suppose the telescope at the center point is tracking the light source. Then the direction of the telescope (as determined by the observer with worldline γ) is represented by a vector field v^a on γ whose value at any point is the component of λ^a orthogonal to τ'^a at that point. It is not difficult to check that the Fermi derivative of v^a along γ vanishes iff $k = k_{crit}(\gamma)$. (For details, see the discussion in Malament (forthcoming). The vanishing of that Fermi derivative serves as a surrogate here for the flatness of the water surface in the bucket.)

34. Here is a rough sketch of a proof (due to Robert Geroch). Let S be the two-dimensional submanifold of axis points. Let $\alpha = (-\tau^a\varphi_a)$ and $\beta = (\varphi^a\varphi_a)$, so that $f = \alpha/\beta$. Let p be a point on S. Given any point q sufficiently close to p, it has a unique orthogonal "projection" q' on S, i.e., there is a unique point q' on S with the property that the geodesic segment running from q' to q is orthogonal to S. So the point q is uniquely distinguished by a pair of objects: (i) the value of β at q, and (ii) its orthogonal projection q' on S. Thus, if we restrict attention to a suitable open neighborhod of p, we can think of α as a function defined on a subset A of the product manifold (with boundary) $[0,\infty) \times \mathbf{R}^2$. We first show that α is smooth, not just as a function on M, but also when construed this way (as a function on A). To do so, we consider a finite Taylor series expansion of α, up to order n, at p, with partial derivatives taken in directions tangent to, and orthogonal to, S. Since α is constant on orbits of φ^a, the coefficients in the series, i.e., the mixed partial derivatives of α at p, have a special, simple structure. Those of odd order in directions orthogonal to S must be 0, and those of even order in those directions can be expressed in terms of derivatives in any one orthogonal direction (and directions tangent to S). This allows us to reinterpret the series as a finite Taylor series expansion (at p) of α *construed as a function on A.*

Next we observe that $\alpha = 0$ and $\nabla_a\alpha = \mathbf{0}$ at p. (The second equation can be proved using clause (v) of lemma 1.) It follows that the terms in the expansion of 0th order in β are 0. So we can divide by β and generate a finite Taylor series expansion for $f = \alpha/\beta$ at p. Since the number of terms n in the original expansion was arbitrary, so is the number of terms in the derived expansion. It follows that all partial derivatives of f (as a function on M) exist and are continuous at p.

35. The computation is straight forward.

$$\tau'_a\nabla_b \, \tau'_c = g\,(\nabla_a h)\nabla_b\,(g\,(\nabla_c h)) = g^2(\nabla_a h)(\nabla_b\nabla_c h) + g\,(\nabla_a h)(\nabla_b g)(\nabla_c h).$$

So, since, $(\nabla_{[b}\nabla_{c]}h) = \mathbf{0}$, and $(\nabla_{[a}h)(\nabla_{c]}h) = \mathbf{0}$, it follows that $\tau'_{[a}\nabla_b\tau'_{c]} = \mathbf{0}$.

36. For a counterexample, it suffices to find a stationary axi-symmetric space-time that is not static, but exhibits "cylindrical symmetry," i.e., in which the axis direction field σ^α is a Killing field. For the latter condition will guarantee that the

function k_{crit} is constant as one moves along the axis. Gödel spacetime is one such. (In terms of standard t, φ, r, y coordinates, σ^a turns out to be—up to a constant— just the translational Killing field $(\partial/\partial y)^a$.) (For a description of Gödel spacetime, see Hawking and Ellis 1973. For further discussion of rotation in the model, see Malament [forthcoming].)

37. It would be nice to have a simpler or more instructive characterization. (I do not have one.)

38. One can verify this with a calculation, but there is a painless way to see that the stated constancy condition cannot hold. Start at a point p on the axis with positive r coordinate, choose a future directed null vector λ^a at p orthogonal to σ^a, and consider the (maximally extended) null geodesic through p that has tangent λ^a at p. It comes in from "past infinity" where, asymptotically, the value of $[-(\tau^a\varphi_a)(\varphi^n\varphi_n)^{-1}]$ is 0. (Recall (8).) Since its value at p is *not* 0, the function cannot be constant on the geodesic.

39. By lemma 2, $\tau_{[b}\nabla_c\tau_{d]} + k_{crit}\tau_{[b}\nabla_c\varphi_{d]} = 0$ on γ. So

$\{\omega_i{}^a\}$ converges to 0 on γ
$\Leftrightarrow \{\tau_{[b}\nabla_c\tau_{d]} + k_i\,\tau_{[b}\nabla_c\varphi_{d]}\}$ converges to **0** on γ
$\Leftrightarrow \{(k_{crit} - k_i)\,\tau_{[b}\nabla_c\varphi_{d]}\}$ converges to **0** on γ.

Since $\tau_{[b}\nabla_c\varphi_{d]} \neq \mathbf{0}$ on γ, the third conditions holds iff $(k_{crit} - k_i)$ converges to 0.

REFERENCES

Ashtekar, A. and A. Magnon. (1975). "The Sagnac Effect in General Relativity." *The Journal of Mathematical Physics* 16: 341–44.

Bardeen, J. M. (1970). "A Variational Principle for Rotating Stars in General Relativity." *Astrophysical Journal* 162: 71–95

Ciufolini, I. and J. Wheeler. (1995). *Gravitation and Inertia*. Princeton: Princeton University Press.

Hawking, S. W. and G. F. R. Ellis. (1973). *The Large Scale Structure of Space-Time*. Cambridge: Cambridge University Press.

Malament, D. (Forthcoming). "On Relative Orbital Rotation in General Relativity." In J. Renn *et al.* (eds.), *Space-Time, Quantum Entanglement, and Critical Epistemology*. Dordrecht: Kluwer Academic Publishers.

O'Neill, B. (1995). *The Geometry of Kerr Black Holes*. Wellesley, Mass.: A.K. Peters.

Page, D. (1998). "Maximal Acceleration is Non-Rotating." *Classical and Quantum Gravity* 15: 1669–1719.

Wald, R. (1984). *General Relativity*. Chicago: The University of Chicago Press.

PART III

General Epistemology
and Philosophy
of Mathematics

[11]

Some Intellectual Obligations of Epistemological Naturalism

ABNER SHIMONY

1. Introduction

The central thesis of epistemological naturalism, as I understand it, is the following:

> Whatever knowledge human beings have, about anything in the universe, can be understood in terms of the natural faculties of human beings and the interaction of these faculties with the objects of knowledge.

The natural faculties are characteristics of human beings as entities subject to natural laws, interacting with other natural entities, and participating in the evolutionary processes of the biosphere. The epistemologically relevant faculties are those for gathering and processing sensory stimuli, selecting and remembering empirical data, introspecting and imagining, formulating concepts and hypotheses and applying these to empirical data, and making inferences. According to epistemological naturalism all of these are conceived to be as natural as human anatomical and physiological features.

There is nothing surprising in epistemological naturalism as just stated, in view of the great prestige of the natural sciences in the general contemporary culture and the great influence in the philosophical profession of Quine's "epistemology naturalized" (see Quine 1960). Nevertheless, we have neither a fully satisfactory formulation of the assertions of epistemological naturalism (which need not accede to Quine's proposal to reconstruct epistemology, or something like it, as a chapter of psychology) nor a fully satisfactory set of answers to a number of serious questions about these assertions. The purpose of this paper is to state what I consider to be the main intellectual obligations of research on epistemological naturalism, if a satisfactory theory is to be achieved. A secondary purpose is to assess the progress made so far in meeting these obligations and the prospects of further progress. My progress report will unavoidably be sketchy, reflecting the great gaps in our present scientific understanding of human cognitive faculties and the much greater gaps in my own knowledge. Nevertheless, I hope to make a good case for optimism regarding the prospects of epistemological naturalism.

An Enumeration of Intellectual Obligations

1. In order to know what one is talking about when one uses the word "naturalism" in a philosophical context, one should say what is meant by "nature." And this obligation seems to require metaphysical investigations—what kinds of things are there in the universe? and what delimits natural things from other things if there are others? If nature is coextensive with the universe, then the central thesis of epistemological naturalism stated in the introduction is surely true, however its detailed content is construed. If nature is not coextensive with the universe, then one needs to investigate how a naturalistic account can be provided for knowledge of that part of the universe that lies beyond nature. Mathematical knowledge may be a case in point. Since numbers and sets are manifestly different in character from entities that are commonly recognized as natural, and since they may have the independent ontological status attributed to them by Platonists, what natural psychological processes enable human beings to achieve mathematical knowledge?

2. The ontology of nature and its causal structure should be characterized sufficiently to understand—not completely, but as much as epistemology requires—the status within nature of the knowing subject. This demand is the central intellectual obligation of epistemological naturalism, and evidently it raises deep questions in various natural sciences and in the philosophy of mind. In my progress report I shall discuss this obligation on two levels. (a) The first is the level of those natural sciences—especially empirical psychology and certain branches of biology—which throw light in a *phenomenological manner* upon the acquisition of knowledge by higher organisms, without pretense of penetration to the greatest depths that can be envisaged for the natural sciences. (b) The other is the *foundational level*, which aspires to depths below phenomenological science. Specifically, it is at this level that the relation of mind to the physical aspects of the world is investigated. There are, in addition, some features of ordinary experience—such as temporal transience—which require treatment at the foundational level because they are at odds with the contemporary physical world view.

3. We seem to have some kind of knowledge of norms. Where, if at all, do norms belong in nature, and if they do not belong or do so idiosyncratically, then how is knowledge of them accommodated by epistemological naturalism?

4. The structure of epistemological naturalism poses two metaphilosophical problems. One is an apparent circularity, arising from the fact that on the one hand epistemological naturalism draws upon the results of the natural sciences, and on the other hand the justification of the methods of the natural sciences—particularly of induction or its surrogate—is itself an epistemological problem. There is an obligation either to show that this apparent circularity is specious, or that it exists but in some clear sense is nonvicious.

There is also a problem of self-reference implicit in epistemological naturalism. This program seeks to characterize human cognitive subjects as entities capable of understanding the workings of the natural world—a world which permits the emergence of cognitive subjects with just such a capacity. But self-reference in conceptual structures can generate inconsistency. There is an obligation to show that this does not occur in the kind of self-reference exhibited by epistemological naturalism.

Comment on (1): Nature and Beyond

I shall not survey the many senses of "nature" proposed in the history of philosophy, but shall simply state the definition which I implicitly used in asserting, in the introduction, the central thesis of my version of epistemological naturalism: that is, *the domain of entities governed by the laws of the natural sciences in their present formulations or, better, in future improvements and continuations of these formulations.* This definition is broad enough to be shared by very different philosophies. It even seems compatible with Kant's characterization of nature as "the connection of appearances as regards their existence according to necessary rules, that is, according to laws" (*Critique of Pure Reason* A216, B263). But I am unconvinced by Kant's arguments that categories such as causality are inapplicable to things-in-themselves (see Shimony 1993, vol. 1, 24–28). I hold instead that there are things-in-themselves to which the laws of physics apply and that cognitive subjects are things-in-themselves to which the laws of psychology apply—so that all of these are entities in nature as characterized above.

In the history of philosophy, various theologies are the predominant doctrines that there is something outside nature. Among these, obviously, are those theologies in which there is a creator God who imposed the laws of nature (interrupted perhaps by miraculous interventions) upon the domain of creation. Although Aristotle did not have a creationist theology, his unmoved mover was outside nature in the sense of being pure actuality and hence not susceptible to change. Nevertheless, the unmoved mover is not in all respects separated from nature, since its mode of operation is by providing a final cause for the rotations of the heavenly spheres, and final causation is one of Aristotle's four causes in his analysis of nature. One comment about theology that I shall make here is that prima facie it is not incompatible with epistemological naturalism. There has traditionally been a well-recognized discipline of "natural theology," which concerned investigations of the existence and attributes of God that rely only upon the natural cognitive faculties of human beings, without reference to faith or authority. (A famous remark of Newton in the *Principia* endorses the legitimacy of this discipline.) However, over and above the well-known objections that have been raised against arguments in natural theology for the existence of God, there seems to me to be a negative generic consideration of some weight: that if God is outside nature, how is the interaction between God and the natural faculties of human beings comprehensible to our natural faculties? Possibly this objection can be handled by postulating a mode of causation that applies both in the natural and the transnatural domains (and Aristotle's doctrine of final causation may be cited to make this answer concrete—but at what intellectual cost?). And if it is said that the mode of interaction is there, but mysterious to us, that answer verges upon a "negative theology" which seems to me little different from agnosticism.

A much discussed candidate for occupancy of the domain beyond nature is the class of mathematical entities—numbers and sets, and perhaps also such things as proofs and decision procedures. Platonists maintain that these entities have an ontological status coordinate with physical and other natural entities, sharing with them the attribute of existing whether or not they are known by human beings, but because of their abstract character not subsumable under the class of entities in nature. Because of limitations of my knowledge, I shall only briefly indicate the cases for and against mathematical Platonism, but I do not wish entirely to omit consideration of the topic because the thesis of epistemological naturalism suggests an interesting question about Platonism that I have never seen explicitly addressed. Kurt Gödel espoused mathematical Platonism in almost all of his writings and oral discussions on philosophy of mathematics (see Wang 1996), despite a few startling expressions of doubt. Here are two of the characteristic statements by Gödel:

> The real argument for objectivism is the following. We know many general propositions about natural numbers to be true . . . and, for example, we believe that Goldbach's conjecture makes sense, must be either true or false, without

there being any room for arbitrary convention. Hence, there must be objective facts about natural numbers. But these objective facts must refer to objects that are different from physical objects because, among other things, they are unchangeable in time. (quoted in Wang 1996, 211)

and

we do have something like a perception also of the objects of set theory, as is seen from the fact that the axioms force themselves upon us as being true, (op. cit., 226).

Tait (2001) objects that

super-realism [his phrase for Platonism] implies an alienation of truth in mathematics from what we actually do: . . . If there are grounds for truth and existence and they are not the axioms, then the axioms could be false.

Gödel is fully aware of this objection and indeed agrees to the second sentence just quoted, but maintains that axioms can be assessed indirectly by the agreement or discord of their consequences with intuition—which is an application of the hypothetico-deductive method to mathematics—and by the cultivation of a Husserlian analysis of the relevant mathematical concepts (Wang 1996, 242–46 and 158–61); but it must be noted that Wang—despite strong sympathies with Gödel—is skeptical of the effectivness of Husserl's phenomenology. Feferman, among various criticisms of Platonism, makes an important relevant remark about objectivity in set theory. He points out that mathematical Platonism entails that the cardinality of the continuum must have a definite place among the alephs, and then says that the fact that the continuum problem "has not been settled by any remotely plausible assumption leads me, for one, to agree with Weyl that it is an inherently indefinite problem" (1998, 72–73).

I hinted above that epistemological naturalism—which is a thesis about understanding human knowledge in terms of human natural faculties and their interaction with the objects of knowledge—seems to have some bearing on the plausibility of mathematical Platonism. If numbers and sets and other mathematical objects have an ontological status beyond nature, just what natural faculties do they interact with and how is this interaction effected? Gödel is fond of comparing intuition of mathematical objects with sensory perception of physical objects of ordinary experience. But regarding the latter we have a large and coherent body of information about the physical transmission of optical and acoustical and other signals from physical objects to the periphery of sense organs, of the sensory response of these organs, and of the selective transmission of neural signals from the sense organs to the perceptual centers of the brain; and we also have a smaller and less coherent, but still impressive, body of information about the deployment of memories, templates, concepts, and conjectures whereby sensation gives rise to perception of objects. If the problem of consciousness is bracketed, then this naturalistic program of understanding the perception of physical objects is very well launched and promises well for the future. Can anything comparable be said about our "perception" of independently existing mathematical entities? The absence of a satisfactory answer to this question after sustained effort (how long?) would surely count as evidence against mathematical Platonism.

Gödel was clearly aware of this difficulty, for he remarked to Wang:

> I conjecture that some physical organ is necessary to make the handling of abstract impressions (as opposed to sense impressions) possible. . . . Such a sensory organ must be closely related to the neural center for language. But we simply do not know enough now, and the primitive theory on such questions at the present stage is likely to be comparable to the atomic theory as formulated by Democritus. (Wang 1996, 233)

Gödel's conjecture has some plausibility, in view of various features of language that have counterparts in mathematical thinking: recursion, generality, freedom in applications, possibility of self-reference, and definiteness of rules combined with flexibility in the selection of rules. There may, however, be something self-defeating in Gödel's conjecture, if it is supposed to bolster his case for mathematical Platonism, for the features of language just mentioned are typical of what must be attributed to the mind in order to formulate a coherent constructivist account of mathematics. Feferman, for example, writes,

> I am in agreement with the constructivist position as to the subjective source of basic mathematical conceptions, but for me these are supposed to be conceptions of certain kinds of ideal worlds, including ones which are not countenanced constructively (such as 'platonistic' worlds of sets). These worlds (or world pictures of mathematical structures) are presented more or less directly to the imagination, from which basic principles are derived by examination. All else (in each picture) is obtained by rational reflection on, and from, basic concepts and principles. (1998, 124)

Is not this non-Platonist interpretation of mathematics as compatible as Platonism with Gödel's conjecture about the "organ" of mathematical thinking? Or—a final thought—might not the attempt to understand mathematical knowledge within a framework of epistemological naturalism have the consequence of narrowing the conceptual gap between Platonism and constructivism?

Comments on (2a): The Place of the Knowing Subject in Nature—Phenomenological Level

In commenting above on the difficulty of giving a naturalistic account of the "perception" of Platonistically conceived mathematical entities, some remarks were made about the remarkable success of the program of accounting naturalistically for our perceptual knowledge of the physical objects. Those remarks were much too brief. The proper placement in the natural world of the organs of perception, and the characterization of their *modi operandi* at a phenomenological level, are the tasks of a large number of overlapping disciplines: geometrical and wave optics, acoustics, anatomy, physiology of the sense organs, neuropsychology, evolutionary biology, developmental psychology, perceptual psychology, cognitive psychology, social psychology, and others. To give a proper progress report of a naturalistic account of ordinary perception one would need to attach (to borrow a bit of computer jargon) a small library, including *Helmholtz's Treatise of Physiological Optics*, the *Handbook of Experimental Psychology*, the works of the Gestalt school, von Bekesy's *Sensory Inhibition* , von Frisch's works on the vision of bees, J. J. Gibson's *The Perception of the Visual World*, U. Neisser's *Cognitive Psychology*, the works of the evolutionary theorists such as K. Lorenz and D. Campbell, Piagetian studies of the cognitive development of children, and so forth. Since a survey of the relevant discoveries is out of the question here, a few examples will have to suffice for the purpose of indicating the richness and progressiveness of the entire program. Four examples illustrate mechanisms and strategies for effective perception which are exhibited by the sense organs and nervous system of contemporary human beings (and probably of other higher animals); one example illustrates the evolution of extraordinarily well-adapted sense organs.

(a) T. Bower (1966) has shown that babies as young as two months track objects, and when a moving object of interest is blocked by a screen the baby shows excitement, as measured by increase of blood pressure, when the object reappears at the farther edge of the screen. It may be an exaggeration to interpret this experimental result as showing an innate conception of object, but it surely indicates a hospitality to the development of that conception.

(b) Physical objects are perceived to be located in a space that is somehow common to all sensory modes. There are innate behavioral patterns in very young infants that contribute to achievement of a multimodal space. Soon after birth a baby tends to shift its eyes toward the source of a sound, which is acoustically accessible to it because sounds coming from the left and from the right are discriminable by difference in arrival time at the two ears; furthermore, a few weeks after birth, when control of neck muscles develops, there is an instinctive turn of the head toward the source of the sound. Visual and tactile spatial perception are integrated by a baby, even before it can crawl, by such playful experiments as grasping the feet and simultaneously watching them.

(c) Although distinct visual images are obtained only from light incident upon the fovea, the edge of the retina responds to a moving stimulus and prompts head and eye movements to bring the image of the moving source to the fovea. Here is a mechanism, quite independent of conceptual attitudes of search and anticipation, for building into the ocular system an exploratory strategy toward objects in the visual field.

(d) Although the mapping of neuronal connections in the brain is very incomplete, there is strong evidence that

> large cortical areas of the frontal lobe . . . are used for new levels of integration. . . . Thus, in the brain, there is not only an almost total centralization and control of processes of the body and input from the outer world, but also the processing of inner information according to the same rules as processing real world objects and still higher processes of comparative and reflexive evaluation. (Seitelberger 1984)

Whatever the mechanism of coordination of neural and mental events may be, this reflexiveness of neural structure is plausibly connected with imagination, envisagement of possibilities, and thought experiments—which immensely enrich our perception of physical objects.

(e) A beautiful example of the adaptation of the sensory capacity of a type of animal to its behavior is provided by von Frisch's (1950) discoveries that the visible spectrum for bees begins at green and extends into the ultraviolet and also that bees use the polarization of light for navigational purposes. Without any commitment as to whether one of these traits was established first, thereby influencing the evolution of the other, or instead the two traits evolved in tandem, von Frisch gives a plausible reason for their correlation. Scattered sunlight incident upon a receptor in a direction making an angle f with the direction of the sun is partially polarized in a way that depends upon f and the frequency of the light. The directional dependence of the polarization increases with the frequency of the light, being greater for blue than for red and for ultraviolet than for blue. Hence, the shift of bees' visible spectrum to the ultraviolet enhances the utility of the polarization of light incident upon their eyes as a compass for flying between hive and food source.

It is noteworthy that examples (a), (b), (c), and (d) above are naturalistic accounts of aspects of "being-in-the world," which is essential to perceptual experience according to the phenomenological theory of Merleau-Ponty (1964). That theory seems to me subtle and observant—

the work of someone who remarkably combines the mentalities of a man of action and a representational painter—and I have no intention to denigrate it by my applications of epistemological naturalism. The "primacy of perception" (ibid.) that Merleau-Ponty insists upon seems to me, however, to be what Aristotle characterizes as "first in the order of knowing," in contrast to "first in the order of being" (Physics 184a 20–21). Epistemological naturalism is unwilling to treat the former "primacy" as something entirely self-contained, but aims instead at illuminating it by means of the latter "primacy"—and has indeed made great progress in doing so.

Comments on (2b): The Place of the Knowing Subject in Nature—Foundational Level

When one tries to carry out at a foundational level the naturalistic program of understanding the knowing subject in terms of the ontology and causal structure of nature one confronts the massive fact that the basic natural science is physics. "Reduction" is a notoriously ambiguous term, but I believe that when sufficiently careful distinctions are made (between laws and contingencies, between detailed prediction and understanding in principle, etc.) then there is a convincing case for the reducibility of astronomy, chemistry, and (setting aside mentality) also of biology to physics (see Shimony 1993b, 191–227). But epistemology is concerned with knowledge and belief, and these are elements of mental states. If a physicalistic theory of mentality were successful, then there would in principle be no obstacle to the program of epistemological naturalism at the foundational level, but only a formidable array of experimental problems of neurology and theoretical problems of the "software" governing the nervous system. But many sophisticated arguments have been given against the possibility of construing mentalistic discourse—concerning sensations, references, volitions, memories, concepts, beliefs, hypotheses, inferences, and so on— as implicit neural discourse, and some naive arguments, as well, that seem to me sufficiently convincing. My own naive argument (elaborated in Shimony 1997) is that physicalism violates the Phenomenological Principle, which asserts that *a minimal condition on ontology is to recognize a sufficient set of realities to account for appearances qua appearances;* physicalism is not capable of accounting for features of mental states like sensory *qualia* as appearances without surreptitiously assuming something to which the appearances appear—and that something is not characterizable by physics. If this is so, does the program of epistemological naturalism run aground in spite of the remarkable successes at the phenomenological level that were noted in the preceding section? Just such a conclusion has been reached by some philosophers who are scientifically very well informed, for example E. Schrödinger (1967) and H. Putnam (1990). The latter, incidentally, had earlier been a sophisticated physicalist, but afterwards arrived at the view that physicalistic (hence naturalistic) descriptions and mentalistic descriptions of higher organisms are complementary, neither dispensable and neither reducible to the other.

In order to hold onto epistemological naturalism at the foundational level without resorting to physicalism a profound metaphysical and scientific revolution is required, or more accurately, a revival of certain views with a long pedigree in the history of philosophy, but—it is earnestly to be wished—a revival with a precision of formulation and a careful linkage to experiment that were missing in the past. In this revolution, physics—in its current formulation or in extrapolations from it that can now be envisaged—would not be the basic natural science. Mentality would have a fundamental status in nature, either coordinate with physical reality or yet more fundamental. A. N. Whitehead's "philosophy of organism" (1929) is a speculation about such a fundamentally mentalistic natural science. There have also been speculations that quantum mechanics points to, or at least is hospitable to, such a mentalistic revolution (e.g., R. Penrose 1989, H. Stapp 1993, A. Shimony 1997, S. Malin 2001). None of them are convincing, but to me physicalism is even less so. We can look forward to a long epoch of uncertainty.

Current physics poses certain other difficulties. It is an obligation of epistemological naturalism to explain naturalistically the phenomenology of ordinary experience, and indeed some strong claims were made at the conclusion of the preceding section, in the comments on Merleau-Ponty, that it has done so with considerable success. But nothing is more pervasive in ordinary experience than *becoming*, or the *transiency* of the present moment, and yet there have been famous arguments that becoming is illusory (J. M. E. McTaggart 1927) or subjective (e.g., A. Grünbaum 1971) and that relativistic space-time theory has no use whatever for transiency. There have been, however, strong defenses of the uneliminability of becoming as an objective fact (Shimony 1993b, 271–87 and 1998, chapter 10, with references to Zeilicovici and Favrhold), and these point either to a modification of current physics so as to accommodate becoming or to a basic natural science that accomplishes this accommodation by mentalism. Finally, quantum mechanics is a fundamental part of physics, and quantum mechanics is plagued with the problem of the reduction of superpositions, sometimes also called "the measurement problem." Briefly, the problem is that linear dynamics of quantum mechanics precludes the occurrence of definite events at the conclusion of measurements—contrary to the obvious phenomenology of the measuring process. There are, of course, serious proposals to resolve this problem without modifying the formalism of quantum mechanics, but there are also students of the subject, among whom I am one, who believe that phenomenology is right and quantum mechanics is wrong, and that a successor to present quantum mechanics will account in a natural way for the occurrence of definite events. Without a solution to the measurement problem, one way or another, there is an embarrassment for epistemological naturalism at the foundational level. I am confident that there will be no such embarrassment.

Comments on (3): Knowledge of Norms

Statements about norms are part of ordinary discourse and to some extent are well understood. We therefore seem to attach a truth value to a statement like "X is right," where X may be, for instance, a rule of inference, or a technique in a factory, or a political strategy, or a rule of behavior. But if "X is right" is true, and we accept Aristotle's characterization of truth or some appropriate refinement thereof, then what it says *is so*. But what *is so* is factual, and that is what is puzzling about a norm. If there is an "is so" associated with a norm, is it not something ideal or aimed at rather than lumpish factuality? And doesn't that peculiarity make knowledge of norms problematic for epistemological naturalism?

Much of the force of this objection is exorcized by eliminating the tendentious word "lumpish." Throughout nature there is a great variety of facts—generalities and singularities, necessities and contingencies, actualities and potentialities. In the biosphere there are ends or goal states—which, as Monod (1972) emphasizes, are exhibited by organisms as primitive as bacteria without any hint of conscious envisagement—and there are facts of achieving and falling short of goals. And with the emergence of conscious organisms in nature there are many new genera of facts: pain, pleasure, exhilaration, happiness, unhappiness, hope, love, hatred, fear, desire, satisfaction, disappointment, envisagement of alternative possibilities, competition among goals, assessment of probabilities, the feeling of imperatives, remorse, sympathy, withholding of sympathy, and many more. All of these are facts, as surely as the density of lead. These facts do not constitute norms (though the feeling of an imperative can masquerade as a norm and has been taken as such by a whole school of ethicists). However, these subtle facts are the elements whereby norms are related to nature and hence are amenable to treatment by epistemological naturalism.

In the first three examples of rightness mentioned above—when X is a rule of inference, when X is a technique in a factory, and when X is political strategy—the sentence "X is right" asserts the fact that X is an effective means to an end Y, where Y is indicated by context even though it is not explicitly spelled out. When X is a rule of inference, then Y is the end of inferring a conclusion from premises of a certain type in such a way that if the premises are true then the conclusion is certainly or probably true; when X is a technique in a factory, then Y is the manufacture of a specified kind of product. If the proposition that X is an effective means to an end Y is true, then it asserts a fact whose knowability by natural faculties is not problematic. In addition, "X is right" usually conveys the additional proposition that the tacitly indicated end Y is something desirable to the speaker, and perhaps to the auditors as well. The desirability of Y to the speaker is surely a fact knowable by the speaker's natural faculties, among which is a certain amount of self-knowledge by introspection; and its desirability to persons other than the self is accessible by the remarkable and natural, though incompletely understood, means whereby human beings make reliable inferences about the mental states of other persons. (I do not wish to underestimate, however, the difficulty of the epistemological problem of knowledge of other minds. It is related to problem (2) above of understanding the ontological status of the knowing subject in nature, and it needs to be treated on the foundational level as well as on the phenomenological level.)

The first three examples of "X is right" are treated by Kant as hypothetical imperatives, and he seems to be right in not regarding them as problematic. The fourth example, in which X is a rule of behavior, is an aspect of the categorical imperative, which is not so easily analyzed. Furthermore, if Kant is right about the absoluteness of the categorical imperative and its sharp separation from any calculation of the satisfaction of human desires, then it is hard to see how this kind of norm can be understood by epistemological naturalism. Just as Kant deploys a transcendental method in the *Critique of Pure Reason* to exhibit the source of synthetic a priori judgments in the faculties of pure intuition and pure understanding he deploys a transcendental method in the *Critique of Practical Reason* to exhibit the source of the ultimate principle of morality. In both *Critiques* the operations of rationality are exhibited—of theoretical rationality in the one case, of practical rationality in the other case—but in neither is the operation of rationality examined as an application of natural law.

Aristotle's ethical theory, in my opinion, does more justice than Kant to human nature, to the immersion of human beings in societies, and to the phenomenology of moral life; and knowledge of behavioral norms as Aristotle views them can indeed be treated as epistemological naturalism. This claim may be criticized by careful readers of the text of the *Nicomachean Ethics* on the ground that Aristotle treats virtues as a kind of habit, and habits are not fixed by human nature but are rather contingencies resulting from education. This objection is not weighty, however, for it rests upon the semantic difference between Aristotle's use of the term "nature" (φυσις), emphasizing the formal cause which characterizes the species universally, and the contemporary use of "nature," which recognizes genetic diversity within a species and allows for acquired diversity of great scope. Aristotle points out a number of considerations relevant to ethics that are in some tension with each other: that all men desire happiness, that what one seeks in the pursuit of happiness differs from person to person according to genetic propensity and upbringing, that what gives satisfaction in the short run can prove to be unsatisfying in the long run, that pleasure is a constituent in satisfying activity and nevertheless a superficial constituent, that human beings are by nature (in his sense) political animals who need a society in order to survive and whose habits are therefore shaped by socially sanctioned education, and that even though habits are youthfully acquired the wisdom for properly judging ethical matters is not acquired until late in life. The sharp separation between the moral law and a calculation for achieving personal satisfaction, which characterizes Kant's ethical theory, is eliminated or at least mitigated in Aristotelian ethics, and Kant's postulation of an imperative imposed transcendentally is replaced by the lasting effects of enculturation by a family and a society that control the educational process. Modern biological and psychological speculations and/or discoveries—for example, Freud's theory of the formation of the superego, the scenario that a sentiment of altruism has survival value in human evolution, the absorption of tribal mores documented by relativistic ethnologists—all can be incorporated into a sufficiently flexible Aristotelian ethics, making it more "naturalistic" in a modern sense (see, for example, W. Rottschaefer 1998). Something of Aristotle that seems to me usually to be missing from much contemporary ethical theory is his recommendation of reflection—for the purpose of adjudicating the tensions within one person's moral experience and among the mores of different societies—that becomes possible after long experience and discipline. Rawls (1993) is exceptional in having a modern version of this idea of Aristotle, emphasizing "reflective equilibrium" as a criterion for achieving objectivity when there are political disagreements. It seems reasonable to extrapolate this concept to moral tensions, both among cultures and within one psyche, as Rawls implicitly does himself. How effective this criterion is for replacing moral relativism by one systematic set of moral principles, or by a small number of such sets (comparable to the "stable states" of Bohr's atomic theory), is a matter on which I have no competence to judge. What I do assert with confidence is that the reflection requisite for reaching such equilibrium is unequivocally a process performed by the natural faculties of human beings: recall the passage from Seitelberger quoted above about the large areas of the brain used for "new higher levels of integration" and "processing of information," which indicates that reflection is part of human mentality. It also seems to me that if the criterion of reflective equilibrium falls short of eliminating moral relativism, that would be evidence of a feature of human nature that we must somehow learn to live with.

Comments on (4): Two Metaphilosophical Problems

The locus of circularity in epistemological naturalism is the treatment of induction. A Bayesian framework for induction can be justified by decision theoretical considerations (Shimony 1993a, chapters 6 and 7 and section 3C of chapter 9), which make no factual assumptions about the natural world. But the Bayesian framework is notoriously tolerant toward an immense range of prior probability assignments. Even supplementary decision theoretical principles—such as the tempering principle, which prescribes that any properly formulated hypothesis be given sufficiently high prior probability to allow it a chance to be favored a posteriori over rival hypotheses if supported sufficiently by data—still leave an immense range of priors after eliminating those which are dogmatic and narrowminded. In order to formulate an effective inductive method within a Bayesian framework, it seems necessary to supplement the framework with factual assumptions to the effect that the *actual* world has such and such characteristics, even though there are possible worlds which do not have them. Two problems are thereby posed. The first is to specify the appropriate factual assumptions or at least to lay down criteria for finding them. The second is to justify the manifest circularity that arises if these factual assumptions cannot be justified a priori but only inductively. Since I have already written almost all that I now have to say about these two problems (Shimony 1993a, chapters 9 and 10) I shall make only a few comments.

The first problem is the more difficult of the two, and it seems to me far from being completely solved. The most promising avenue of research is to combine careful considerations of Bayesian probability theory with careful attention to preeminent historical examples of inductive reasoning. Franklin (1986) has analyzed some important experiments in elementary particle physics from a Bayesian point of view and has thrown much light on the reasoning, but he has not paid attention to the identification and deployment of factual assumptions of the reasoning. Stein (1967, 1970, and elsewhere) has noted the subtlety of Newton's reasoning about universal gravitation in the *Principia* showing how complex is Newton's "deduction from experiment," and his work has been valuably continued in the present volume by Harper and Smith. But none of these analyses of Newton's methodology has been explicit either about the fertility (or absence thereof) of a Bayesian framework or about the precise use (if needed) of auxiliary factual assumptions. The vein which they have already explored is certain to yield much more treasure.

The second problem seems to me the easier of the two, and its solution is known at least in outline. A circularity of procedure is nonvicious if, in Peirce's phrase, it does not "block the road of inquiry." If tentative factual assumptions are made in order to allow the Bayesian machinery to proceed to a relative assessment of relevant hypotheses in an investigation, there is no blockage of the road to inquiry if the possibility is kept open—seriously, not just pro forma—that the factual assumptions themselves could be assessed and possibly rejected or refined. If careful attention will be paid to the history of scientific methodology, I am confident that this openness of inquiry will be found to have prevailed, in spite of episodes of blockage; and examples will be found of all the possibilities regarding tentative factual assumptions: confirmation, rejection, and refinement. The circularity in a naturalistic treatment of induction is actually a dialectic between scientific methodology and scientific results. In view of the human condition, it is hard to see how anything better in our search for knowledge can be expected.

As to the second metaphilosophical problem of epistemological naturalism, that of self-reference, I believe that it evaporates upon examination. There is, of course, an exemplary situation in mathematical logic in which self-reference in connected with inconsistency: Gödel's theorem that no consistent formal system in which deduction is precisely representable can yield a proof of its own consistency. The existence of Gödel's theorem and the incomplete understanding of it have engendered a widespread fear of self-reference. The self-reference implicit in epistemological naturalism is not analogous to that involved in Gödel's theorem. Epistemological naturalism aims to show that the constitution of nature is knowable in its principles by human cognitive faculties, but makes no pretense that human beings are capable of chronicling all the contingent particularities of nature; furthermore, epistemological naturalism aims to show not the certainty of the emergence of organisms with those faculties but only the compatibility of their evolution with the general principles governing nature. There is nothing in the "closure of the circle" of this program which remotely resembles a precise representation of deduction in a formal system. Indeed, many of the steps in the envisaged closing of the circle are qualitative and others are inductive, and all of these are hard to formalize. In addition, a crucial part of the entire scheme is the *evolution* of organisms capable of scientific discoveries—and evolution is an extraordinarily complex stochastic process whose detailed history is evidently beyond our powers of reconstruction and whose gross character can be understood only globally and qualitatively. There is no inconsistency on the surface in the kind of self-reference exhibited by epistemological naturalism, nor is there likely to be any beneath the surface.

REFERENCES

Aristotle. (1941). *Physics.* In *The Basic Works of Aristotle*, ed. R. McKeon. New York: Random House.

Aristotle. (1941). *Nicomachean Ethics. The Basic Works of Aristotle.*

Bower, T. G. R. (1966). "The Visual World of Infants." *Scientific American* 215, no. 12: 80–92.

Feferman, S. (1998). *In the Light of Logic.* New York and Oxford: Oxford University Press.

Franklin, A. (1986). *The Neglect of Experiment.* Cambridge: Cambridge University Press.

Grünbaum, A. (1971). "The Meaning of Time." In E. Freeman and W. Sellars (eds.), *Basic Issues in the Philosophy of Time.* La Salle Ill.: Open Court.

Harper, W. This volume.

Kant, I. (1929). *Critique of Pure Reason.* Translated by N. Kemp Smith. New York: Humanities Press.

Kant, I. (1997). *Critique of Practical Reason.* Translated by M. J. Gregor. Cambridge: Cambridge University Press.

Malin, S. (2001). *Nature Loves to Hide.* Oxford: Oxford University Press.

McTaggart, J. M. E. (1927). *The Nature of Existence.* Cambridge: Cambridge University Press.

Merleau-Ponty, M. (1962). *The Primacy of Perception and Other Essays.* Evanston, Ill.: Northwestern University Press.

Monod, J. (1972). *Chance and Necessity.* New York: Random House.

Penrose, R. (1989). *The Emperor's New Mind.* Oxford: Oxford University Press.

Putnam, H. (1990). *Realism with a Human Face.* Cambridge, Mass.: Harvard University Press.

Quine, W. V. O. (1969). "Epistemology Naturalized." In *Ontological Relativity and Other Essays.* New York: Columbia University Press.

Rawls, J. (1993). *Political Liberalism.* New York: Columbia University Press.

Rottschaefer, W. (1998). *The Biology and Psychology of Moral Agency.* Cambridge: Cambridge University Press.

Schrödinger, E. (1967). *What is Life?* Cambridge: Cambridge University Press.

Seitelberger, F. (1984). "Neurobiological Aspects of Intelligence." In F. Wuketits (ed.), *Concepts and Approaches in Evolutionary Epistemology.* Dordrecht: Reidel, 123–48.

Shimony, A. (1993a,b). *Search for a Naturalistic World View.* Vols. 1 and 2. Cambridge: Cambridge University Press.

Shimony, A. (1997). "Comments." In R. Penrose, *The Large, the Small, and the Human Mind.* Cambridge: Cambridge University Press.

Shimony, A. (1998). "Implications of Transience for Spacetime Structure." Chapter 10 of S.A. Huggett, L. Mason, and P. Tod (eds.), *The Geometric Universe.* Oxford: Oxford University Press.

Smith, G. This volume.

Stapp, H. P. (1993). *Mind, Matter, and Quantum Mechanics.* Heidelberg: Springer Verlag.

Stein, H. (1967). "Newtonian Space-Time." *The Texas Quarterly* 10, no. 3 (autumn): 174–200.

Stein, H. (1970). "On the Notion of Field in Newton, Maxwell, and Beyond." In R. Steuwer (ed.), *Historical and Philosophical Perspectives of Science*. Minneapolis: University of Minnesota Press.

Tait, W. (2001). "Beyond the Axioms: The Question of Objectivity in Mathematics." *Philosophia Mathematica* 3, no. 9: 21–36.

Wang, H. (1996). *A Logical Journey*. Cambridge, Mass.: MIT Press.

Whitehead, A. N. (1929). *Process and Reality*. New York: Macmillan.

[12]

Maximizing and Satisficing Evidential Support

ISAAC LEVI

In the context of any given inquiry, a distinction is made between propositions taken for granted as being certainly true, their negations that count as certainly false, and propositions whose truth values are unsettled. Propositions taken for granted as being certainly true (false) are propositions whose truth values are settled beyond doubt—at least for the time being. The propositions whose truth-values are in doubt might be true and might be false—again, at least for the time being. When such serious possibilities are potential answers in the context of bona fide inquiries, they are conjectures or hypotheses.

The purpose of appraising hypotheses with respect to how well they are confirmed is to help determine whether new information ought to be added to the current state of full belief. Should the initial state of full belief, relative to which assessments of evidential support are being made, be modified by rejecting potential answers that are disconfirmed to a sufficiently high degree and adding the negations of these rejected answers to the full beliefs? Should the best-confirmed potential answer be adopted? Whether confirmation is satisficed or maximized, once information is added to the full beliefs, the new assumptions become so settled that they may be used as resources in subsequent deliberation. One way or the other, confirmation or evidential support is relevant to changing states of full belief—that is to say, to altering the distinction between full beliefs and serious possibilities or between certainty and hypothesis.

Not all serious possibilities are conjectures or hypotheses relevant as potential answers to the question under study. Real and living doubts do indeed concern serious possibilities but they must also be about propositions that are potential solutions to the problems of interest to the inquirer or to the community in which the inquirer participates.

315

But even if attention is focused on potential answers to a specific question, some potential answers or relevant conjectures carry more information of value relevant to answering the question than others do. If we ask who will win the Presidential Election in 2000, "A Democrat" and "Al Gore" are potential answers. But "Al Gore" carries more valuable information than "A Democrat" relative to the question as to who will win. This may not be true if the question is: "Which party will win the Presidency?" To respond that "Al Gore will win" is either considered irrelevant to the question or equal in informational value to the answer that a Democrat will win. This is so even though one potential answer is weaker than another. Weaker answers should never carry more valuable information than stronger ones; but there is no requirement that they carry strictly less. In choosing between rival theories that are not comparable with respect to logical strength, theories can be compared with respect to their explanatory power or, perhaps, their power in facilitating unification of some domain in order to assess value of information. Considerations such as these may have relatively poor correlation with truth or probability of truth. But answers having such properties offer an incentive to risk error in adding such theories to background assumptions.

I shall represent an inquirer X's system of commitments to full belief or certainty, insofar as this commitment is expressible in a suitably regimented language $\underline{L}$, by a set $\underline{K}$ of sentences in $\underline{L}$ closed under deductive consequence.

Let $\underline{B}$ be another deductively closed set included in $\underline{K}$. It represents a potential state of full belief that is maximally uninformative with respect to the problem identified by X for investigation.

Let $\underline{V}$ be a set of hypotheses exclusive and exhaustive relative to $\underline{B}$ and each consistent with $\underline{B}$. V is the *basic partition* for the problem identified by X when it represents the strongest consistent potential answers to X's question (insofar as X has been able to identify such answers) relative to the minimal or background corpus $\underline{B}$. Whether V is a basic partition depends in part on $\underline{B}$ and in part on the demands for information occasioned by X's problem as X understands it.

$\underline{K}$ is supposed to represent the result of adding information to $\underline{B}$ through experimentation and observation as well as collateral inferences. $\underline{K}$ may be incompatible with some elements of V. The hypotheses in V that survive rejection constitute the *ultimate partition* U relative to $\underline{K}$.[1] The elements of U as well as Boolean combinations of these consistent with $\underline{K}$ and not implied by it are the conjectures that remain open questions when $\underline{K}$ has replaced $\underline{B}$ as the corpus of full beliefs.

The inquirer X, I suppose, is interested in coming to a more definite resolution of his question than X has achieved beforehand. X seeks to add new information to X's full beliefs in $\underline{K}$ on the basis of the evidence available to X—that is to say, on the basis of $\underline{K}$. Relative to $\underline{K}$, X seeks to reject some nonempty subset of U and to add the disjunction of the survivors to $\underline{K}$ along with the logical consequences of $\underline{K}$ and this disjunction.

Thus, potential answers may be represented in one of three equivalent ways:

1. by the subsets of U whose elements are rejected;
2. by the disjunction d of those elements of U that survive rejection (or the assertion that one of the unrejected elements is true);
3. by the logical consequences of $\underline{K}$ and the disjunction d.

No matter which mode of representation is deployed, the intention is to represent a change from one state of full belief (the one represented by the deductively closed corpus $\underline{K}$) to another state of full belief (represented by another deductively closed corpus $\underline{K}^+_d$. d is the disjunction of some subset of U. $\underline{K}$ does not entail d unless d is the disjunction of all elements of U. In this setting d is not information received by observation, experiment, or testimony but is an answer selected from the various potential answers generated by U on the basis of the information in $\underline{K}$. The transition from $\underline{K}$ to $\underline{K}^+_d$ manifests a decision to adopt one potential solution to the problem rather than another on the basis of the information available in $\underline{K}$. Because implementing the decision adds new information to $\underline{K}$ not already entailed by it, the decision may also be considered to be an ampliative or inductive inference.

The conjectures in U might be predictions about who will win the next world series, how the troubles in Kosovo will develop in the next month, the estimation of the value of some parameter, or a choice of some theory to adopt as a correct systematization of a given domain. What I wish to say should be relevant to all such contexts.

The entire point of making appraisals of conjectures prior to expanding $\underline{K}$ (or refusing to do so) with respect to how well the evidence supports these conjectures is to form the basis for making ampliative or inductive inferences that eliminate some elements of such a U. Elimination of elements of U in virtue of information received via observation and experiment or from the testimony of others is not contemplated. Such information as well as information from other ampliative inferences is already embodied in $\underline{K}$. Given $\underline{K}$ and other features of the context to be mentioned later, the inquirer judges that he or she is justified in eliminating elements of U and becoming absolutely certain of the truth of the disjunction of the survivors even though that disjunction is not entailed by $\underline{K}$. The evaluations of the potential answers represented by assessments of support or confirmation are supposed to reflect how the inquirer should decide between competing strategies for inductive expansion. Assessments of evidential support for potential answers are supposed to inform the choices to be made.

According to *satisficing* measures of evidential support, Boolean combinations of elements of U are added to $\underline{K}$ via induction if and only if their evidential support reaches a certain threshold. The support afforded a conjecture by evidence should be high enough to entitle us to convert the conjecture to a full belief.

Maximizing measures of evidential support insist in choosing among potential answers by adding the potential answer carrying maximum support to the network of full beliefs together with the deductive consequences.

The notions of evidential and maximizing support involved cannot be the same without denying the legitimacy of ampliative reasoning.

To see this, consider that the set of sentences added to $\underline{K}$ must yield the set of deductive consequences of $\underline{K}$ and the disjunction of all and only unrejected elements of U. Satisficing support (relative to $\underline{K}$) must assign all and only these sentences support at or above the threshold.

If the threshold is set at the maximum, the disjunction of all and only unrejected elements of U must be the disjunction of all elements of U; for at the maximum threshold, the inquiring agent should suspend judgment by refusing to reject any elements of U. Hence, the maximum degree of satisficing support assigned to a disjunction of a subset of U is the support assigned the disjunction of all members of U or to any sentence entailed by $\underline{K}$. No other sentence can be assigned maximum support.

Consequently, equating maximizing support with satisficing support is equivalent to embracing the view that the sentence d should be the disjunction of all members of U. In effect, one should never expand ampliatively. Probability cannot be both a maximizing and a satisficing measure of evidential support unless antiinductivism is embraced.[2] In that case the notion of evidential support is trivialized.

Can probability be a measure of evidential support in one of these respects?

It may appear surprising that probability is ever considered to be an index of evidential support in a maximizing sense. As Popper pointed out a long time ago (1962, 54–55), if an inquirer were to favor a potential answer carrying maximum probability, he would restrict himself to answers carrying probability 1 on the evidence. Indeed, probability 1 here should be understood as absolute certainty so that the inquirer should come to believe only the logical consequences of what he already believed. That is to say, he should not make any inductive expansion at all. This is a result that antiinductivists who are also antiprobabilist like Popper can embrace with equanimity. The conception of evidential support in the maximizing sense is well defined but has no nontrivial application. If probability is a measure of evidential support or confirmation in the maximizing sense, that application cannot be the one to which much philosophical importance may be attached.

Quine and Ullian (1970) espouse a popular variant of antiinductivism closely resembling Popper's. These authors undermine inductivism by claiming (unlike Popper) that induction is but a species of hypothesis. Hypothesis, according to these authors, is "guesswork" or conjecturing. This understanding of hypothesis agrees well with Peirce's view of abduction. But, according to Peirce, induction is not guesswork. Given a set of conjectures or guesses and some sort of experimental design for testing the guesses, induction is the outcome of implementing the tests and converting some conjecture and its implications into certainties. In general, the outcomes of tests and the initial information do not entail the conclusion converted from conjecture to full belief. To the extent that this transition is inference it is ampliative in precisely the way induction is understood to be.

Unfortunately in both the 1870s and 1880s Peirce declared induction and hypothesis to be species of synthetic or ampliative reasoning. He illustrated this by citing as a form of hypothetic reasoning, inferences that he later recognized as qualitative inductions. In his later years (when Peirce began speaking of "abduction" rather than "hypothesis"), Peirce clearly recognized his mistake. (See Peirce 1901, 106.) And it is a mistake. Ampliative reasoning involves obtaining new beliefs not already implied by what is known. Hypothesis does not yield new beliefs unless conjectures are confused with beliefs. But to confuse conjecturing with belief in the sense of judging true is akin to confounding fantasy with belief.

There is no ampliative reasoning at all in hypothetic reasoning in the sense in which ampliative reasoning yields new belief. On this point, Popper is clearer than Quine and Ullian. Although Peirce is guilty of treating hypothetic reasoning as a species of ampliative reasoning, Peirce never held that induction is a species of hypothesis. The conclusion of hypothetical reasoning or abduction is a *conjecture*. The initial state of full belief K does not entail the conjecture and, in that sense, the reasoning is ampliative. But the reasoning is not ampliative in the sense in which ampliative reasoning yields new *belief*. By way of contrast, induction is ampliative in this latter sense.

Popper insists that it is never legitimate to acquire new belief in laws or theories ampliatively. In 1972, chap. 1, Quine and Ullian resolutely insist on claiming that induction is a species of hypothesis as do those who favor some form of inference to the best explanation. It seems to me that in spite of their use of a different rhetoric, the views of Quine and Ullian represent a form of antiinductivism quite close to Popper's.

Carnap was also an antiinductivist.[3] In contrast to his fellow antiinductivist, Popper, Carnap was interested in replacing what he acknowledged to be the customary view of inductive reasoning with assessments of probability on the evidence. Carnap muddied the waters by insisting that the sort of probabilistic reasoning he considered is inductive and that his two notions of confirmation are notions of inductive support or confirmation. The probabilistic reasoning, with which Carnap sought to replace ampliative, inductive reasoning, is not at all ampliative.[4] Consequently, his notions of degree of confirmation are not intended to be used as confirmation in either the maximizing or the satisficing sense.

Inductivists think that the distinction between full belief and doubt is often subject to modification. One way in which the distinction may be modified is by removing information and by adding information to a body of full belief when the information is obtained by the testimony of the senses when they are judged reliable and experts when they are considered reliable and honest. However, the distinction may also be changed by ampliative inference. In contrast to antiinductivist probabilists who can make no sense of inductive support in the maximizing sense, inductivists can. But inductivists cannot embrace the thesis that probability is evidential support in the maximizing sense.

In spite of the fairly broad consensus among inductivists and antiinductivists that probability is not a measure of inductive support in a maximizing sense, some well known authors do seem to endorse the opposite view.

Consider the reports of experimental results published by A. Tversky and D. Kahneman (1983) concerning the so called "conjunction fallacy." The probability of the "conjunction" A&B of two events is no greater than the probability of either one of the conjunctions. If by probability, one means relative frequency, this means that the relative frequency of A's among the C's should be no less than the relative frequency of things that are both A's and B's among the C's. Yet Tversky and Kahneman report that experimental subjects violate the requirement when asked to rank hypotheses with respect to probability after having been given some background information. Here is a well-known scenario that Tversky and Kahneman report having offered to experimental subjects.

> Linda is 31 years old, single, outspoken and very bright. She majored in philosophy. As a student, she was deeply concerned with issues of discrimination and social justice, and also participated in anti-nuclear demonstrations.

The experimental psychologists then invite the experimental subjects to rank a series of propositions about Linda with respect to probability including the following.

Linda is active in the feminist movement. (f)
Linda is a bank teller. (b)
Linda is both a bank teller and active in the feminist movement. (b&f)

Tversky and Kahneman report that most experimental subjects regard b&f as more probable than b.

In this exercise and in problems concerning medical diagnosis where a similar result occurs, the experimental subject is invited to rank the results with respect to probability. But there is no reason here to suppose that the experimental subjects understand "probability" in the sense of the calculus of probabilities.

Indeed, there is substantial reason to doubt that they do so. The setting of the problem invites them to ascertain which of the rival hypotheses is better supported on the basis of the "evidence" supplied in the brief description of Linda. It is a reasonable conjecture that they interpret the question to be: Which of the sentences should be the strongest added to the background $\underline{K}$ via induction? That is to say, the experimental subjects might well interpret the request to rank the propositions with respect to probability as asking for a ranking with respect to evidential support in the maximizing sense.

It is arguable, however, that probability cannot be inductive support in the maximizing sense *precisely because probability ranks weaker answers at least as well supported as stronger ones.* Experimental subjects in the Linda and cognate cases may have some presystematic understanding of this. They may not be suffering from cognitive illusions, making a mistake, or committing a fallacy except the linguistic mistake (if it is one) of not using "probability" in the sense of the calculus of probabilities.

My aim is not to defend experimental subjects as free of all fallacies. Tversky and Kahneman appear to have a convincing case to make that experimental subjects commit the conjunction fallacy in some tasks. This seems so in estimating frequencies and assessing fair betting rates.

It may be that experimental subjects use the same "heuristics" for making probability judgments in estimating frequencies and assessing fair betting rates as they do when making judgments of evidential support in the maximizing sense in cases of medical diagnosis and in problems like the question of Linda. As Kahneman and Tversky have themselves observed, heuristics often give correct results. They are, after all, rules of thumb for rough-and-ready answers to questions that one uses in lieu of exact rules. That they will go wrong in some cases is to be expected. It is important, however, to identify the conditions where they go wrong and where they do not.

Tversky and Kahneman have disavowed any prescriptive or evaluative intent when they charge the presence of a fallacy. But it seems quite clear that they are claiming that experimental subjects are making mistakes or are subject to cognitive illusions in the Linda case. They and their admirers have urged that steps should be taken to develop methods for preventing such illusions from impacting on more serious contexts of judgment. Kahneman and Tversky urge that attention to these illusions and how they may be avoided be paid in the training of medical researchers. If experimental subjects are *not* making mistakes in the medical diagnosis cases and are being urged to think that they are, listening to Kahneman and Tversky may do more harm than good.

In any case, Kahneman and Tversky seem to think that the right approach to their questions should be to treat probability as a measure of evidential support in the maximizing sense. On that assumption, the experimental subjects are committing a fallacy in the Linda problem. I have argued that probability cannot be a measure of evidential support in the maximizing sense precisely because in this kind of problem, the hypothesis that Linda is a feminist bank teller is better supported in the maximizing sense than Linda is a bank teller.

A favorite (though not the sole) measure of inductive support among Bayesians whether they are antiinductivist as Carnap was or not is the difference between a posterior probability relative to the current evidence $\underline{K}$ and some prior measure of probability relative to a standard state of ignorance. Let us write that as $P_K(x) - P_B(x) = P_B(x/B) - P_B(x)$.

Utilizing such measures of evidential support, we can compare $P_K(b\&f) - P_B(b\&f)$ with $P_K(b) - P_B(b)$. If P_K is the probability judgment based on the story about Linda, we might well conclude that $P_K(b\&f) > P_K(b\&{\sim}f)$ whereas $P_B(b\&f)$ might equal $P_B(b\&{\sim}f)$. Then the evidential support for b&f would be greater than the evidential support for b. An experimental subject who ranked the propositions in terms of evidential support so conceived would rank Linda is a feminist bank teller over the claim that she is a bank teller. Bayesians who favor such measures of support seem to be committed to what Tversky and Kahneman are calling a fallacy.

I do not mean to suggest that all Bayesians officially regard such measures as measures of support in the maximizing sense. How could the antiinductivist Carnap do so? To be sure Carnap is willing to call the difference between a posterior and a prior a measure of confirmation. He is also ready to think that his measures of logical probability are confirmation measures in another sense. But he leaves unclear why it should matter to anyone interested in ampliative reasoning whether a hypothesis is highly confirmed or highly disconfirmed.

Good Bayesians could also be inductivists. They could then have some use for a measure of confirmation. Still differences between posteriors and priors cannot serve as measures of maximizing support. They are not. A good measure of inductive or evidential support ought to meet the following condition of adequacy.

If $\underline{K}$ is the total evidence available and h (h' is entailed by $\underline{K}$, then the evidential support for h afforded by $\underline{K}$ ought to be equal to the evidential support afforded by $\underline{K}$ for h'.

The difference between a posterior and a prior (and other measures like the ratio of the posterior and the prior, the difference between log of the posterior and the log of the prior, and the like) fail to satisfy this condition. Of course, Bayesians who are also antiinductivist would reject my condition of adequacy. The requirement makes sense only if the inductive conclusion h is to be added to $\underline{K}$ along with the logical consequences. If $\underline{K}$ entails that h and h' share the same truth value, then adding h' to $\underline{K}$ along with the logical consequences yields the same state of full belief. If ampliative reasoning is proscribed and with it evidential support in the maximizing sense, the Bayesian notion of inductive support whatever its import might be has no need to meet the condition of adequacy. Inductivists must, however, continue to embrace it.

Thus Popper and Miller (1983) claimed that universal generalizations can be factored into ampliative and explicative factors and that the ampliative part can never be inductively supported positively by data. Jeffrey (1984) pointed out that factoring into ampliative and explicative factors is not unique. There are factorings where the Popper-Miller claim fails. But the ampliative component according to any one factorization is equivalent given $\underline{K}$ to any other such factorization. According to the condition of adequacy, the degrees of inductive support ought to be the same (Levi 1984b). Popper and Miller were right to claim that the difference between a posterior and a prior could not be a good measure of inductive support, but for the wrong reason.

It is not too difficult to find a modification of $P_K(x) - P_B(x)$ that satisfies the condition of adequacy (Levi 1984a, chap. 5). Replace $P_B(x)$ by a probability measure $M_K(x)$ relative to $\underline{K}$ that need not be the same as $P_K(x)$. The reformed measure of evidential support in the maximizing sense then looks like this:

(4) $P_K(x) - M_K(x)$

The proposal works well enough formally but it lacks motivation as it stands. Fortunately, there is a motivation ready to hand. According to Bayesian decision theory, expected utility is to be maximized in choosing among the available options in practical deliberation. I have long favored the view that seeking an answer to a specific question relative to the available evidence $\underline{K}$ ought to be rationalized as a decision problem. To the extent that inquirers seek to maximize expected utility relative to cognitive goals, it is entertainable that maximizing indices of evidential support may be equated with evaluations of expected utility. Keeping in mind the tradition of Popper, Carnap, and Bar Hillel according to which informational

value varies inversely with probability, we may consider interpreting the
M–function as that probability measure such that $1 - M_K(x)$ represents the
informational value adopting potential answer x adds to $\underline{K}$. The proposal
deviates from the practice of Popper, Carnap, and Bar Hillel by insisting
that the probability measure M_K need not be equal to the credal probability measure P_K. With this understood, (1) can be derived as a special case
of a more general formula I proposed some time ago as an index of
expected epistemic utility.

(1) $P_K(x) - qM_K(x)$ where $0 < q \leq 1$.

(1) is the special case where $q = 1$.

The index of expected epistemic utility (2) is a quantity to be maximized in choosing between different inductive expansion strategies
according to the widely endorsed principle urging maximization of
expected utility. If this is right, any function of the form (2) is a candidate
for consideration as an index of evidential support in the maximizing sense
provided that P_K and M_K are given the appropriate interpretations as
expectation determining and informational value determining probabilities
relative to $\underline{K}$.

In order to understand the role of the parameter q some of the details
of the utility function used to obtain (2) need to be fleshed out.

The key idea is similar to the voluntarist view of how we ought to
change our points of view expressed in William James's *Will to Believe* by
the slogan "Seek Truth, Shun Error!" In a somewhat more careful
moment, James modified the injunction to seek Truth by the remark that
it is possibility of Truth that is to be sought. And in the context of his essay,
it is clear he intended that the possibility be important. I suggest that
"Truth" be replaced by "Important Information" as determined by an
informational value determining probability. I am inclined to think that
James would agree. But like Peirce, I would object to James's willingness
to give total priority to "Truth" or "Important Information" when it
comes to matters of religious faith. Still I do think that the pragmatists
were on the right path when they thought of specific inquiries as goal
directed so that arguments rationalizing the solutions we adopt to problems we face ought to be practical or decision theoretic arguments. The
difference between scientific deliberation and other forms of deliberation
lies not in a difference between practical and theoretical rationality but in
the character of the ends we seek.

James's idea was not merely that important information and avoidance of error are desiderata in fixing belief but that because the values
are multidimensional, there should be some trade off between them—
some method of weighing the rival desiderata. In the case of religious

belief, he thought the weight attached to avoidance of error should be set at 0.

We may say that the utility or value of adding x to $\underline{K}$ when it is true is equal to the following:

(2) $\alpha T(x,t/f) + (1 - a)[1 - M_K(x)]$

In (3) $0 \leq a \leq 1$. When x is true, $T(x,t) = 1$. When x is false $T(x,f) = 0$. This leads to an expected utility function given below:

(3) $\alpha P_K(x) + (1 - \alpha)[1 - M_K(x)]$

In contrast to James, I suggest that no case of adding false information is ever strictly preferred to avoiding error in adding information. This entails that $a \geq 0.5$.

Let $q = (1 - \alpha)/\alpha$. Divide (4) by a and subtract q. The result is (1). This is a positive affine transformation of (4) and hence is suitable for maximization if (4) is.

Thus, we have an interpretation of the parameter q. The higher q the greater the relative importance the inquirer attaches to obtaining valuable information rather than minimizing risk of error.

If x is a disjunction of elements of a finite ultimate partition U, $P_K(x)$ $- qM_K(x)$ is itself a sum of components of the form $P_K(x_j) - qM_K(x_j)$. To be a maximum, all of these components must be nonnegative. The weakest of the potential answers for which the inductive support is a maximum is precisely the disjunction of all those x_j's for which the components are nonnegative. If this weakest maximum is adopted, an element of U is rejected if and only if $P_K(x_j) < qM_K(x_j)$.

If this rejection rule is deployed and $\underline{K}$ is expanded by rejecting elements of U, both the P_K and M_K – functions may be updated and the procedure reiterated. Eventually a "fixed point" will be reached at which the conclusion is "stable."

At this point, the importance of q—the index of boldness—becomes apparent. If q equals 1, then with finite U, the stable conclusion rejects all elements of U except those for which P_K/M_K is a maximum.

Thus, if q = 1, someone concerned to predict the outcome of tossing a fair coin a thousand times (where tosses are probabilistically independent) should conclude that the coin will land heads exactly five hundred times— provided all elements of U carry equal M_K-value.

I do not think most of us are willing to be so rash. For values of q less than 1, we obtain more plausible estimates. The estimate would be that the coin would land heads approximately five hundred times with the approximation depending on the value of q.

Thus, we can give an account of why it is to be expected that the coin would land heads approximately five hundred times but not why it is to be expected that the coin would land heads exactly five hundred times. We can distinguish between explanations of why it is not surprising that the coin would land heads exactly five hundred times and why it is surprising that the coin would land heads every time. The former is not surprising because it would not be rejected no matter how bold one is; but the latter would be rejected by all but the virtually skeptical where q is near 0. Yet, we may show how it is possible that the coin landed heads every time.

All of this and more can be achieved by appealing to a notion of inductive support in the maximizing sense along the lines that I have suggested. Of course, the model I have just sketched exploits quantitative assessments of belief probability and informational value far too determinate to be realistically applied in many contexts. But I believe that a more general account of how to make decisions with indeterminate probabilities and utilities can be deployed to address this point.

On the view I have been developing, probability is not a measure of inductive support in the maximizing sense. Expected epistemic utility is. Consequently, inductive support in the maximizing sense is partially but not entirely determined by assessments of probability.

Let us now turn to the concept of evidential support in the satisficing sense.

That probability is a measure of satisficing support is more widely endorsed than the idea that it is a measure of maximizing support. Henry Kyburg built an account of induction on this idea. C. G. Hempel's approach to inductive acceptance was similarly constructed. Ernest Adams's account of indicative conditionals is based on the idea. And it has become the canonical account of indicative conditionals among those who think (wrongly, I believe) that we should embrace some variant of the closest worlds semantics for subjunctive conditionals.

Probability as satisficing support implies that in adding new information to the corpus of absolute certainty $\underline{K}$, it is necessary and sufficient that the new information carry probability relative to $\underline{K}$ at least as great as some threshold value.

All we need to do is to remind ourselves of the infamous lottery paradox to appreciate what is wrong here. If the satisficing measure of evidential support is to serve its purpose, the set of sentences added to $\underline{K}$ must, if $\underline{K}$ is consistent, constitute a deductively closed and consistent set of absolute certainties. Otherwise, we cannot say that the new set of sentences represents a new state of full belief. If we consider a finite but large number of tickets, then for some probability $1 - k$ with k small, X is obliged to conclude that none of the tickets will win in contradiction to the claim that at least one will. If the lottery has a countable infinity of tickets, each ticket carries 0 probability of winning. This is no contradiction as long as countable additivity is not assumed. But we may still conclude that none of the tickets will win. And if one is engaged in estimating the value of some real valued parameter, each point estimate carries 0 probability and so must be rejected.

With the aid of the measure of maximizing support already proposed, it is relatively easy to sketch the shape of measures of satisficing support that avoid the difficulties in a fairly plausible manner (Levi 1984a, chap.14).

The measures of maximizing support I have proposed are *boldness dependent* in the sense that they contain a boldness parameter. Of course, to apply such a measure, one has to fix on a value of q. But it is not necessary always to state such a value when evaluating hypotheses. Given $\underline{K}$, U, P_K, and M_K, it is always possible to determine for each x in U the threshold value $q(x)$ such that x is rejected (or stably rejected) for values of q greater than or equal to $q(x)$ unless there is no such value in which case $q(x) = 1$.

The degree $d(x)$ to which x is surprising is equal to $1 - q(x)$.

If h is a disjunction of elements of U, the degree $d(h)$ to which h is surprising is the minimum degree of surprise attached to a disjunct in h.

$q(x)$ itself is a measure of degree of possibility carrying formal properties similar to those favored by Lotfi Zadeh and other aficionados of fuzzy set theory.

The degree of belief or evidential support $b(h)$ in the satisficing sense that h is $d(\sim h)$. $b(x \& y) = \min[b(x), b(y)]$.

Of course, one can also introduce a measure $1 - b(x)$.

All of these measures are determined once one of them is. Since to my knowledge G. L. S. Shackle (1949, 1961) is the first author to explore measures with these formal properties, I call such measures "Shackle measures." Critics of the use of such measures have often complained that neither Shackle nor any of the other reinventors of these measures have given a clear idea of their intended application.

Some progress may be made in answering this question once it is recognized that fixing on a value for the index of boldness q is equivalent to fixing on a threshold value such that h is added to $\underline{K}$ if and only if $b(h)$ is at least as great as the threshold. The set of potential answers added in this way to $\underline{K}$ form a deductively closed set.

In deliberation aimed at terminating inquiry by adding some erstwhile conjecture to the settled assumptions to be used, as Dewey would put it, as a "resource" in subsequent deliberation, one may wish to take a look at what sorts of conclusions would be reached at various levels of boldness.

This sort of appraisal can be reformulated as an assessment as evaluation of *b*–values relative to $\underline{K}$. This, I suggest, is evaluation of potential answers to a question with respect to evidential support in the satisficng sense (Levi 1967, 1980).

Thus, in the case of the million ticket fair lottery, the hypothesis that ticket j will be drawn carries probability 10^{-6} and informational value $1 - 10^{-6}$. No matter what value of q is adopted, no element of *U* will be rejected. So for each j, q(ticket j will be drawn) = 1. d(ticket j will be drawn) = 0. b(ticket j will not be drawn) = 0. There is no lottery paradox. That is because we have abandoned the idea that one should "accept" hypothesis if and only if its probability is above a given threshold. Evidential support in the satisficing sense is not a probability. It is a Shackle measure.

The proposal I have made is distinguished from the others I have mentioned in that it derives evidential support in the satisficing sense from a probability based notion of inductive support in the maximizing sense without reducing evidential support in the satisficing sense to probability. In addition, Shackle-like measures play an important role in some recent reconstructions of default reasoning (Gärdenfors and Makinson, 1993). Because students of default reasoning have been at loggerheads over the relevance of probabilistic and statistical thinking to default and nonmonotonic reasoning, the introduction of models that provide a clear explication of the relation between probability and satisficing measures of inductive support may prove of value.

Thus, probability in the sense of subjective or credal probability is not a measure of evidential or inductive support in either the maximizing or satisficing sense. It does not help if one derives the subjective probabilities from logical or statistical probabilities.

Applications of the calculus of probabilities are relevant to the assessment of expectations and in statistical explanation and prediction. Probability measures can be used in assessing informational value. These and cognate applications suffice to make probability an important concept.

The twentieth century has witnessed an obsession with the idea of replacing notions of ampliative inference with accounts of probabilistic inference and conceptions of support or confirmation that are not motivated by concern with ampliative reasoning at all. The net effect has been to promote a form of antiinductivism as extreme as the antiinductivism of Popper while claiming (in contrast to Popper) to be characterizing inductive reasoning.

I do not know how to demonstrate the untenability of any form of antiinductivism whether it is Popperian, Quinean, or probabilistic. My opposition to antiinductivism is predicated on certain assumptions that ought to be but unfortunately are not uncontroversial.

Peirce and Dewey were right to maintain that inquiry begins with an effort to remove doubt and terminates with the removal of doubt. As Peirce insisted, conjecturing plays an important role in this endeavor. Popper and Quine were right to emphasize the importance of hypothesizing and subjecting hypotheses to test. But the results of testing will lead to the eliminating of conjectures by ampliative or inductive reasoning from the data obtained as Peirce so clearly understood. Conceptions of evidential support or confirmation have no clear relevance in inquiry understood as concerned to remove doubt unless they are tied to the evaluation of potential "inductive leaps" to potential answers to serious questions. I have pointed out that such evaluation can be made by appealing to two distinct kinds of assessments of evidential support: a kind that one maximizes and a kind that one satisfices. Probability, so I claim, is relevant to both kinds of assessment; but neither can be equated with probability nor be explicated exclusively in terms of probabilistic concepts.

Antiinductivists seek to identify notions of corroboration or of evidential support consonant with their visions. A well-corroborated conjecture is, according to Popperians, one that has survived severe testing. But why should we care whether a conjecture is severely tested if such testing can never support conclusions reached by ampliative reasoning?

Probabilists can offer a somewhat better answer for collecting data. Collection of new data can improve the expectations of decision-makers under certain conditions. These conditions include absolute certainty on the part of the experimenter that data to be collected will be error proof and that probability judgments are maximally determinate. Unfortunately for the probabilists these conditions are rarely satisfied in real life. Probabilists have sought to avoid being absolutely certain of anything even of death and taxes. So how can they be certain that the procedure for collecting data is error proof? In addition, reasonable inquirers recognize that probability judgment should often be indeterminate. So the probabilist answer is not much better than the Popperian one.

Quineans, like Popperians, make much of conjecturing and worry about the "virtues" of good hypotheses. As noted before, they differ from Popperians because they think of induction as a species of hypothetic inference—inference that places great emphasis on conservatism and modesty. Conservatism urges us to change current beliefs as little as possible and modesty favors logically weaker over logically stronger conjectures among potential answers consistent with old beliefs. Before I complained that if the conclusion of an induction is but a conjecture no induction has taken place. Nothing has been removed from the serious possibilities. If we do think of induction along Quinean lines as relieving doubt by removing serious possibilities from that status, then conservatism and modesty—

counter to Quine's intentions—favor complete suspension of judgment. That is the most conservative and modest move one can make. Simplicity, generality, and refutability may allow for more boldness. What is lacking is any sense that risk of importing error matters. Ignoring risk of error is acceptable in the case of hypothesis or abduction; for the conclusions of such reasoning are not the acquisition of new belief. Conjecturing does not *pace* Quine and Ullian change beliefs. So there is no error to risk.

But induction involves change in belief. Quine and Ullian admit this much. They also claim, however, that induction is a species of hypothesis. They cannot have it both ways. If induction is a species of hypothesis, there is no change of belief and no risk of error can be incurred. Probability plays no greater role in evaluating conclusions of such inductions than it does for Popperian conjecturing.

If induction does involve a change of belief and incurs risk of error, induction cannot be a species of conjecturing. Perhaps all that Quine and Ullian mean to claim is that inductions are evaluated taking into account the same virtues as are taken into account in evaluating conjectures. If that is so, avoidance of error cannot matter in induction any more than it does in abduction or hypothesis. Peirce complained that the methods of tenacity, authority, and a priori reason fail to take into account the question of truth. All that matters is relief from doubt. He would object to the Quinean view according to this second inductivist reading and so do I.

Ultimately the trouble with the Popperian, Quinean, and probabilist points of view is that if doubt is ineradicable, there does not seem to be much point to inquiry. And if there is no point to inquiry, there is no point to evaluating conjectures with respect to one's favorite measure of inductive support or corroboration. I cannot claim to have demonstrated to the resolutely skeptical that there is any point to inquiry. But if inquiry is taken seriously, the two main types of inductive support should have structural features along the lines I have indicated.

In recent years a new form of antiinductivism has emerged that has relevance to our discussion of evidential support.

Inductive reasoning is a species of what is now called nonmonotonic inference. Consider some initial corpus $\underline{B}$ and let $\underline{B}^i$ be the result of inductive expansion of $\underline{B}$ (relative to V, M_B. P_B, and q). Let $\underline{B}$ be consistent with e. If one compares $\underline{B}^i$ with $[\underline{B}^+_e]^i$, the latter will not, in general, be a superset of the former. Consider a case where e is inconsistent with $\underline{B}^i$. Many authors assume that in such cases, $[\underline{B}^+_e]^i$ should be the same as $[\underline{B}^i]^*_e$ where the * represents the revision transformation for theories proposed by Alchourrón, Gärdenfors, and Makinson (AGM) in 1985. AGM revision is a nonmonotonic transformation. Its nonmonotonicity arises, however, from the fact that it transforms an initial consistent theory or belief state into another such theory as long as e is logically consistent. It allows for consistency preserving belief contravening transformations. Nonmonotonicity does not in this case involve any ampliative reasoning. To reduce ampliative or default reasoning to AGM revision is, in effect, to eliminate ampliative reasoning.

I do not have the opportunity here to demonstrate what I shall now dogmatically and flatly assert. The reduction that is claimed in a paper by Gärdenfors and Makinson (1993) and is endorsed tacitly in work by Lehmann and Magidor (1992) rests on two assumptions (Levi 1996):

1 Degrees of belief or evidential support in the satisficing sense also measure degrees of entrenchment or vulnerability to being given up.

2 When e is consistent with $\underline{B}^i$, $[\underline{B}^i]^+_e = [\underline{B}^+_e]^i$. (Permutability)

Permutability obtains if and only if $\underline{B}^i$ is derived from an inductive rejection rule of the sort I describe on p. 408 for the case where q = 1 and the rule is reiterated to a fixed point. Keep in mind that each application of the inductive rejection rule may be rationalized as maximizing expected epistemic utility relative to the question under consideration.

Adopting q = 1 is questionable. If it is known that a coin has been tossed a thousand times, it seems unreasonable to conclude that it lands heads exactly half of the time.

Assumption (1) is just as questionable. A strong case can be made that measures of entrenchment should satisfy the formal requirements imposed on Shackle measures just as degrees of evidential support in the satisficing sense should. But just as M_K should not be confused with P_K, entrenchment should not be confused with evidential support in the satisficing sense.

Assumption (1) is a vestige of the tendency to conflate certainty with incorrigibility or invulnerability to being given up. This equation, I believe, also fuels Popperian and probabilistic forms of antiinductivism. Many probabilists have, for example, wondered how one could legitimately withdraw a judgment of certainty once it is made.

As Howard Stein has emphasized, one of the respects in which Isaac Newton counts as a better philosopher than he has been credited with being has been his sensitivity to the distinction between certainty and conjecture. The tendency among the epigones of Descartes to deemphasize that distinction seems to be, in part, a response to a conviction that what is certain is immune to revision. Inductivists should beware of such prejudices in any guise in which they appear.

NOTES

1. I have used "Ultimate Partition" since Levi 1967 even though calling the partition "ultimate" is misleadingly hyperbolic. The elements of U represent the strongest consistent potential answers meeting the demands for information imposed by the question under investigation and consistent with K.

2. Antiinductivists may, of course, propose their own intended applications for what they choose to call measures of confirmation. Calling such assessments measures of inductive support or confirmation does not convert such antiinductivists into inductivists.

3. See Carnap 1960, where, like Popper, Carnap claims to solve Hume's problem by disallowing inductive inferences.

4. Carnap (1962, introduction to the second edition) had two understandings of degrees of confirmation: (a) as the degrees of subjective probability justified by the available evidence and (b) as the degree to which evidence e added to state of full belief K increased (decreased) the degree of subjective probability of a target proposition. Given any hypothesis h consistent with K, h is both possibly true and possibly false no matter what probability is assigned to h relative to K. Assigning a degree of probability to h relative to K cannot represent an inductive inference in the sense of an ampliative inference. This is so even if h is assigned probability 0 or 1. As long as h is consistent with K, Carnap's view would appear to be that the inquirer for whom K is the total evidence should not be absolutely certain that h is true or absolutely certain that h is false. Carnap's misleadingly labeled "inductive logic" licenses no alteration in the inquirer's distinction between serious possibility and impossibility. *A fortiori* no ampliative inference is warranted. Changes in probability judgment can be licensed by changes in state of full belief. But if the change in state of full belief is the product of observation or expert testimony, ampliative reasoning is not involved in the change. I am not suggesting that study of changes in probability judgment due to changes in states of full belief is uninteresting and has no application. I am objecting to the antiinductivist suggestion that the study of inductive or evidential confirmation in a sense relevant to inductive inference can be simulated or replaced by confirmation in one of Carnap's two senses.

REFERENCES

Alchourrón, C., P. Gärdenfors and D. Makinson. (1985). "On the Logic of Theory Change: Partial Meet Functions for Contraction and Revision." *Journal of Symbolic Logic* 50: 510–30.

Carnap, R. (1960). "The Aim of Inductive Logic." In *Logic, Methodology and Philosophy of Science,* edited by E. Nagel, P. Suppes and A. Tarski. Stanford University Press, 302–18.

Carnap, R. (1962). *Logical Foundations of Probability.* 2d ed. Chicago: University of Chicago Press.

Gärdenfors, P. and D. Makinson. (1993). "Nonmonotonic Inference Based on Expectations." *Artificial Intelligence* 65: 197–246.

Jeffrey, R. C. (1984). "The Impossibility of Inductive Probability." *Nature* 310: 433.

Lehmann, D. and M. Magidor. (1992). "What Does a Conditional Knowledge Base Entail?" *Artificial Intelligence* 55: 1–60.

Levi, I. (1967). *Gambling with Truth*. New York: Knopf. Reprinted by MIT Press in 1973.

———. (1980). "Potential Surprise: Its role in Inference and Decision Making." In *Applications of Inductive Logic*, ed. by L. J. Cohen and M. Hesse. Oxford: Clarendon and Oxford, 1–27. Reprinted with additions in Levi 1984a, chap.14.

———. (1984a). *Decisions and Revisions*. Cambridge: Cambridge University Press.

———. (1984b). "The Impossibility of Inductive Probability." *Nature* 310: 433.

———. (1996). *For the Sake of the Argument*. Cambridge: Cambridge University Press.

Peirce, C. S. (1901). "On the Logic of Drawing History from Ancient Documents, Especially from Testimonies." First half of the manuscript is published in *The Essential Peirce*, vol. 2, ed. by N. Houser et al., Bloomington: Indiana University Press.

Popper, K. R. (1962). *Conjectures and Refutations*. New York: Basic Books.

Popper, K. R. and D. Miller. (1983). "A Proof of the Impossibility of Inductive Probability." *Nature* 302: 687–88.

Quine, W. V. and J. Ullian. (1970). *The Web of Belief*. New York: Random House.

Shackle, G. L. S. (1949). *Expectations in Economics*. 2d ed. Cambridge: Cambridge University Press. 1952.

———. (1961). *Decision, Order, and Time in Human Affairs*. 2d ed. Cambridge: Cambridge University Press. 1969.

Tversky, A. and D. Kahneman. (1983). "Extensional versus Intuitive Reasoning: The Conjunction Fallacy in Probability Judgment." *Psychological Review* 90: 293–315.

[13]

Maximality vs. Extendability: Reflections on Structuralism and Set Theory

GEOFFREY HELLMAN

1. Introduction

In a recent paper, while discussing the role of the notion of *analyticity* in Carnap's thought, Howard Stein wrote:

> The primitive view—surely that of Kant—was that whatever is trivial is obvious. We know that this is wrong; and I would put it that the nature of mathematical knowledge appears more deeply mysterious today than it ever did in earlier centuries—that one of the *advances* we have made in philosophy has been to come to an understanding of just how deeply puzzling the epistemology of mathematics really is. (1993, 283)

Although our principal concern here is not with analyticity but rather with competing visions of the very subject matter of mathematics, the present essay can certainly be read as an extended illustration of this poignant remark.[1]

There is a recurring emphasis in Stein's writing on the importance of theoretical insights of our leading intellectual forebears in science and mathematics for the healthy practice of philosophy. In this spirit, I can think of no better way to introduce the subject of this essay than by quoting from the concluding remarks of Zermelo's great, yet underappreciated, 1930 paper, *"Über Grenzzahlen und Mengenbereiche."* Having formulated

---

I am grateful to audiences at the Steinfest, University of Chicago, May 21–23, 1999, and at the Philosophy of Mathematics Conference at the University of California, Santa Barbara, Feb. 4–6, 2000, and especially to Stewart Shapiro and Tony Anderson, for helpful comments on earlier drafts of this paper.

essentially what we now would call the Zermelo-Fraenkel axioms of set theory (in second-order form), and having established some of the most fundamental facts about structures satisfying these axioms, including their ordinal characteristic or limit numbers as strongly inaccessible cardinals, Zermelo wrote:

> The 'ultrafinite antinomies of set theory', which the scientific reactionaries and anti-mathematicians eagerly and delightedly call on in their campaign against set theory, these specious 'contradictions' arise solely from a confusion between the non-categorical axioms of set theory and the various particular models of them: What in one model appears as an 'ultrafinite un- or super-set' is in the next higher domain a perfectly good 'set' with a cardinal number and order type of its own, which serves as the foundation stone for the construction of the new domain. The boundless series of Cantor's ordinal numbers gives rise to an equally boundless series of essentially different models of set theory, in each of which the whole classical theory can be expressed. The two polar opposite tendencies of the thinking mind, the idea of creative progress (*"schöpferischen Fortschrittes"*) and that of all-embracing completeness (*"zusammenfassended Abschlusses"*), which lie at the root of Kant's 'antinomies', find their symbolic expression and resolution in the concept of the wellordered transfinite number-series, whose unrestricted progress comes to no real conclusion, but only to relative stopping-points, the 'boundary numbers' that divide the lower from the higher models. And so the 'antinomies' of set theory, properly understood, lead not to a restriction and mutilation, but rather to a further, as yet unsurveyable, unfolding and enrichment of mathematical science. (Zermelo 1930, 47, trans. J. Burgess and the author.)

While the formal accomplishments of this paper of Zermelo's are well known to set theorists (it is considered the locus of the discovery of large cardinals), little attention has been given to these closing remarks which sum up Zermelo's own assessment of the significance of his results. Indeed, far from being merely a rhetorical flourish, Zermelo's summation has here, in its reference to "two polar opposite tendencies of the human mind, creative progress and all-embracing completeness," identified one of the most fundamental problems in the foundations and philosophy of mathematics, going to the heart of our conception of mathematical reality and truth. As is evident from the passage, Zermelo was confident that he had in fact hit upon the resolution of that problem, but, as will emerge below, further steps are necessary before this can be maintained. Indeed, from our current vantage point—which is, of course, unlike Zermelo's, a post-Gödelian one—a proper resolution appears impossible without transcending the extensional logical frameworks standardly appealed to in formalizing set theory, and therewith most of mathematics.

To anticipate, the central problem reflecting "the two polar opposite tendencies of the thinking mind," in a nutshell, is this: Over what totality should we take our set variables to range (in other words, what is the range of unrestricted quantifiers, $\exists x, \forall x$ of set theory)? We know that it cannot be taken to be a set, on pain of contradiction, but even if we treat it as a collection of another type, we run up against the strongly held intuition that it can be conceived to be properly extended, that is, to be treated as part of a more extensive domain of mathematical objects (of whatever higher type has been reached). True, we can consistently avoid ever speaking of "all sets" as forming a collection of any type (the first-order solution, also the putative solution via so-called "plural quantifiers"), but the problem persists then in another guise: although we don't *have to* recognize such an object, still we *can*, and it is this very possibility of embracing a new totality that seems intimately bound up with our mathematical capacities and indeed the very "collecting" capacities that lead naturally to set theory in the first place. On the other hand, we also do seem to have available in our language genuinely unrestricted quantifiers, implying a capacity to speak unrestrictedly of "all sets," indeed "all objects (period)." Such quantifiers can be used to speak of "all set-like objects," or "all collection-like objects," and this can be thought of as a unique, fixed, determinate totality. Indeed, this "all embracing tendency," although it appears to limit the "creative progress" of conceiving proper extensions of *any* given domain, may even be invoked to shore up the full *truth-determinateness* of set theory, the "holy grail" of the set-theoretic realist. A rather elaborate argument making precisely this move in fairly rigorous detail has, in fact, recently been presented by McGee, and, curiously, it rests on results on models of set theory formally very close to those of Zermelo's 1930 paper. But the philosophical vision McGee's "categoricity theorem" (as he calls it) supports is, I believe, quite deeply at odds with Zermelo's, and yet this opposition seems to be the very one that Zermelo took himself to have resolved. It will be our task below to attempt to disentangle this and to propose a resolution of our own in the spirit of Zermelo's.

The plan of the rest of this paper is as follows. In the next section, we shall call attention to the complex dual role of set theory, as a unifying "foundational" framework for mathematics, on the one hand, and as a part of mathematics in its own right, on the other, ripe for a structuralist interpretation along with other mathematical theories. This dual role is correlated with the problem of "one vs. many" models or worlds, in turn correlated with the polar opposite tendencies highlighted by Zermelo. In section 3, we shall review the principal results of Zermelo (1930), bringing out his *Extendability Principle* and its central role in his "resolution" of the problem of set-theoretic antinomies. Next, in section 4, we shall summarize McGee's main result on behalf of full set-theoretic determinateness and examine its relation to Zermelo (1930). Then, in section 5, we shall attempt to formulate a Zermelian response to McGee's work. Here we will see clearly the problem of remaining within an extensional, second-order logical framework. In the final section, we shall summarize how modal logic can be mobilized to give a proper expression to Zermelo's resolution, and we shall make some remarks on the question of determinateness vs. open-endedness of set theory.

2. The Dual Role of Set Theory: As Universal Framework and as Part of Mathematics

As it actually evolved historically, set theory served as a universal framework in which all of standard, classical mathematics could be carried out. As such, it can be viewed as an extensional, streamlined alternative to the type theory of Russell and Whitehead's *Principia Mathematica*. When put in rigorous, axiomatic form (as either first- or second-order Zermelo or Zermelo-Fraenkel set theory), it serves both as a standard of classical mathematical proof and as a general theory of mathematical existence. Surely it must count as one of the great achievements of modern mathematical logic to have developed such systems, adequate to all ordinary mathematical purposes, even despite the impossibility of a combinatorial consistency proof, as implied by Gödel's second incompleteness theorem.

It is interesting, however, that even prior to Gödel's remarkable discovery, set theory had also begun to be viewed as part of mathematics in its own right, dealing not with just one fixed universe of sets—*the* proposed sufficient ontological framework for classical mathematics—but with *multiple universes*, indeed with boundlessly infinitely many, as described in Zermelo (1930). As such, it resembles abstract algebra rather than number theory, and it can be approached in a structuralist spirit along with other mathematical theories (including number theory as well, it should be noted). Moreover, this perspective arose in the context of what we today call higher-order (at least second-order) logic, in its standard, semantical sense (in which full power sets are available as the ranges of higher-order quantifiers); that is, only mathematically standard models are being considered, not the nonstandard ones that arose later in the aftermath of Gödel's work, and notably in the work of Cohen and successors. (In particular, all the models considered by Zermelo agree on all mathematical questions pertaining to natural numbers, real numbers, sets and functions of real numbers, set and functions of these, and so on without end.) The differences among the many models are intrinsically set-theoretic, having to do with large transfinite ordinals and cardinals, not with any questions of classical analysis, or even modern functional analysis for that matter.[2]

Now there is a serious tension between these two roles or views of set theory. The first role, as universal framework for mathematics, surely encourages the idea of a fixed background ontology, adequate to all mathematical purposes. Thus, even if one attempts to carry out a structuralist interpretation of set theory within set theory itself, via model theory, as is standardly done—thereby accommodating Zermelo's many models—one confronts the limitation that the intended "real universe of sets" cannot itself be a set, so that either it is taken as a collection of another order, leading either to proper classes or hyperclasses (and so *ad infinitum*) or it is not recognized at all. In either case, a structuralist interpretation of set theory within set theory will have failed. Moreover, the role of "universal framework" for ordinary mathematics—or at least the role of "unifying framework"—can be served by many set theories. We already have had to take into account the distinction between Zermelo set theory (with the axiom of Separation but not the axiom of Replacement) and Zermelo-Fraenkel set theory (with both). Then there are the choices between set theory with or without Regularity (or well foundedness), between set theory with or without the Axiom of Choice, a plethora of choices over large cardinals, determinacy hypotheses, etc. Which of these many choices is to serve as the universal framework? The staunch realist-platonist may reply, "The true one," meaning that there is a single real world of sets to be described. But, apart from its other defects—notably its failure to do justice to mathematics' freedom to explore any of these alternatives *without* having to claim that one is uniquely correct—this stance is clearly at odds with a structural interpretation. For such an interpretation sees the subject matter of any set theory as "set theoretic structure (of the sort corresponding to that theory)," exemplified by any objects whatever standing in the right structural relations. (Just what "exempflication" here comes to needs to be spelled out, of course. Different versions of structuralism do this differently.) No set theory—as a list of axioms as usually presented—is, on the structuralist view, true *simpliciter* in virtue of correctly describing a part of the world in the way that, say, chemistry is true. At best such a theory is true *in* or *of* structures of the appropriate type. The plethora of alternative set theories surely adds to the motivation for developing a structuralist perspective. (Clearly, here we are siding with Hilbert and against Frege in their famous debate over the nature of mathematical axioms and reference.)

On the other hand, adopting such a perspective does not require sacrificing or denying the unifying role of set theory (better theories), although claims of absolute universality become suspect. One of the lessons of modern proof theory is that only very weak set theoretic assumptions are needed in order to carry out the vast bulk of ordinary mathematics. (The programs of Friedman, Simpson, and others—see, for example, Simpson

1999 and Feferman 1989—teach us this.) The power and generality of set-theoretic language and weak axioms of set existence (relative to the full ZF or even Z) stand ready to be incorporated in any structuralist account of set theory. Even if it should turn out that answers to intrinsically set-theoretic questions, such as the existence of large cardinals of various sorts, are genuinely required in order to solve combinatorial problems or problems of real or complex or functional analysis, etc. (as Friedman's program predicts), that will just reveal a further unity among mathematical disciplines which a structural interpretation of set theory should accommodate. In sum, set theories do not lose their unifying power *vis-à-vis* ordinary mathematics by themselves being interpreted along structuralist lines as *part* of mathematics.

3. Zermelo 1930

Here we present a brief overview of the main results of this landmark paper, concentrating on those needed for a comparison with those of McGee 1997.[3]

Having articulated in ordinary mathematical language what we would recognize as the ZF axioms (minus the Axiom of Infinity), Zermelo goes on to prove basic results characterizing the "domains of sets" that satisfy these axioms.[4] These turn out to be (up to isomorphism) what we recognize as the "natural models" of the axioms, with or without urelements, that is, those models built from an urelement basis (or just the null set) by iterating the power set operation at every successor stage and unioning at limit stages "sufficiently far" into the transfinite. Formally speaking, the two main achievements of the paper are, first, to prove that all models of the axioms are of this structure, that is, in modern terms, isomorphic to a natural model V_α or the analogous structure built on urelements, and second, to specify exactly what "sufficiently far" into the transfinite means, namely, that a must be a regular, strong-limit cardinal, known as a strongly inaccessible cardinal.

The first result, which I have called the "quasi-categoricity" of ZF^2—where the superscript indicates that the axioms are taken to be the second-order ZF axioms—is broken down into two theorems:

Theorem 1 *If M_1 and M_2 are any two models of $ZF^2(i)$ with urelement bases of the same cardinality and (ii) having the same ordinal height, then M_1 and M_2 are isomorphic.*

Theorem 2 *For M_1 and M_2 as above satisfying (i) but with different heights, either M_1 is isomorphic to an initial segment of M_2 or vice versa.*

The second result can be stated as

Theorem 3 *The ordinal height of any model of ZF² is a strongly inaccessible cardinal.*

(Here 'ordinal height' means supremum of the ordinals of the model in the sense of the order type of the well-ordered "sets" serving as ordinals in the model.)[5]

Two remarks are in order concerning these theorems. The first is that they are presented as ordinary, informal mathematics, without any indication of a formal system in which they are to be thought of as rigorously proved. Moreover, the axioms themselves are presented informally, not in symbolic logical notation. Second, regardless of just where, formally speaking, the theorems are proved, it should be emphasized that they hold only if the system ZF is formulated as a second-order one, in other words, the axiom of Replacement must be formulated as a statement about arbitrary functions from sets to sets, not as a first-order schema with denumerably many instances corresponding to first-order definable functions. Analogous remarks apply to Separation, if it is stated separately. (In a footnote, Zermelo explicitly states that the propositional function of Separation and the replacement function of Replacement "can be quite arbitrary [*beliebig*]," and asserts that "none of the consequences of restricting these functions to a particular class are relevant for the point of view taken here." (Zermelo 1930, 30, n.1; trans. Hallett 1996, n. 21). The multiplicity of models possible for first-order ZF far outstrips that which Zermelo insists upon for his own ZF, and none of the above theorems holds good for first-order ZF. Not only do models of first-order ZF(C) differ on first-order undecidables at the levels of natural numbers, reals, functions of reals, etc., but even models of the form V_λ, in which the power-set operation is genuine ("full"), may have λ singular. (See Drake 1974, 111, ex. 1.6[4]; also Montague and Vaught 1959.) Zermelo's theorem that λ must be strongly inaccessible is correct, however, for ZF², and this can be proved in standard model theory, formalized either in ZF¹ or a formalized fragment of ZF², or in a formalized fragment of second-order logic, as indicated in Hellman 1989.

So far—as Zermelo recognizes at the outset of the final section 5 of his 1930 paper, entitled "Existence, Consistency, and Categoricity"—the existence of different models (in the second-order sense) of the axioms, ZF² (minus Infinity), and the very consistency of these axioms have simply been assumed. He writes, "We shall not attempt here to *prove* this consistency logically and formally. On the contrary, under the general *assumption* of the freedom from contradiction of set theory, we shall examine the (mathematical, i.e., ideal) *existence* of the different model types considered here.

Thus, we assume the existence of set domains with an arbitrary basis which satisfy the ZF axioms" (Zermelo 1930, sec. 5; trans. Hallett 1996.) He then sketches a construction of the least normal domain (modeling ZF' = ZF with Regularity but without Infinity), with characteristic ω, which he takes to demonstrate at least the consistency of ZF'. However, Zermelo recognizes that this by itself does not take one past the hereditarily finite sets, and if one adds the Axiom of Infinity, the most one can say so far is that the smallest "normal domain" of these axioms (ZF²) must have "characteristic π_1," in other words, ordinal height = the first strongly inaccessible cardinal. He continues,

> But are there, following ω, such "boundary numbers" at all? Certainly, in so far as there is an "infinitistic" set theory at all, that is, in so far as there are normal domains with infinite sets. For the *totality* of all the "basic sequences" [equivalently ordinals] of such a domain itself has an ordinal type π, even if there is no set of this type π *inside* the domain. And if there are "boundary numbers" $\pi > \omega$ at all, then among them is a *smallest* π_1. Certainly neither their existence nor their non-existence can be "proved," i.e. be derived from the general ZF' axioms [ZF² without Infinity], because, for example, the boundary number ω exists in the "Cantorian" domain (the smallest model of ZF² including Infinity) but not in the "finitistic" domain. In other words, the question can be answered *differently* in different "models" of set theory, thus it is not yet settled by the axioms alone. Our axiom system is *non-categorical*, which in this case is not a disadvantage but rather an advantage, for on this very fact rests the enormous importance and unlimited applicability of set theory. *Naturally one can always force categoricity artificially by the addition of further 'axioms', but always at the cost of generality.* (Zermelo 1930, sec. 5; trans. Hallett, my emphasis.)

This passage will serve us well in connection with McGee's results, to be presented and discussed in the next section.

Rather than attempt to enforce categoricity, Zermelo proceeds to postulate the following principle, which I call the *"Extendability Principle"*:

> *Every categorically determined domain* [i.e., determined via its basis and its ordinal characteristic, in accordance with the preceding theorems] *can also somehow be conceived of [aufgefasst] as a 'set'*, i.e. can occur as an element of a (suitably chosen) normal domain. (Zermelo 1930, 46)

The latter will be a proper extension of the first with a greater "boundary number," with ordinal height a greater inaccessible cardinal, and so on without end. Furthermore, Zermelo postulates an *"Extended Extendability Principle"* as follows:

> In the same way, from each infinite sequence of different normal domains with a common basis, which are such that of any two one always contains the other

as canonical segment, there arises, through union and fusion [*Verschmelzung*], a categorically determined domain of sets which again can be extended [*ergänzt*] to a normal domain of higher characteristic. (Ibid.)

Now Zermelo does not specify here the source of any infinite sequence, nor even whether such a sequence can be a transfinite one. But we may assume that it can be, and that its transfinite type can come from an ordinal already lying in any normal domain obtained so far (that is already postulated). In any case, Zermelo asserts that "the series of 'all' boundary numbers [the original quotes around 'all'] is unbounded in the same way as the [ordinal] number series itself" (Ibid.). (Set theorists will recognize in this the "Axiom of Inaccessible Cardinals." We shall have more to say about this below.) Again, he recognizes that this cannot be "proved" from the ZF' axioms: "Rather, the *existence of an unbounded sequence of boundary numbers* must be postulated as a new axiom of 'meta-set theory', and in so doing the 'consistency' question needs to be looked at more closely. If I here limit myself to just the foregoing sketch and must refer to its subsequent working out ['*Ausführing*'], still the following, which can be regarded as the essential result of the present investigation, should already be clear" (Ibid.). Here follows directly the final paragraph with its ringing defence of higher set theory, quoted at the outset of this paper (and which the reader may now wish to reread). This completes our overview.

4. McGee 1997

In this thought-provoking paper, McGee sets out to extend to set theory what we are already able to say generally in the case of number theory in answer to the question, how we are able (in principle) to learn mathematical vocabulary so that its meanings provide a determinate truth value for each sentence of (the relevant) mathematical language, in accordance with a realist conception. The general answer we can already provide in the case of number theory runs thus: in addition to activities of counting, measuring, etc.—by themselves insufficient for more than a fragment of the language—we can in principle learn a body of theory, namely, axioms, which serve as "meaning postulates"; these axioms implicitly define a unique isomorphism type of structure, so that, however referents are assigned to the terms of the language, a unique truth value is determined for each sentence. Realism in truth values is thereby detached from uniqueness of reference of individual designators, and this is seen as an advantage in meeting Benacerraf's (1965) well-known challenge to mathematical realism. Moreover, a psychologically realistic answer (in the vernacular sense) to the question of the paper's title is not sought, but rather an answer to the rather different "in principle" Kantian style question, "How is it possible for beings with humanlike cognitive capacities to 'learn mathematical language' in the required sense, i.e., to sustain mathematical realism?"

The axioms that suffice for number theory are well known; of course they are the (second-order) Peano-Dedekind axioms, and this brings up the question of the status of second-order logic. The strategy adopted is, basically, to accept it provisionally as genuine logic, and later to argue that an open-ended understanding of *schemata* can suffice for the same ultimate purposes (in connection with *set theory*), without the ontological commitments of second-order machinery. Regarding first-order logic, however, McGee emphasizes that he takes universal quantification as absolute and completely unrestricted, that is, unrelativized in any way: "everything" in the intended sense means "absolutely everything that (actually) exists," and this is taken to be unproblematic and fully determinate, regardless of any epistemological limitations we may confront over cases.[6] This stance is absolutely crucial to McGee's results on the "categoricity" of set theory, as we shall see momentarily.

The set theory with which McGee begins is ZFCU, in second-order form, that is, ZF^2 with the Axiom of Choice and with provision for *Urelemente* (reflected in adjustments in the axioms of Extensionality and Regularity). Appealing to Tarski (and perhaps also to Boolos 1985), we can make sense of the notion of a sentence in this language being a logical consequence of these axioms. If we could somehow achieve a categorical system of axioms which we could regard as "meaning postulates" on the primitives, $Set(x)$ and $\in$, then we could appeal to our grasp of "logical consequence" to account for (humanly comprehensible) truth-determinateness of arbitrary set-theoretic sentences, in exact analogy with the familiar, in-principle account we can give for number theory: the axioms are determinately true; and any logical consequence of them is also; in light of categoricity (and the bivalence of classical logic), any *other* sentence of the language in question is equivalent to the negation of a logical consequence, and so is determinately false.

As McGee recognizes, however, the theory cannot be ZFCU, even in second-order form (call it ZF^2CU, for that theory has multiple models in accordance with Zermelo's theorems (although Zermelo's 1930 paper is not cited by McGee), unless inaccessible cardinals (past ω) are ruled out *a priori*, a suggestion which McGee dismisses, acceding to majority opinion among set theorists. Moreover, even if one restricts attention to models with *un*restricted domain, in other words, concentrates on models in which the universe of discourse contains "absolutely *everything*" (taking this uncritically now, for the sake of argument), one still confronts the phenomenon of nonisomorphic models of ZF^2CU disagreeing on some purely set-theoretic sentences, by the device of counting as *Urelemente* of a new model items of a (given) big model with at least one inaccessible—call the least such κ—which are not accessible sets in that model, having cardinality $\geq \kappa$. The new model so constructed will recognize as sets only the accessible sets of the original and will simply recognize a great many more *Urelemente* than were originally given. Thus, some new axiom(s) must be added to ZF^{2}CU.[7]

The essence of McGee's results can now be easily stated: the method just sketched of upsetting isomorphism between the pure sets of models of ZF^{2}CU which have the same universe of discourse—that is, the union of what are counted as sets and what are counted as *Urelemente*—turns out to be essentially the *only* way. That method is ruled out if to ZF^{2}CU is added one new axiom, the *Urelement-Set Axiom*, which asserts that the *Urelemente* form a set:

$$(\exists x)(Set\,(x) \wedge (\forall y)\,(\neg Set\,(y) \to y \in\, x)).$$

Clearly this suffices to rule out the above construction, since the expanded collection of *Urelemente* for the new model is "too big" to be a set in that model (or in the original, for that matter, though this is not relevant), not having accessible cardinality in the original model. Relative to the new model (and the old, as well), it cannot have any cardinality; the *Urelement-Set Axiom* is violated. Through a series of lemmas, McGee shows how to prove what he calls a

> **Categoricity Theorem.** [General Part] Any two models of second-order ZFCU + the *Urelement*-Set Axiom with the same universe of discourse have isomorphic pure sets. (McGee 1997, 55. I have supplied the bracketed designation.)

This is a good set-theoretic result. Below we will comment on how Zermelo could recover it.

By itself, however, this is not enough for McGee's purposes, for different models of ZF^{2}CU + the *Urelement*-Set Axiom may have *different* universes of discourse. Which of these—and, according to Zermelo's theorems and his Extendability principles, there are boundlessly infinitely many—is to serve as the standard of (pure) set-theoretical truth? The answer is given by McGee in the continuation of the very Categoricity Theorem whose first, general part we have just quoted:

> **Categoricity Theorem, continued.** [Application] In particular, any two models of second-order ZFCU + the *Urelement*-Set Axiom in which the first-order variables range over everything have isomorphic pure sets. (Ibid.)

If we restrict attention to just those models whose universe of discourse contains exactly everything there in fact happens to be, we find that they must agree up to isomorphism on what they call the "pure sets." This doesn't tell us whether in fact there are exactly *seventeen* strongly inaccessible cardinal numbers, for example, or whether there are more or fewer, etc., but it does tell us that, in reality, there is a determinate answer, regardless of how we may fiddle with the boundary between "sets" and "*Urelemente*" in our interpretations of ZFCU + the *Urelement*-Set Axiom. (In effect, no matter what we say the cardinality of the "*Urelemente*" is, so long as that cardinality exists *as a set* in the real world, all interpretations must agree on the *purely* set-theoretical sentences, in other words, those making no reference to *Urelemente*. Just as the categoricity of second-order Dedekind-Peano arithmetic doesn't tell us the truth value of any unsolved problem of number theory but merely guarantees that there is a well-defined, determinate answer, regardless of our abilities to find it and regardless of how we may fiddle with particular designations of number-theoretic vocabulary (numerals, operation symbols, relation symbols), so too the "categoricity" in McGee's sense of ZF^2CU + the *Urelement*-Set Axiom guarantees truth-determinateness apart from any questions of epistemic access to truth values. (The reader is invited to extend the analogy to second-order real analysis and the determinateness of the Continuum Hypothesis, for example.)

Let us turn now to a Zermelian assessment of these results.

5. Zermelo* 2000

It would be most interesting to be able to ask Zermelo directly about McGee's paper, but in lieu of that, we can attempt to formulate a response ourselves based on the views expressed in his 1930 paper, recognizing that in the process we may well be imposing our own views to some extent. We trust that the '*' will satisfy the legal community in this regard (without detracting from the simultaneous expression of admiration of this true "superstar" of mathematical foundations!).

Put in the terms of Zermelo's closing summation, it seems clear that he would say that the striving for "all embracing completeness" has gotten the upper hand over the polar opposite tendency of "creative progress," and at the undue expense of the latter. For, as Michael Hallett puts it in his introduction (1996), "the paper [Zermelo 1930] sets out to show that the standard set-theoretical axioms, the ZF axioms (minus the axiom of infinity) have an unending sequence of different models" (2), and the great significance Zermelo attaches to this is hardly reconcilable with an admission that, "in the end" (so to speak), the "unending sequence" is to be conceived as occurring within—ending after all in—a single universe of sets, "*the* Cantorian cumulative hierarchy," itself unambiguously picked out (if only plurally and if only "up to isomorphism") as part of "absolutely everything (period)." Indeed, according to Tait, reference to a totality of "all sets" or "all ordinals" makes sense for Zermelo only relative to a given model of set theory.[8] It seems to me that Putnam expressed a view closer to Zermelo's (although without describing it as such) in his 1967 paper, when he wrote, "Even God could not make a model of [Zermelo] set theory that it would be *mathematically* impossible to extend" (Benacerraf and Putnam 1983, 310). "All the sets existing among all the things there are" is no exception.

Now I said above that the first "general part" of McGee's "Categoricity Theorem" is a perfectly good, intelligible piece of mathematics, one that Zermelo could recover. Indeed, he could do this in either of two ways, formalizing the theorem either in pure second-order logic, as McGee himself would prefer, or formalizing it within ZF set theory itself. For the first option, it is easy to write out a second-order logical formula expressing that a class X, a subclass S, and a binary relation E serve as a model of ZFCU + the *Urelement*-Set Axiom (X the domain, S the "pure sets" thereof, and E the "membership" relation), and then to say that any two such models, X, S, E and X', S', E', are such that there is a "membership" isomorphism between S and S'. And McGee's proof of this much can be carried out in second-order logic. On the alternative formalization, one could employ Tarskian model theory (i.e. introducing a relation of satisfaction between sets and formulas, etc.). The proofs would look very similar, but, as we shall see, Zermelo would have good reasons for preferring Tarskian model theory, for then all the results leading up to McGee's "Categoricity Theorem" can be interpreted as holding in any model of the axioms, including results which speak of "the pure sets" unrestrictedly constituting a maximal totality. But we are anticipating. What deserves to be emphasized just now is that the second part of McGee's "Categoricity Theorem"—what we labeled "Application" above—does not even belong to mathematics, appealing as it does to a clearly nonmathematical idea, the idea of "everything there is, mathematical or otherwise." This is clearly a term of metaphysics, not mathematics, and I do not see why Zermelo should have to rely on it. In particular, the idea that the determinateness of *pure mathematics (i.e., set theory)* should depend on the determinateness of this metaphysical notion is one that strikes me as quite alien to a perspective which sees mathematics as quite capable of standing on its own, and it would be surprising, to say the least, if that were not Zermelo's perspective. This is not to say that Zermelo is committed to the truth-determinateness of arbitrary set-theoretic sentences. Whether he is or not is not entirely clear, at least from his 1930 paper. But surely what determinateness there is should not derive from a nonmathematical idea such as "everything in reality."

So the task for Zermelo (or Zermelo*) on being confronted with McGee's results is to sift out the mathematical content, and then to see whether that content coheres with Zermelo's own proposed resolution of the tension between "all-embracing completeness" and "creative progress."

We have already indicated that there is no problem in recovering the first general part of McGee's "Categoricity Theorem." But I suggested that Zermelo (*) should prefer a formalization within set theory itself, rather than within pure second-order logic. To see why, let us consider a related result of McGee's, what he calls the

> **Completeness Principle.** The pure sets are not isomorphic to a proper initial segment of the "pure sets" of any other model of second-order ZFCU. (McGee 1997, 54)

McGee says that this is "entailed by" the *Urelement*-Set Axiom together with ZF^2CU , and it is clear that he intends this in the sense that the Principle is to be derived from those axioms in second-order logic, not that the Principle is to be understood merely as *holding in any model of* ZF^2CU + *the Urelement-Set Axiom*. This is a subtle but important difference. For, on the first understanding, one takes the reference to "the pure sets" in the Completeness Principle quite literally and unrestrictedly: it means "all the pure sets there are in reality." Since models can only be constructed out of what actually exists (presumably), the Principle says that there is no way of redrawing the boundary between sets and nonsets in such a way that the axioms of ZF^2CU would still be satisfied and so that the pure sets (all the real ones, you know) could be embedded in the new "pure sets" as an initial segment. In other words, the pure sets together with the *Urelemente* (in reality, assuming they form a set) form a maximal model of ZF^2CU, in direct violation of Zermelo's Extendability Principle! Clearly, on this interpretation, McGee's Completeness Principle and Zermelo's perspective and proposed resolution of the "central tension" are directly at odds with one another.

On the alternative reading, however, the situation is very different. For as a constraint on models of ZF^2CU , the Completeness Principle is harmless. Its opening reference to "the pure sets" now effectively means "the pure sets of the given model"; and now what the Principle says is that if the (i.e., any given) model also satisfies the *Urelement*-Set Axiom, then *its* boundary between "pure sets" and "nonsets" cannot be redrawn so as to violate the "maximality" of the model in the appropriate *relative* sense, that is in comparison with any other model constructed from the domain of the original one. (In effect, all the quantifiers in the statement of the Completeness Principle become relativized to the domain of a model satisfying it.) Obviously, this is entirely compatible with the Extendability Principle, and one could even imagine Zermelo continuing: "And just as the proper classes of one model become (or correspond to) sets in the next higher domain, so the 'completeness' of each model is only of the character of a 'relative stopping point' and 'no true completion in the unrestricted advance [of the endless ascending series of models]'." (See the closing paragraph of Zermelo 1930, quoted above.)

So far so good. Unfortunately, however, this is not the end of the matter, for there lies lurking a contradiction which must now be exposed. Even if Zermelo interprets McGee's results as set-theoretic theorems, hence as constraining models of (the relevant) set theory, still it is most natural and reasonable to render that theory—as understood by Zermelo—in second-order form. This means, however, that second-order logic is still in the background, and, as usually presented, that in turn includes a comprehension scheme for classes and relations implying the existence of a universal class (that is to say, of *all* first-order objects, concreta, sets, etc.). A straightforward, naive reading of this would have Zermelo implicitly committed to a maximal totality of sets after all, contravening the Extendability Principle, and contravening the relativity of "all sets," "all ordinals," etc., to a given model (see the last note quoting Tait). The fact that Zermelo does not *have to* interpret McGee's results in these terms does not evade the problem that, on a straightforward reading, he seems committed to the intelligibility of doing so, and, as already emphasized, this seems contrary to the whole spirit of his enterprise! Indeed, even if, on independent grounds, Zermelo (or anyone) found the *Urelement*-Set Axiom objectionable—read not merely as a constraint on (possible) models (*à la* Hilbert) but as a direct statement about reality (*à la* Frege)—that would only mean that he could not accept McGee's theorems as establishing "truths of the real world," so, for example, full determinateness for pure set theory might still be questioned. That in itself, however, would not alleviate the conflict with Extendability.

There are essentially three different ways out of this impasse. One is the one implicit in standard set-theoretic practice, namely to retreat to first-order logic as the background logic, giving up on the second-order framework for set theory. This solves the problem of implicit commitment to any universal collections, since one simply has no logical comprehension principle to respect, merely the restricted comprehension principles of set theory itself (instances of Separation or Replacement). The fact that one employs unrestricted quantifiers over sets, ordinals, even models of set theory, and so forth, does not in itself carry any implication of existence of any higher-order collections of these things. As Quine has repeatedly emphasized, speaking of all red things does not imply recognition of a property of redness nor even of a set of all red things. Set theory of course is in the business of recognizing sets of given objects, but it balks at any such thing as the class of all sets or all ordinals, etc., except as a convenient *façon de parler* in the form of a conservative extension of ZFC, such as NBG (von Neumann-Bernays-Gödel's set theory with proper classes).

This resolution comes however at a steep price. First of all, one loses the greater expressive power of second-order logic. As Kreisel, Shapiro 1985, and others have emphasized, one cannot even state the general principle of Replacement which motivates the scheme for infinitely many first-order instances. (This is, of course, in direct analogy with Mathematical Induction of number theory.) This is symptomatic of the fact that first-order set theories inevitably countenance many nonstandard and unintended models which are ruled out at a stroke by second-order formulations (even if the latter still permit a multiplicity of models, for example, the many natural models of ZFC discovered by Zermelo). It is perfectly true that one can carry out the model theory of second-order axiom systems within a first-order background set theory; indeed, that is the standard way of presenting Zermelo's 1930 results. Second-order statements are then merely *mentioned*, not used. But one has then failed to take advantage of second-order expressive resources in stating one's official assumptions.

Secondly, straightforward adoption of the first-order axioms of ZFC, say, carries direct commitment to those axioms and their consequences as genuine truths (Frege's view of axiomatics, in contrast to Hilbert's), and one seems to be giving up on a structuralist approach to set theory as the study of certain kinds of mathematical structures on a par with other branches of mathematics. One seems thereby committed to a naive platonist view of a unique, real world of sets correctly described by such and such axioms. And this, then, immediately raises a third problem: which axioms *do* describe the real world of sets, and how on earth (or in heaven's name) do we tell? Must the set theorist, qua mathematician, really confront such skeptical questions as "How do you know that there aren't really non-well-founded sets, or that there aren't really exceptions to such elementary axioms as Pairing, let alone Power Set and Replacement?" As Shaprio (1997) has brought out nicely, the traditional, naive platonist must confront the prospect of a kind of gross error or mismatch between axioms and reality that seems an utter embarrassment to the view. Surely pure mathematics does not turn on such a prospect!

The second way of attempting to resolve the difficulty faced by the Zermelian in using second-order logic while maintaining the Extendability Principle is implicit in Boolos's proposal (1985) to interpret the second-order quantifiers of set theory *plurally*. Thus, the usual argument based on examples such as Geach's—"Some critics admire only one another"—that many ordinary plural constructions build in a hidden commitment to classes of individuals, is stood directly on its head, and it is claimed that in fact such plural constructions can be used to provide a class-free interpretation of second-order quantification. Thus, as Lewis (1991) has persuasively argued, we can say, without any threat of contradiction, that there are exactly the non-self-membered classes (or sets). We can even invoke direct expression of universal plural quantification to gain the advantages of, for example, second-order mathematical induction as a defining condition on natural-number-type structures: Any numbers whatever which include 0 and which include the successor of any among them also include all numbers. Making use of a pairing function, one can reduce polyadic second-order quantification to monadic and so use plural quantifiers to express the second-order Replacement axiom of ZF and a great deal more.

So far, this looks promising. But how does appeal to plural quantification as a reading of second-order set theory treat the Extendability Principle? Insofar as a model is required to be a set, as in first-order set theory, Extendability can be respected by adopting the Axiom of Inaccessibles,

$$\forall\alpha(\text{``}\alpha \text{ an ordinal} \rightarrow \exists\beta(\beta > \alpha \ \& \ \beta \text{ a strongly inaccessible cardinal''})).$$

But once we have the power of second-order language, we can get the effect of speaking of a "model whose domain is absolutely all the sets." How is it a resolution of Zermelo's problem to manage to do this without ontological commitment to an object called "the class of all the sets"? It still seems we have commitment to a multiplicity—a plurality, "absolutely all the sets"—which is, by construction, as it were, not extendable. It is little more than a play on words to say that Extendability is respected because the plurality is not counted as a collection in its own right. Mathematically speaking, we can always *introduce* talk of such a collection without contradiction (of course, we won't call it a "set," or a "hyperset" of any order already recognized—it will have to be of a new order), and then we can speak mathematically of proper extensions by set-like operations. And, if it is replied that this just takes us to higher and higher types of actual collections, we will naturally ask for a theory systematizing talk of all of them; such a theory will look very much like set theory all over again with many inaccessibles; and we will appear to be regenerating all our puzzles about "absolutely all sets" all over again. Whatever our most universal of universal quantifiers may be, treating its range plurally will not alter the fact that, mathematically speaking, we can entertain a still new, still higher-order collection of "all the things in that range" (by hypothesis, we have such unrestricted quantification available), with further imaginable extensions. It seems that any mathematically possible world can be properly extended. Plural quantifiers, as employed by Boolos and others range over, or within, only a single such world, and so they do not really address the problem of avoiding inextendability. Is there, then, any way of retaining second-order logic in the background, with its expressive power and advantages for a structuralist view of set theory, while at the same time avoiding commitment to inextendability? If one is prepared to speak openly of *mathematical possibility*, indeed there is. For, as I have elaborated elsewhere (1989), if one recognizes assertions of purely mathematical existence as assertions of mere conceptual coherence obeying an appropriate logic of possibility, one then has all the resources one needs for charting a safe course between the Scilla of a first-order background and the Charybdis of inextendability.

6. On the Possibility of a Modal Resolution

It is commonplace nowadays, under the influence of Quine, to understand "mathematical existence" as simply the existence of mathematical objects, in a putatively ordinary, univocal sense of "existence," obeying a universal, classical quantifier logic. That, apparently, was not Zermelo's understanding, for in the last section of his 1930, entitled "Existence, Consistency, and Categoricity," he speaks of "*mathematische d.h. ideelle Existenz*" ("mathematical, that is ideal existence") of models of set theory (43), which is presumably something different. The core idea of a modal resolution is to take this "ideal existence" as a kind of logical possibility, obeying a suitable modal logic of its own.

Such a resolution begins with an ordinary, "actualist" understanding of modal discourse. Modal operators function perfectly well without any commitment to a realm of possible objects.[9] To say that there might have been a second planet in our solar system that supported intelligent life is to say just *that*—based for example on an appeal to the possibility of the right kind of planet formation out of planetesimals in condensing, swirling stardust, together with lots of cooperating biological conditions, etc.; *not* that there *is* a *possible planet* with the entertained properties. Of course, ordinary discourse has its vagaries, and we might well say that there is (or once was) the *possibility* of such a planet. But that is to be understood in the first way, not as a reification of possibilities. The fact that, until quite recently, perhaps, it has not been seen how to present a formal semantics of quantified modal logic without appeal to possible objects (worlds and their contents) should not distract us. First, it seems perfectly cogent to *use* modality in our background language to present such a semantics (as Chihara 1998, for example, has done). Second, in any case, ordinary understanding precedes formal semantics (of necessity, for the latter is presented in language which must already be understood), although this is not to say that formal semantics may not "feed back to" or inform ordinary understanding.[10]

The next step is to integrate this actualist understanding of modal discourse with talk of sets of objects. Now sets may well be thought of, at least initially, as reifications of possibilities in their own right, namely—as Putnam (1975) put it, paraphrasing Mill—as "permanent possibilities of selection" (71). (I say "at least initially" to concede that impredicative set theory, with its power-set axiom applied to infinite sets, goes well beyond any ordinary notion of "selectability," that is to say, by intelligent organisms using countable languages. That is a subsidiary matter here.) But even so, it is only possible to select that which is available to be selected. We may well entertain the possibility of making selections under counterfactual circumstances, but such possibilities are limited to what is (entertained as) available in those circumstances. If we entertain the possibility of a world (in the ordinary sense) with exactly a trillion horses, we may also envision selecting a trillion horses; but we could not, in *those* circumstances, select more than a trillion, even though a world with more than a trillion is also—and would also be—possible. Moving to set-theoretic talk, we can say that, if there were exactly a trillion horses, there would also be the set of them; but in *those* circumstances there would not also be a set with *more* than a trillion horses, although in *other* entertainable circumstances, there would be. Sets are determined by, and so limited by, their members. Relative to any possible circumstances, sets can contain only what exists in those circumstances; "that which" would merely have been possible relative to those circumstances could not have been elements of anything in those circumstances. This is simply a generalization of the actual situation: there can be no actually existing set of not-actually-existing but merely possible objects.[11] And we take this to be a linguistic-conceptual limitation, reflecting our understanding of modal discourse, prior to set theory: no not-actually-existing but merely possible object has—actually has, that is—any property whatever, for such talk of "possible objects" is strictly a *façon de parler* to be paraphrased away.

Before moving on, let us pause to note one implication of the foregoing. It is often said that abstracta, especially mathematical abstracta, are *necessarily existing* (although the consistent set-theoretic platonist who insists on the just noted amodal or metamodal usage should regard even this as a kind of category mistake). Fregean numbers count as archtypical. The possibility that there might have been nothing whatever is thereby ruled out, counter to widespread intuition (and certainly counter to the desert-landscape aesthetic, which must yearn to eliminate even those billions of metric tons of sand!). Setting Fregean numbers and their ilk to one

side (metaphorically speaking, of course), and working with sets as a sufficient realm of mathematical abstracta (thus, setting thoroughgoing nominalism to one side as well, for present purposes), we can see that the
possibility of nothing is quite compatible with a rich set-theoretic ontology. Along with some leading set theorists (including Zermelo 1908, 202;
Fraenkel, Bar-Hillel, Levy 1973, 24; and perhaps Gödel 1944, 459), we
may regard the null set as a convenient fiction, so that indeed all hereditarily pure sets are fictional as well. But any existing object(s) can serve as
Urelement(e) at the base of a cumulative hierarchy of sets. (Slight modifications of the ZF axioms, especially Extensionality and Regularity, to
accommodate this are readily available. See, for example, Maddy 1990,
appendix.) But if all *Urelemente* are contingent, so are all sets, ultimately,
and this may be generalized to the possible, that is, if only contingent
objects might have existed as *Urelemente*, then only contingent sets are
possible. Ordinary mathematical truths (set-theoretically rendered), however, can still count as necessary, holding in any (sufficiently rich) cumulative structure that there might be or have been.

How then can all this help resolve Zermelo's dilemma? The key idea is
to allow second-order modal logic as the background framework but to
restrict the comprehension principle for second-order objects (classes and
relations) in such a way that only classes and relations of (first-order)
objects hypothesized as possibly existing in given circumstances (or "at a
world") are recognized as existing in those circumstances. Formally this
amounts to allowing instances of the scheme

$$\Box \exists P \forall x_1, \ldots, x_n \, [R(x_1, \ldots, x_n) \leftrightarrow \varphi],$$

where φ lacks free 'R' and contains no modal operators. (This may be
called the "Extensional Comprehension Scheme.") Note that no boxes
immediately precede the initial universal quantifier(s). This rules out, for
example, existing collections of objects that merely might have existed,
exactly as desired. In particular, there will be no commitment to a class of
all possible sets or of all possible ordinals or of all possible set-theoretic
models, etc. This is precisely what allows free reign of the Extendability
Principle, but now, of course, that is read literally as a modal principle: Any
model there *might* be *could* be extended. What then of "the universe of all
sets," guaranteed by the ordinary second-order comprehension principle,
of which the displayed is just the necessitation? As just indicated, in the
modal setting this cannot be "the universe of all possible sets," for that is
not countenanced. Instead, each world or possible model (i.e., model
there might be) will have its universe, a proper class relative to its world or
model, exactly as Zermelo envisioned. Moreover, the Extendability
Principle guarantees that, as he also envisioned, each such (relative) proper

class corresponds to a collection which is treated as a set in proper extensions of the given (possible) model, hence which is but an element of a vast new hierarchy in which "the whole classical set theory can be expressed," and so on, without conceivable end. In sum, the modal framework provides just the flexibility needed to guarantee unrestricted validity of the Extendability Principle while at the same time preserving the expressive advantages of second-order logic. Unwanted violations of Extendability are blocked by natural restrictions on second-order logical comprehension, restrictions grounded on our prior (actualist) understanding of modal operators and their interactions with set-theoretic concepts.

This resolution, however, brings with it a new puzzle: What sense are we to make of the idea that certain sets, say highly inaccessible ones, are merely possible but not actual, that they *might have existed* but in fact do not? The problem has nothing to do with the status of the *Urelemente*; these may be taken to be actual. But it is essential to the modal resolution of Zermelo's problem that not all possible set-theoretic structures built on these *Urelemente* be actual, for whatever ones are taken as actual can be accumulated into a further structure, itself properly extendable *ad infinitum.* "Which ones then *are* actual?" one may well ask. Where is the line between actual and merely possible to be drawn?

As long as one retains an absolutist conception of sets as unique, fully determinate abstract objects, I see no non-arbitrary answer to this question. It may indeed be pointed out that mainstream mathematics, pure and applied, is indifferent to the answer (so long as a few transfinite levels are allowed), but that hardly resolves the philosophical, interpretive difficulty. What is the situation, however, if one gives up the absolutist conception in favor of a structuralist one? According to this, set theory may be understood as investigating *any* structures meeting specified defining conditions, such as the ZFC axioms. Any objects whatever may serve as items of such a structure so long as they are interrelated in the right ways—we cannot say, by *the* membership relation, but by a "membership-type" relation, that is, a binary relation fulfilling the axioms. (This was the course spelled out formally in my 1989, chap. 2.) Now a principled answer to the question, "Which sets are actual?" becomes available. For now the question *means,* "Which set-theoretic structures are actual?" And, if this in turn means full ZF or even Z structures, then the answer that suggests itself is simply that, as far as we know, none of them is, although a great many are possible, but that this is all that mathematics really needs. That this is indeed a principled answer depends on the case for (a component of) *epistemic nominalism*, the case for the view that we really have no good evidence for the actuality of such vast structures,[12] although we may have some reason to accept their possibility, the strength of which depends on the particular type of structure described. That case cannot be entered into here in any detail, but let me suggest that it is bolstered by the viability of several nominalist programs for ordinary mathematics, including the predicativist program, Chihara's 1990 modal linguistic type theory, modal-structuralism applied to number theory and analysis (including functional analysis), and variants of these. Their effect, individually and cumulatively, is that set theory (with the axioms of Infinity and Power Sets) is quite dispensable for scientific purposes, so that, whatever force indispensability arguments may have for low-level mathematical assumptions, such arguments so far appear impotent on behalf of transfinite set theory.[13] Furthermore, the fact that current scientific practice acquiesces in a face-value reading of some low-level mathematical assumptions counts, for me, as pretty weak evidence for platonism, even as regards low-level mathematical objects. The empirical sciences are not in the business of mathematical epistemology, at least not yet, and it is perfectly understandable that they would simply reach for whatever is available on the shelf. I see no overarching principle of "naturalism" that would force mathematical epistemology to take this current practice as providing "the true semantics" of mathematical discourse.[14] Finally, the notion that reference to sets in the absolutist sense is somehow guaranteed by grammar, via a "syntactic priority thesis" attributed to Frege (see Wright 1983), seems to me ultimately unconvincing for a variety of reasons, including circularity arising from the need to dismiss a host of apparent exceptions to the thesis. But a fuller treatment of this will have to await another occasion.

Finally, what can be said about the determinateness of set theory? As we have seen, results such as McGee's achieve this by uncritically assuming "all-embracing completeness" in our reference to sets, but, we have suggested, unsuccessfully. Does the alternative sketched, enforcing the Extendability Principle without restriction via a modal-structural interpretation, perforce renounce determinateness in the interests of "creative progress"? If this means "full determinateness," the idea that every set-theoretic question, including all those unlimited to sets of given rank, has a determinate answer, then I think it must be conceded that we have yet to find a non-arbitrary way to achieve this. Surely, appeal to modality only highlights the difficulty. Even if we are convinced that infinite structures are possible, even if we are convinced that full structures for Zermelo set theory or even Zermelo-Fraenkel set theory are possible, this still leaves open a host of questions concerning possible extensions that we have no clue how to pin down. Note that these questions about possibility are not merely questions of formal consistency of axioms—all such questions can be cast in purely number-theoretic terms and are therefore without question determinate. The questions concern the possibility of structures of genuinely large uncountable power, *inconceivably* large by ordinary lights. We have, to be sure, learned a great deal since Zermelo 1930 about the transfinite through the study of large cardinals, determinacy principles, constructibility, and so forth. Still, we confront an "as yet unsurveyable unfolding and enrichment of mathematical science," and perhaps that shall always be so. Perhaps at the higher reaches of transfinite set theory, our mathematical knowledge does become merely conditional; perhaps it would be hubris for finite beings to pretend to more. That still leaves standard, mainstream mathematics fully determinate—interestingly, without in any way limiting its own creative unfolding and enrichment!—and perhaps that is as much determinateness as we could have hoped for.

NOTES

1. Indeed, there is a more specific link between what follows here and Stein's paper, concerning his understanding of Carnapian "frameworks" as "constitutive of alternative notions of *possibility*" (1993, 285, my emphasis), but it would be premature to enter into that in this essay.

2. This is not to say, of course, that proof-theoretically, extensions of ZFC, say, do not have consequences for mathematics at the level of number theory and analysis, perhaps even mathematically interesting consequences. It was precisely this prospect that inspired Gödel's program and continues to inspire it through work of Harvey Friedman on "necessary uses of abstract set theory." We are here commenting on model-theoretic determinateness, as established by Zermelo's theorems on second-order ZFC, which is, of course, fully compatible with proof-theoretic incompleteness, as we already know from Gödel's theorems applied to formal number theory.

3. For a more extensive overview of the content of Zermelo 1930, the reader is referred to Michael Hallett's (1996) introductory comments preceding his translation of this paper.

4. For purposes of comparison with related results, it should be noted that Zermelo presupposes also the Axiom of Choice, although he regards this as part of the background logic rather than as a set-theoretic axiom.

5. First, Zermelo requires that the ordinal index π, to be a "boundary number," corresponding to a level of the cumulative hierarchy constituting a model or "normal domain" as he calls it, must be an infinite, regular ordinal, that is one which is not equal to the limit of a sequence of smaller ordinals indexed by a strictly smaller limit ordinal. (In other words, if $\pi = \lim \xi_\alpha \alpha$ indexed by a limit number β such that each $\xi_\alpha < \pi$, then $\beta = \pi$. π is not cofinal with a smaller limit ordinal.) Such an ordinal must also be an initial number, that is, not equinumerous with any smaller ordinal, and so a cardinal by von Neumann's definition, of the form $\omega\alpha$ (identified with $\aleph_\alpha$).

But as Zermelo recognizes, this is not enough. It must also be the case (in order that the Replacement Axiom be satisfied) that π be a "critical number" or fixed point of the monotone, continuous function $\alpha \to \omega_\alpha$ (Cantor's aleph function), which will be the case for regular ω_α just in case α is a limit ordinal. ω_α is then called a *limit cardinal*. So π must be an infinite, regular limit cardinal, thereby satisfying $\pi = \omega_\pi$. (In modern parlance, π must be a weakly inaccessible cardinal. Note, incidentally, that regularity must be imposed; many singular cardinals are also fixed points of the aleph function.)

If Cantor's generalized continuum hypothesis is true, this would suffice, but, not being able to rely on this, Zermelo further requires, in effect, that π be a fixed point of the beth function, defined by $\beth_0 = \aleph_0, \beth_{\beta+1} = 2^{\beth_\beta}$, and $\beth_\lambda = U_{\alpha<\lambda}\beth_\alpha$ for limits λ. Thus, π must be a strongly inaccessible cardinal. This guarantees that the power set axiom is satisfied in V_π (or isomorphic structure, or corresponding structure built on urelements).

6. A famous paper of Quine's (1961), begins thus: "A curious thing about the ontological problem is its simplicity. It can be put in three Anglo-Saxon monosyllables: 'What is there?' It can be answered, moreover, in a word—'Everything'—and everyone will accept this answer as true. However this is merely to say that there is what there is. There remains room for disagreement over cases; and so the issue has stayed alive down the centuries" (1).

7. Indeed, it should be noted that Menzel (1986) proposed models of set theory in which the *Urelemente* form a proper class, serving as the ordinals themselves. This is in diametric opposition to McGee's proposed new axiom.

8. As Tait puts it: "We can speak of a universe of sets as a model of set theory only if we have a categorical determination of it; but then it would not be the universe of all sets. We may 'define' the totality V of all sets or the totality $\Omega = \{x|x$ is a von Neumann ordinal$\}$ of all ordinals, etc.; but these are just formal terms which have no meaning in themselves: what is the 'x' in the expression $\{x|...\}$ supposed to range over? It is only with respect to a given structure $\langle M, \varepsilon_M \rangle$ that they have meaning" (1998, 472).

9. This is the usage of Kripke's seminal work (1972).

10. It is interesting to compare the present stance with that of W. D. Hart 1976. In this interesting paper, Hart considered a modal explication of "the potential infinite" but, in his (very reasonable) unwillingness to commit himself to possibilia, abandoned it, lacking "a sound and complete possible worlds semantics for a plausible quantified modal logic without interpreting possible existence as the existence of possibilia" (254). In its place, Hart pursued a set-theoretic explication, but was naturally led to the very problems highlighted in the present paper concerning the indefinite extendability of whatever cumulative hierarchy we care to "construct" (Hart's term). Thus (according to Hart's proposal), what appeared as a merely potential infinity relative to a given hierarchy (in effect by being cofinal in the ordinal levels of that hierarchy) appeared as actual in an extension of that hierarchy. As he recognized, "we have now come full circle: we have returned to the intuitive picture of processes we rejected at the outset as not sufficiently theoretical to yield an account of the potential infinite adequate to decide basic questions in a decisive way" (263). While we have not been concerned with the problem of explicating "potential infinity" in this paper, but rather with the tension between set-theoretic realism and extendability, the parallels are striking.

11. One may object here that, just as there is an atemporal sense of 'exists' appropriate to sets—even sets of dated particulars—so there is an amodal, or meta-modal, sense of 'exists' appropriate to sets—even sets of objects governed by

modality. On this understanding, one should not, for instance, speak of sets of actual objects as themselves "actually existing" *qua sets*, for that would be strictly a category mistake, just as it may be regarded as a category mistake to speak of the set of today's raindrops as "existing today." (Remember that set-theoretic talk allows, for example, for a set with some of yesterday's, today's, and tomorrow's raindrops. Such a set is said to exist atemporally.) Does it also make sense to speak amodally of sets of "objects that merely might have existed"? On a *possibilist* view, such as Lewis' (1979), yes, but not on the actualist view adopted here. It seems that the modal-structuralist must simply forgo the amodal usage, for it is of the essence to bring set-theoretic existence under the general rubric of possible existence.

12. It is useful, I think, to distinguish this claim concerning set-theoretic structures, for ZF or even Z, say, from the more general (*Goodmanian*) nominalist claim that we have no compelling reason to accept sets at all, even low-level sets of material objects, although it should be noted that, even for Goodman, countenancing sets *meant* countenancing the means of "generating" endlessly (if not transfinitely) many levels "above" the initially granted ones. (Goodman's nominalism itself should, of course, be distinguished from the still more general, *traditional* nominalist claim that there are no abstract entities at all, or no compelling reason to accept any.) See Goodman 1977, chap. 2.

13. A similar conclusion can be based on results of the reverse mathematics program of Friedman, Simpson, et al., although this does not address the residual ontological questions concerning the status of the weak set-theoretic assumptions that are still needed for scientifically applicable mathematics. Here the nominalist reconstruction programs have something to add. See my 1999.

14. For an attempt to articulate a "naturalist" objection to nominalist reconstruction programs, see Burgess and Rosen 1997. For a nominalist-inspired response, see my review 1998.

REFERENCES

Benacerraf, P. (1965). "What Numbers Could Not Be."*Philosophical Review* 74: 47–73. Reprinted in Benacerraf and Putnam 1983, 272–94.

Benacerraf, P. and H. Putnam. (1983). *Philosophy of Mathematics: Selected Readings.* 2d ed. Cambridge: Cambridge University Press.

Boolos, G. (1985). "Nominalist Platonism." *Philosophical Review* 94: 327–44.

Burgess, J. and G. Rosen. (1997). *A Subject with No Object: Strategies for Nominalistic Interpretation of Mathematics.* Oxford: Clarendon Press.

Chihara, C. (1990). *Constructibility and Mathematical Existence.* Oxford: Clarendon Press.

Chihara, C. (1998). *The Worlds of Possibility.* Oxford: Clarendon Press.

Drake, F. (1974). *Set Theory: An Introduction to Large Cardinals.* Amsterdam: North Holland.

Feferman, S. (1989). *"Infinity in Mathematics: Is Cantor Necessary?" Philosophical Topics* 17, 2: 23–45.

Fraenkel, A. A., Y. Bar-Hillel and A. Levy. (1973). *Foundations of Set Theory.* 2d edition. Amsterdam: North Holland.

Gödel, K. (1944). "Russell's Mathematical Logic." Reprinted in Benacerraf and Putnam 1983, 447–69.

Goodman, N. (1977). *The Structure of Appearance.* 3d edition. In *Boston Studies in the Philosophy of Science,* vol. 53, ed. R. S. Cohen and M. W. Wartofsky. Dordrecht: Reidel.

Hallett, M. (1996). Introductory comments and translation of Zermelo 1930. In W. Ewald, ed., *From Kant to Hilbert: Readings in the Foundations of Mathematics,* 2 *vols.* Oxford: Clarendon Press, 1208–1233.

Hart, W. D. (1976). "The Potential Infinite." *Proceedings of the Aristotelian Society* New Series 76: 248–64.

Hellman, G. (1989). *Mathematics without Numbers: Towards a Modal-Structural Interpretation.* Oxford: Clarendon Press.

———. (1998). "Maoist Mathematics?" Review of Burgess and Rosen 1997, *Philosophia Mathematica,* series 3, vol. 6: 334–45.

———. (1999). "Some Ins and Outs of Indispensability: A Modal-Structural Perspective." In A. Cantini, E. Casari, and P. Minari, eds., *Logic and Foundations of Mathematics.* Dordrecht: Kluwer, 25–39.

Kripke, S. (1972). "Naming and Necessity." In D. Davidson and G. Harmon, eds., *Semantics of Natural Language.* Dordrecht: Reidel, 253–355.

Lewis, D. (1979). "Counterpart Theory and Quantified Modal Logic." In M. J. Loux, ed., *The Possible and the Actual.* Ithaca, N.Y.: Cornell University Press, pp. 110–28.

Maddy, P. (1990). "Physicalist Platonism." In A. Irvine, ed., *Physicalism in Mathematics.* Dordrecht: Kluwer, 259–89.

McGee, V. (1997). "How We Learn Mathematical Language." *The Philosophical Review* 106: 35–68.

Menzel, C. (1986). "On the Iterative Explanation of the Paradoxes." *Philosophical Studies* 49: 37–61.

Montague, R. and R. Vaught. (1959). "Natural Models of Set Theories." *Fundamenta Mathematicae* 47: 219–42.

Putnam, H. (1967). "Mathematics without Foundations." *Journal of Philosophy* 64: 5–22. Reprinted in Benacerraf and Putnam 1983, 295–311.

———. (1975). "What Is Mathematical Truth?" In *Mathematics, Matter and Method: Philosophical Papers,* Vol. 1. Cambridge: Cambridge University Press, 60–78.

Quine, W. V. (1961). "On What There Is." In *From a Logical Point of View,* 2d edition. New York: Harper and Row, 1–19.

Shapiro, S. (1985). "Second Order Languages and Mathematical Practice." *Journal of Symbolic Logic* 50: 714–42.

Simpson, S. G. (1999). *Subsystems of Second Order Arithmetic.* Berlin: Springer.

Stein, H. (1993). "Was Carnap Entirely Wrong After All?" *Synthese* 93: 275–95.

Tait, W. W. (1998). "Zermelo's Conception of Set Theory and Reflection Principles." In M. Schirn, ed., *Philosophy of Mathematics Today.* Oxford: Clarendon Press, 1998, 469–83.

Wright, C. (1983). *Frege's Conception of Numbers as Objects* . Aberdeen: Aberdeen University Press/Humanities Press.

Zermelo, E. (1908). "Investigations in the Foundations of Set Theory I." In J. van Heijenoort, ed., *From Frege to Gödel.* Cambridge, Mass.: Harvard University Press, 1967, 200–215.

———. (1930). "Über Grenzzahlen und Mengenbereiche: Neue Untersuchungen über die Grundlangen der Mengenlehre." *Fundamenta Mathematicae* 16: 29–47.

[14]

Beyond Hilbert's Reach?

WILFRIED SIEG

. . . historical reflection serves, in the end, to shape the present and the future.

—Ernst Cassirer[1]

0. What Is at Issue?

Work in the foundations of mathematics should provide systematic frameworks for important parts of the practice of mathematics, and the frameworks should be grounded in conceptual analyses that reflect central aspects of mathematical experience. The Hilbert School of the 1920s used suitable frameworks to formalize (parts of) mathematics and provided conceptual analyses. However, its analyses were mostly restricted to finitist mathematics, the programmatic basis for proving the consistency of frameworks and, thus, their instrumental usefulness. Is the broader foundational quest beyond Hilbert's reach? The answer to this question seems simple: "Yes and No." It is "Yes" if we focus exclusively on Hilbert's finitism; it is "No" if we take into account the more sweeping scope of Hilbert and Bernays's foundational thinking. The evident limitations of Hilbert's "formalism" have been pointed out all too frequently; in contrast, I will trace connections of Hilbert's work, beginning in the late nineteenth century, to contemporary work in mathematical logic. Bernays's reflective philosophical investigations play a significant role in reinforcing these connections.

It is a fact of intellectual history, perhaps a curious one, but nonetheless a fact, that the *Grundlagenstreit* of the 1920s colors even now our perspectives on the foundations of mathematics and beyond. In those early debates, we find dramatically formulated stances, and we tend to interpret them as being substantively and starkly opposed to each other. Minimal historical awareness should have undermined that tendency a long time ago, as the finitist program, first formulated in Hilbert's Leipzig talk of September 22, 1922, was explicitly intended to mediate between constructivist and logicist, set theoretic positions. Weyl recognized that point in papers[2] from 1925 and 1928, and even Brouwer's polemical *Intuitionistic Reflections on Formalism*, published in 1927, contain these remarks:

> The disagreement over which is correct, the formalistic way of founding mathematics anew or the intuitionistic way of reconstructing it, will vanish, and the choice between the two activities be reduced to a matter of taste, as soon as the following insights[3] . . . are generally accepted. The acceptance of these insights is only a question of time, since they are the results of pure reflection and hence contain no disputable element, so that anyone who has once understood them must accept them.[4]

From Hilbert's perspective, there was every principled reason to view the mathematical substance of Weyl's and Brouwer's foundational work as part of broadly conceived axiomatic investigations; in lecture notes from the summer term of 1920 he had already emphasized:

> But what these two researchers [Weyl and Brouwer] have achieved in terms of positive and fruitful results through their investigations on the foundations of mathematics, that fits very well with the axiomatic method, indeed, it is exactly in the spirit of this method. For it is being investigated, how a part of analysis can be delimited by a narrower system of assumptions.[5]

One is tempted to think that the *Grundlagenstreit* could have given way to a calmer discussion; but it did not.[6] I will not try to disentangle aspects of personality, professional aspirations, or ideological judgments. Rather, I intend to describe the broad foundational context and attempt to understand what the programmatic constructivist Hilbert defended against his flamboyant fellow constructivist Brouwer, how he tried to do so, and how he got there.

The logicophilosophical community has focused on Hilbert's finitist means for securing "classical" mathematics and on the epistemological distinctiveness of those means—as viewed in the twenties; and I think that a deepened mathematical and philosophical analysis of finitism remains an important issue. However, we have not been equally concerned with the substance of what Hilbert strove to secure—over a lifetime. And that is not *classical* mathematics as it evolved until the nineteenth century, but rather *modern* mathematics as it resulted from a radical transformation during the second half of that century. Howard Stein called it a transformation "so profound that it is not too much to call it a second birth of the subject"; he argued, and I agree, that it was effected mainly by the work of Gauss, Dirichlet, Riemann, and Dedekind.[7]

Hilbert was intimately connected to this part of mathematical tradition (in Göttingen), but also to a second significant aspect. I am alluding to the free use of mathematical concepts in, and indeed their invention or free creation for, applications in the sciences. (I should point out that Weyl, in 1928, viewed intuitionistic mathematics as inadequate for the sciences!) In the introductory remarks of his Paris Lecture Hilbert described most vividly the rich interplay of mathematical thought and experience. Discussing the central importance of problems for mathematics, he commented on their origins as follows:

> Surely the first and oldest problems in every branch of mathematics spring from experience and are suggested by the world of external phenomena. . . . But, in the further development of a branch of mathematics, the human mind, encouraged by the success of its solutions, becomes conscious of its independence. It evolves from itself alone . . . by means of logical combination, generalization, specialization, ... and appears then itself as the real questioner. . . .
> . . . while the creative power of pure reason is at work, the outer world again comes into play, forces upon us new questions from actual experience, opens up new branches of mathematics. . . .
> And it seems to me that the numerous and surprising analogies and the apparent harmony which the mathematician so often perceives in the questions, methods, and ideas of the various branches of his science, have their origin in this ever-recurring interplay between thought and experience.[8]

One basic condition has to be met, however, in order to safeguard creative freedom within mathematics and within contexts of applications: the introduced notions must be consistent. That was clearly expressed in the Paris Lecture, and Hilbert reiterated this view four years later in his Heidelberg talk. About the underlying *creative principle* he wrote then: "in its freest use [that principle] justifies us in forming ever new notions with the sole restriction that we avoid a contradiction."[9] Thus, the central methodological issue is, how we can rationally assess whether this restriction has been met.

The methodological issue was more concrete and limited already at this point: Hilbert sought to establish the consistency of axiom systems, for example, in 1900 for the real numbers and in 1904 for the natural numbers. Such a proof was to ensure the existence of the set or, in Cantor's terminology, of the consistent multiplicity of the real and natural numbers. The issue can be traced back to Dedekind and is, according to Bernays (and in harmony with my earlier remarks), most closely connected to the "transformation the methodological approach of mathematics underwent towards the end of the nineteenth century."[10] One characteristic feature of that transformation is the emergence of _existential axiomatics_ described in the first part of my paper. That part is entitled "Logical Models" and examines Dedekind's and Hilbert's attempts to secure the consistency of analysis by logical means, before 1900. The second part, "Direct Proofs," presents in detail Hilbert's attempt to solve the consistency problem for elementary arithmetic in 1904, Poincaré's critical objections, and the impact of _Principia Mathematica_ in Göttingen. "Proof Theoretic Strategies" is the title of the third part; here finitist mathematics moves to center stage, and the methodological perspective for Hilbert's proof theory is described. An analysis of the informal ideas underlying this approach leads to the fourth part, "Accessible Domains," and serves as the motivating background for a programmatic formulation of _reductive structuralism_. My goal there is twofold, namely, to describe a global, integrating perspective of foundational work on the one hand, and to formulate some more local, focused problems for mathematical work on the other hand.[11]

1. Logical Models

Hilbert attempted to secure analysis from contradictions at the close of the nineteenth century. His formulation of a theory for real numbers in 1899 was inspired by Dedekind's and is distinctly modern. Recall that Kronecker, a mere decade earlier, had still been trying to banish the general notion of irrational number from mathematics; and Hilbert's lecture notes from the period between 1894 and 1899 show how difficult it was for him to obtain a proper perspective on the notion of number (_Zahlbegriff_). In the end, Hilbert associated all the central foundational issues with the _axiomatic method_. To proceed axiomatically means for Hilbert to think with consciousness, but also with critical awareness. The method allows the rigorous investigation of independence and completeness issues, and it is needed for securing, completely and logically, the content of our knowledge.[12] Already Dedekind had most explicitly aimed at grounding—by logic—our arithmetical knowledge!

1.1. Existence

A rather direct interpretation of the essay _Was sind und was sollen die Zahlen_ and of the later explanatory letter to Keferstein shows that Dedekind strove to give a consistency proof relative to a logic that allowed the construction of models, here, for simply infinite systems. In other words, a logical proof of the existence of such a system was to secure that the very notion did not contain an "internal contradiction." Dedekind wrote to Keferstein:

> After the essential nature of the simply infinite system, whose abstract type is the number sequence N, had been recognized in my analysis . . . , the question arose: does such a system *exist* at all in the realm of our ideas? Without a logical proof of existence, it would always remain doubtful whether the notion of such a system might not perhaps contain internal contradictions. Hence the need for such a proof (articles 66 and 72 of my essay).[13]

These observations can be extended in a natural way to cuts and complete ordered fields, treated in the earlier essay *Stetigkeit und irrationale Zahlen*. Dedekind viewed his broad methodological considerations not as specific for the foundational context of these essays, but rather as paradigmatic for the sound introduction of axiomatically characterized notions.[14]

In his proof of the existence of a simply infinite system, Dedekind had used however the (in)famous "system of all objects of my thinking." Hilbert learned from Cantor as early as 1897 that this gave rise to a contradiction and, thus, undermined the logical basis of Dedekind's essay. The very title of Hilbert's historically sweeping lectures from 1894, *Quadratur des Kreises*, was consequently expanded in 1897 to *Zahlbegriff und Quadratur des Kreises*. Hilbert emphasized there the importance of the "fixation of the (real) number concept," and he defined the reals as fundamental sequences taking for granted the natural numbers. Two years later, when finishing *Grundlagen der Geometrie*, Hilbert formulated axioms for the reals including the completeness axiom in yet another version of these lectures on the quadrature of the circle. His paper *Über den Zahlbegriff* was completed for publication on October 12, 1899 and summarized these early investigations; Hilbert had presented it already on September 19, 1899 to the Munich meeting of the German Association of Mathematicians.[15]

A neglected, but most significant link to Dedekind should be noted. Hilbert followed Dedekind in formulating the central axiomatic conditions for real numbers, as well as in setting up the very framework by assuming the existence of a system of things satisfying the conditions, "We think a system of things . . . " (Wir denken ein System von Dingen . . .). He proceeded methodologically in exactly the same way for *Grundlagen der Geometrie*, where the existential framework for the axioms of geometry is introduced by "We think three different systems of things . . . " (Wir denken drei verschiedene Systeme von Dingen . . .). Thus, as in Dedekind's case, there is an explicit existential assumption that has to be secured or discharged in some way. To emphasize this crucial aspect of Hilbert's method, both Hilbert and Bernays called it *existential axiomatics* (existentiale Axiomatik). In the case of geometry Hilbert discharged the existential assumption for (parts of) the axiom system by appropriate analytic models. But how could the problem be addressed for analysis? How could the existence of the system of reals be secured? In his answer, Hilbert referred to Cantor's distinction between consistent and inconsistent multiplicities that presented the former as the proper objects of set theory. Hilbert was critical of this distinction. His critical attitude, only implicit here, was made explicit in the Heidelberg talk of 1904 where he claimed that Cantor's conception "leaves latitude for subjective judgement and therefore affords no objective certainty."

1.2. A Partial Syntactic Turn

Dedekind had given a logical existence proof of a simply infinite system in order to guarantee that the very notion of such a system does not contain "internal contradictions." Hilbert recast consistency as a syntactic property of axiom systems, demanding that no contradiction be provable from the axioms in a finite number of steps.[16] That allowed him to attack the problem of arriving at a consistent multiplicity of real numbers in *Über den Zahlbegriff* and his Paris lecture from a new viewpoint: the existence of sets is to be guaranteed by *consistency proofs* for appropriate axiom systems. Hilbert had shifted from consistent multiplicities to consistent theories[17] and suggested to give objective content to Cantor's notion through consistency proofs for theories. In the Paris lecture he thought that a *direct proof* should be possible:

> I am convinced that it must be possible to find a direct proof for the consistency of the arithmetical axioms [proposed in "Über den Zahlbegriff" for the reals] by means of a careful study and suitable modification of the known methods of reasoning in the theory of irrational numbers.[18]

It is quite obscure from the published papers I referred to what would constitute a direct proof. A reasoned, though by no means unproblematic conjecture can be based on earlier lecture notes. Hilbert thought that the construction of the reals, and also of the natural numbers, could be given directly and be exploited as a blueprint for a Dedekindian consistency proof. This seems to be supported also by the (one-sidedly preserved) correspondence with Cantor at the time of the Munich and Paris talks. Cantor insisted in his letters on two points: (i) the consistency even of finite multiplicities (i.e., the existence of finite sets) has to be postulated, and (ii) Dedekind's considerations are fundamentally flawed. I conjecture Hilbert believed, despite the Cantorian admonitions, that Dedekind's logicism with suitable restrictions might after all provide the means for a principled consistency proof.[19]

Hilbert changed his views dramatically after Zermelo and Russell discovered their elementary contradiction, a contradiction that had according to his own testimony "a catastrophic effect in the mathematical world."[20] It had undoubtedly a catastrophic effect on Hilbert himself: Bernays reported to Constance Reid that Hilbert believed at the time, even if only for a very brief period, that Kronecker might have been right in demanding a radical restriction of mathematical notions and methods. In lecture notes from the summer of 1904, just before the Heidelberg talk in August of that year, one finds these illuminating and revealing remarks on Dedekind and Kronecker:

> He [Dedekind] arrived at the opinion that the standpoint of viewing the integers as obvious cannot be sustained; he recognized that the difficulties Kronecker saw in the definition of irrationals arise already for integers; furthermore, if they are removed here, they disappear there. This work [*Was sind und was sollen die Zahlen*] was epochal, but it did not yet provide something definitive, certain difficulties remain. These difficulties are connected, as with the definition of the irrationals, above all to the concept of the infinite.[21]

These difficulties were plainly stated at the very beginning of the Heidelberg lecture, where Hilbert described in detail alternative foundational approaches and remarked about Dedekind:

> R. Dedekind clearly recognized the mathematical difficulties encountered when a foundation is sought for the notion of number; for the first time he offered a construction of the theory of integers, and in fact an extremely sagacious one. However, I would call his method *transcendental* insofar as in proving the existence of the infinite he follows a method that, though its fundamental idea is used also by philosophers, I cannot recognize as practicable or secure because it employs the notion of the totality of all objects, which involves an unavoidable contradiction.[22]

What could be done?—Hilbert shifted, first of all, his efforts from the theory of real numbers to that of integers; he proposed, secondly, to give a genuinely direct proof of the existence of "the smallest infinite," and that was to be done by establishing the consistency of an axiom system that reflected Dedekind's conditions for a simply infinite system.

2. Direct Proofs

The elementary Zermelo-Russell paradox had convinced Hilbert, as we saw, that there *was* a problem with his earlier considerations and that difficulties had to be faced at a more fundamental level. In the Heidelberg

Lecture, Hilbert reasserted most strongly his view that the problems for the reals are resolved once matters are resolved for the natural numbers.

> The existence of the totality of real numbers can be demonstrated in a way similar to that in which the existence of the smallest infinite can be proved; in fact, the axioms for real numbers as I have set them up . . . can be expressed by precisely such formulas as the axioms hitherto assumed. . . . the axioms for the totality of real numbers do not differ qualitatively in any respect from, say, the axioms necessary for the definition of the integers. In the recognition of this fact lies, I believe, the real refutation of the conception of arithmetic associated with L. Kronecker.[23]

Hilbert actually claimed, "In the same way we can show that the fundamental notions of Cantor's set theory, in particular Cantor's alephs, have a consistent existence." Let us come back to the more modest goal of establishing the existence of the smallest infinite.

2.1. Turning Further

In section 1.2., I described the partial syntactic turn Hilbert had taken by recasting consistency as a syntactic notion. However, he had neither specified inference steps nor had he presented other than semantic arguments. That was remedied here at least in a broad programmatic sense: he developed logic and arithmetic simultaneously and inferred the consistency of the joint system from elementary syntactic observations. The methodological starting-point was formulated in this way:

> Arithmetic is often considered to be a part of logic, and the traditional fundamental logical notions are usually presupposed when it is a question of establishing a foundation for arithmetic. If we observe attentively, however, we realize that in the traditional exposition of the laws of logic certain fundamental arithmetic notions are already used, for example, the notion of set and, to some extent, also the notion of number. Thus we find ourselves turning in a circle, and that is why a partly simultaneous development of the laws of logic and of arithmetic is required if paradoxes are to be avoided.[24]

The theory Hilbert proposed consists of axioms for identity and Dedekind's requirements for a simply infinite system, except that induction is not explicitly formulated; in modern notation:

(1) $x = x$
(2) $x = y \;\&\; A(x) \to A(y)$
(3) $x' = y' \to x = y$
(4) $x' \neq 1$

The rules, extracted from Hilbert's description of "consequence," are modus ponens and a substitution rule that allows the replacement of variables by arbitrary sign combinations. Other modes of logical inferences are mentioned later, but neither formally stated nor incorporated into the consistency proof. The idea of the consistency proof is this: formulate a property P and show by induction on derivations that all provable equations have P. The property Hilbert considers is homogeneity: an equation a = b is called *homogeneous* if and only if a and b have the same number of symbol occurrences; it is easily seen that all equations derivable from axioms (1)–(3) are indeed homogeneous. But a contradiction can be obtained only by establishing an unnegated instance of (4) from (1)–(3). Such an instance is necessarily inhomogeneous and, consequently, not provable.

Hilbert saw his considerations as answering, for the first time, the earlier call for a direct proof. He commented:

> The considerations just sketched constitute the *first case* [my emphasis, WS] in which a direct proof of consistency has been successfully carried out for axioms, whereas the method of a suitable specialization, or of the construction of examples, which is otherwise customary for such proofs—in geometry in particular—necessarily fails here.[25]

Hilbert had emphasized the need to develop logic and mathematics simultaneously, but the actual work had significant shortcomings: there is no calculus for sentential logic, there is no proper treatment of quantification, and induction is neither rigorously formulated nor incorporated into the argument. In sum, there *is* an important shift from "semantic" arguments to a "syntactic" one, but the set-up is utterly incomplete as a formal framework for arithmetic.

2.2. Critical Analysis

The presumed foundational import of Hilbert's talk was not left unchallenged. On account of the inductive character of the consistency proof Poincaré criticized Hilbert's considerations severely; this critique is well-known and absolutely to the point. Toward the end of Hilbert's paper there is a peculiar "uncertainty" that reveals underlying methodological problems; they too were pointed out by Poincaré.[26] It becomes very clear how penetrating Poincaré's considerations were when one reads in parallel the 1935 remarks on Hilbert's Heidelberg talk by Bernays: they give a précis of Poincaré's critique. We should keep in mind, however, that the existence of sets was the central issue and was to be guaranteed by the consistency of an appropriate axiom system, a viewpoint shared explicitly and strongly by Poincaré. "If therefore," Poincaré wrote, "we have a system of postulates, and if we can demonstrate that these postulates imply no contradiction, we shall have the right to consider them as representing the definition of one of the notions entering therein."[27] In any event, as to the critical aspect Bernays wrote:

> the systematic standpoint of Hilbert's proof theory is not yet fully and clearly developed. Some places indicate that Hilbert wants to avoid the intuitive conception of number and replace it by its axiomatic introduction. Such a procedure would lead to a circle in the proof theoretic considerations.[28]

Bernays emphasized also that Hilbert had not articulated distinctions central to the later finitist program:

> Also, the viewpoint of restricting the contentual application of the forms of existential and general judgments is not yet put forth explicitly and completely.[29]

Hilbert's own views of his objective accomplishments were formulated in lectures from the summer term of 1905.[30] These lectures contain additional technical details, but point out basic shortcomings as well; Hilbert bemoans the unsatisfactory state of logic, in particular, the state of quantification theory. Hilbert had a distinctive approach already in the Heidelberg lecture, clearly recognized and applauded by Poincaré, as Hilbert's "all" ranged only over the limited domain of combinations of thought-objects, not as Russell's over everything whatsoever; see Poincaré 1905, 1040.

2.3. Proper Formalisms

During the period from 1905 to 1917 Hilbert gave almost annually lectures on the foundations of mathematics, but these lectures did neither break new ground, nor did they return to a proof theoretic study;[31] another approach opened up, however. Around 1913 Hilbert started to become familiar with some of Russell's writings. The official lecture notes from the winter term 1914–15 contain brief remarks about type theory, and the notes from a student, serendipitously preserved in the Institute for Advanced Study at Princeton, more extended ones. There was even some correspondence between Hilbert and Russell, reported in appendix B of Sieg 1999. A number of relevant talks on the foundations of mathematics were given in Göttingen during this period by Behmann, Bernstein, Hilbert, and Zermelo; most significantly, Hilbert directed Behmann's dissertation of 1918, *Die Antinomie der transfiniten Zahl und ihre Auflösung durch die Theorie von Russell und Whitehead*. A detailed description of this work is found in Mancosu 1999, narrowing the real gap in our historical understanding of the details of the Russellian influence on Hilbert.[32] How strongly Russell influenced Hilbert has been clear from the notes for his course on *Set Theory* (summer term 1917) and his Zürich talk *Axiomatisches Denken* given on September 11, 1917; they reveal renewed logicist tendencies in Hilbert's work. Hilbert wrote in the essay on which his talk had been based:

> The examination of consistency is an unavoidable task; thus, it seems to be necessary to axiomatize logic itself and to *show that number theory as well as set theory are just parts of logic.*

If we try to achieve such a reduction to logic, Hilbert said at the very end of the set theory notes, "we are facing one of the most difficult problems of mathematics."[33]

Russell and Whitehead provided not only the stimulus for this programmatic redirection, but also powerful technical tools.[34] The latter were ingeniously adapted and mathematically analyzed in the winter term 1917–18, when Hilbert offered lectures under the title *Prinzipien der Mathematik* with the assistance of Paul Bernays. Hilbert had finally a proper formalism for the development of mathematics: a language for capturing the logical form of informal statements and a calculus for representing the structure of logical arguments. The presentation is carried through with focus, elegance, and directness. The logical work is complemented by real metamathematical considerations; the latter are certainly inspired by the (perspective underlying the) work that had been done at the turn of the century on the foundations of geometry. These beautifully written, detailed notes include all the basic material that is contained in the 1928 book by Hilbert and Ackermann; thus, they are the real beginning of modern *mathematical* logic. For my purpose the main points can be summarized as follows: (i) there is a full development of (the syntax and semantics for) sentential, monadic, first-order logic, and ramified analysis, (ii) independence and completetness problems are formulated and partly solved, and (iii) theories are always presented with appropriate domains or, more precisely, many-sorted structures. The last point brings out in this setting the crucial aspect of existential axiomatics that had been so important in Hilbert's early investigations (see section 1.1.).

Absolutely no proof theoretic considerations are presented in these notes, though consistency is a real issue. The consistency of pure logic is examined, and both sentential and first-order logic are semantically shown to be consistent, the latter by considering a one-element domain. A footnote warns the reader, however, not to overestimate the significance of this result for first-order logic, because "[i]t does not give us a guarantee that the system of provable formulas remains free of contradictions after the symbolic introduction of contentually correct assumptions."[35] After all, these assumptions may force the domain to be infinite.

At the beginning of 1920, having abandoned for good reasons the logicist route and responding in part to the contemporaneous investigations of Brouwer and Weyl, Hilbert and Bernays pursued a radically constructive redevelopment of arithmetic. This took up a recurring Kroneckerian theme in Hilbert's foundational reflections. However, it was realized quickly that this could not provide a foundation for classical forms of reasoning, as the law of excluded middle does not hold for constructively understood quantified statements. Having recognized this fact, Hilbert and Bernays mentioned Brouwer for the first time when closing with: "This consideration helps us to gain an understanding for the sense of the paradoxical claim, made recently by Brouwer, that for infinite systems the law of the excluded middle (the 'tertium non datur') loses its validity."[36]

To us it may seem as if Hilbert had available all the mathematical and logical means for the formulation of *the* program. Yet it took some more time before he had gained the appropriate methodological perspective, and before finitist mathematics and proof theory emerged in a programmatically coherent alignment.

3. Proof Theoretic Strategies

Hilbert had argued for a *theory of proofs* in his 1904 Heidelberg talk; he had mentioned it also in his 1917 Zürich talk, but without any programmatic direction. The suggestion was finally taken up again in the summer semester of 1920: the notes from that term contain a consistency proof for a restricted part of elementary number theory. Indeed, it is (almost exactly) the system of the Heidelberg Lecture.[37]

3.1. Turning Further, Ctd.

The syntactic turn in treating the consistency problem, from choosing a syntactic formulation of consistency to developing logic and arithmetic simultaneously, is pursued further in these notes. The description of the system of elementary arithmetic is given in a more coherent way and is evidently informed by the logical work of the prior years. Attention is paid to the mathematical means used in the proof theoretic arguments, but the formalism that is being investigated is semi-constructive; cf. the end of this subsection. The formalism is almost exactly that of the Heidelberg Lecture, but the argument for its consistency is quite different, mainly through the introduction and use of the notion "kürzbar":

If one considers a proof with respect to a particular concrete property it has, then it is possible that the removal of some formulae in this proof still leaves us with a proof that has that particular property. In this case we are going to say that the proof is *kürzbar* with respect to the given property.[38]

Hilbert establishes three lemmata: the first claims that a theorem can contain at most two occurrences of ->, the second asserts that no statement of the form (A -> B) -> C can be proved, and the third expresses that a formula a = b is provable only if a and b are the same term. To recognize the distinctive character of the arguments, let me look at the proof of the first lemma. Hilbert proceeds indirectly and assumes that there is a theorem with at least three occurrences of ->. Without loss of generality he further assumes that the theorem has a proof that is not "kürzbar" (with regard to the property of having an end formula with at least three occurrences of the ->). The theorem cannot be an axiom, as axioms contain at most two occurrences of the ->. Thus, it must have been obtained by modus ponens. The major premise of that inference contains at least one more occurrence of -> than its conclusion, i.e., the given theorem; we have consequently a contradiction to the "Nicht-Kürzbarkeit" of the given proof.

On its surface, Hilbert's new proof does not use the induction principle. It is structured in analogy to the standard proof of the fact that $\sqrt{2}$ is not rational. Hilbert frequently asserted, not only here, that consistency proofs should be of the same character as the proof of the irrationality of $\sqrt{2}$.[39] The analogy plays even on the double meaning of "kürzbar"; on the one hand, "kürzbar" means, when applied to proofs, "can be shortened," but on the other hand it also applies to fractions and then means "(a common factor) can be canceled." Recall that the standard argument proceeds also indirectly, assuming that $\sqrt{2}$ is rational, i.e., equals p/q, q ≠ 0; without loss of generality it is then assumed further that p/q is not "kürzbar." In his 1922 publication, based on lectures he had given in the spring and summer of 1921 in Kopenhagen and Hamburg (and submitted for publication not before November of 1921), Hilbert makes explicit the strategic point of the modified argument:

> Poincaré's objection, that the principle of complete induction cannot be proved but by complete induction, has been refuted by my theory.[40]

Is this to be taken in the strong sense that induction is not used at all? Or is it to be understood, perhaps, just in the weaker sense that a special procedure is being used—a procedure based on the construction and deconstruction of numerals and that, by its very nature, is different from the induction principle?[41] From a mathematical point of view Hilbert used the least number principle in an elementary form, namely, applied to a purely existential statement. The work in this first of Hilbert's foundational articles of the twenties is evidently transitional. It does have a major problem in not recognizing clearly necessary metamathematical means, but also in not fixing appropriately the very logic of the formal system to be investigated: Hilbert tried to keep it constructive by using, for example, negation only in a restricted way.[42]

3.2. Principled Formulation

Proof theoretic considerations were pursued with novel metamathematical means and with a principled foundational perspective in the lectures from the winter term 1921–22. For the first time, Hilbert and Bernays used the terms *finitist mathematics* and *Hilbert's proof theory* and made explicit the domain of mathematical (finitist) objects appealed to in proof theoretic investigations. They pointed out:

> We have to extend the domain of objects to be considered; i.e., we have to apply our intuitive considerations also to figures that are not number signs. Thus, we have good reason to distance ourselves from the earlier dominant principle according to which each theorem of pure mathematics is in the end a statement concerning integers.

With a jibe at such distinguished mathematicians as Dirichlet and Dedekind, they continued, "This principle was viewed as expressing a fundamental methodological insight, but it has to be given up as a prejudice."[43] After all, formulas and proofs of formal theories are the direct object of proof theoretic investigation, and appropriate definition and proof principles (analogous to those for numbers) have to be used.

Hilbert proved the consistency of a fragment of number theory with *full classical sentential logic* and free variable statements. The very elaborate and detailed proof given in the notes was sketched in Hilbert's Leipzig talk of September 22, 1922.[44] A treatment of quantifiers is indicated there, and genuine transformations of formal proofs are used to carry out the argument. Equally striking is the underlying idea that expresses in a novel, precise way the nineteenth-century methodological maxim that elementary statements should be provable by elementary means. *Elementary* statements are those formulas, also called *numeric*, that are built up solely from =, ≠, numerals, and sentential logical connectives.

The central steps of the proof theoretic argument are described easily: (i) formal proofs with a numeric end formula are transformed into configurations that are not necessarily proofs, but consist only of numeric formulas; (ii) formulas in these configurations are all effectively brought into normal forms; (iii) the resulting normal form statements are all recognized to be "true." Given a formal proof of $0 \neq 0$, the transformations leave the end formula unchanged. From (iii) and the fact that $0 \neq 0$ is not true, it follows that $0 \neq 0$ is not provable. Clearly, these considerations are preliminary in the sense that they concern a theory that is part of finitist mathematics and thus need not be secured by a consistency proof. The next step is crucial with regard to the real issue of securing parts of mathematics that properly extend finitist mathematics.

Hilbert treats quantifiers with the τ-function, the dual of the later ε-operator; τ associates with every predicate $A(a)$ a particular object $\tau a(A(a))$ or simply τA. The *transfinite axiom* $A(\tau A) \rightarrow A(a)$ expresses, according to Hilbert, "if a predicate A holds for the object τA, then it holds for all objects a." The τ-operator allows the definition of the quantifiers:

$$(\forall a)\ A(a) \leftrightarrow A(\tau A),$$
$$(\exists a)\ A(a) \leftrightarrow A(\tau(\neg A)).$$

Hilbert extends the proof theoretic considerations to the "first and simplest case" of going beyond the finitist system. The technique used will become the ε-substitution method, allowing the elimination of quantifiers from proofs of quantifier-free statements. Thus, finitist proof theory is given not only its principled formulation, but also its guiding idea (reflection principle)[45] and a dual version of its technical tool (ε-calculus). Bernays writes in 1935: "With the presentation of proof theory as given in the Leipzig talk the principled form of its structure had been reached."[46] Ackermann's thesis, published in 1925, is a direct continuation of Hilbert's paper.

3.3. Uniform Projection

Instead of pursuing the all-too-well-known sequence that starts with Hilbert and Bernays, goes through Ackermann, von Neumann, Herbrand, and then ends with Gödel, I turn to the question: What is the informal idea underlying the proof theoretic work, including the very idea of formalizing mathematical theories? An answer to this question is found most directly in papers by Bernays from 1922 and in the related lecture notes from 1921–22 and 1922–23.[47] As we saw, Hilbert's *existential axiomatics* assumed always a system of objects satisfying the axiomatic conditions, and Bernays remarked:

> In the assumption of such a system with particular structural properties lies something so-to-speak transcendental for mathematics, and the question arises which principled position with respect to it should be taken.

An intuitive grasp of the completed sequence of natural numbers, for example, or of the manifold of real numbers should not be excluded outright. However, taking into account tendencies in the exact sciences, Bernays suggested a different strategic direction, namely, to try "whether it is not possible to give a foundation to these transcendental assumptions in such a way that only primitive intuitive knowledge is used." This suggestion is supplemented by a wonderful image of how to exploit the formalizability of axiomatic theories for this goal: their formalization serves to *project* the associated structures uniformly into the proper mathematical, finitist domain. Even fifty years later Bernays used that image and emphasized the epistemological significance of such projections:

> In taking the deductive structure of a formalized theory . . . as an object of investigation the [contentual] theory is projected as it were into the number theoretic domain. The number theoretic structure thus obtained is in general essentially different from the structure intended by the [contentual] theory. But it [the number theoretic structure] can serve to recognize the consistency of the theory from a standpoint that is more elementary than the assumption of the intended structure.[48]

The reader may consider this image as merely playful or as genuinely helpful. I choose the latter view, because the substantive point can be recast, and was recast by Bernays in 1930, as an explication of Hilbert's existential axiomatics that reveals a thoroughly structuralist perspective. "Structuralist" is here to be taken in the modern philosophical sense as described so masterfully in Parsons 1990 and discussed extensively in Shapiro 1997. Parsons states there that views of this kind can be traced back to the end of the nineteenth century, but attributes clear general statements only to Bernays in 1950 and to Quine somewhat later.[49] However, Bernays points already in 1930 to a characteristic aspect of the *newer mathematics* and describes the subject repeatedly as the study of structures (e.g., on p. 32). He presents concisely the standard account of if-then-ism or deductivism, and raises—as the starting point of his systematic philosophical investigations—the vacuity issue for that account. He takes this problematic as arising from two moments of modern axiomatics, namely, (i) the purely hypothetical connection between axioms and theorems, abstracting from the content and truth of the axioms, and (ii) the existential formulation of mathematical theories, assuming a given and from the very beginning fixed system of things and relations pertaining to them.[50] Let me present Bernays's discussion (on pp. 20–21) of the central point in greater detail.

Given the perspective on modern axiomatics I just sketched, the axioms and theorems of an axiomatic theory are statements that concern the relations occurring in them, and the relations pertain to the things of an assumed system. The knowledge provided by a proof of a theorem (Lehrsatz) L from axioms $A_1, \ldots, A_k$ consists in the determination (Feststellung) that, if the statements $A_1, \ldots, A_k$ hold for the relations, then the statement L also holds for these relations. Here we have, as Bernays puts it, a general theorem on relations, i.e., a theorem of pure logic: the results of an axiomatic theory present themselves as theorems of logic. However, these theorems have significance only if the axiomatic conditions can be satisfied at all:

> If such a satisfying structure is unthinkable, i.e., logically impossible, then the axiom system does not lead to any theory at all, and the only logically meaningful statement concerning the system [of axioms] is thus the determination (Feststellung) of the contradiction following from the axioms. For this reason there is for every axiomatic theory the requirement of a proof of the *satisfiability*, i.e., the *consistency* of its axioms.[51]

Bernays observes further that such proofs are usually given by providing arithmetical models, unless one can get by with the construction of finite ones. Thus, Bernays has retraced Hilbert's motivation for a consistency proof or, perhaps better, isolated the methodological core of his considerations. What is surprising after Herbrand's and Gödel's dissertations is what seems to be an unapologetic identification of satisfiability and consistency. In an unpublished note (found in the appendix), Bernays describes this connection properly and in harmony with the principles guiding proof theoretic investigations. In any event, Bernays formulates here first a position of (what Parsons calls) *eliminative structuralism* in a concise way. That position is complemented by principled, philosophical reflections and programmatic, mathematical efforts to obtain finitist consistency proofs. It is this additional reflective perspective that allows us to see Hilbert and Bernays's structuralism as being of the noneliminative variety; see the beginning of section 4.2.

We saw how Hilbert and Bernays tried to exploit the special epistemological status of finitist mathematics for consistency proofs. After the discovery of Gödel's Second Incompleteness Theorem, however, the fundamental status of finitist mathematics had to be given up, and finitist considerations had to be expanded by considerations in stronger theories. To avoid a threatening vicious circle, as in Hilbert 1905, these stronger theories had to be constructively motivated. The first consistency proof satisfying such an informal demand was obtained independently by Gödel and Gentzen and established the consistency of classical arithmetic relative to its intuitionistic version: the latter theory was indeed based on an extended constructive viewpoint, ironically, the intuitionistic one. Bernays

(1954) called this *sharpened axiomatics* (verschärfte Axiomatik) and formulated as a minimal requirement that "the objects [making up the intended model of the theory] are not taken from a domain that is thought as being already given, but are rather constituted by generative processes."[52] There is no indication in Bernays's 1954 or in his later writings what kind of generative processes should be considered, and why that particular feature of domains should play a distinctive, foundational role. These two issues are at the center of the considerations in the next section.

4. Accessible Domains

When Gödel considered Platonism still as a doctrine "which cannot justify any critical mind and which does not even produce the conviction that they [the axioms of set theory] are consistent," he analyzed also different layers of constructive mathematics in most informative ways. The lowest layer, identified with finitist mathematics, had one important characteristic:

> The application of the notion "all" or "any" is to be restricted to those infinite totalities for which we can give a finite procedure for generating all their elements (as we can, e.g., for the totality of integers by the process of forming the next greater integer and as we cannot, e.g., for the totality of all properties of integers).[53]

Directly associated with the procedure for generating the integers are the principles of proof by induction and definition by recursion. There is no further analysis of this direct association; the principles are simply taken to have a high degree of evidence. Can one go beyond such a brief, purely descriptive account and, perhaps, extend the considerations to other classes of mathematical objects?

4.1. Finitist Objects and Processes

Gödel considered the totality of integers as just one example of totalities whose elements are generated by a finite procedure. That a greater class of such totalities has directly associated principles had been emphasized already by Poincaré (1905). After a discussion of the induction principle for natural numbers, he remarked:

> I did not mean to say, as has been supposed, that all mathematical reasonings can be reduced to an application of this principle. Examining these reasonings closely, we should see applied there many analogous principles, presenting the same essential characteristics. In this category of principles, that of complete

induction is only the simplest of all and this is why I have chosen it as a type. (1025)

Modern expositions and critical examinations of Hilbert's considerations, for example, those of Parsons and Tait, focus on natural numbers. As a matter of fact, so did Bernays (1930), but he viewed the case of numbers as paradigmatic and embedded it into broader reflections on the nature of mathematical knowledge; the latter was to be captured in a principled way, independently of the current inventory of mathematical disciplines. Bernays viewed a *certain kind of abstraction* as distinctive for the nature of mathematical thought:

> This abstraction may be called formal or mathematical abstraction. It consists in emphasizing the structural moments of an object, i.e., the way it is composed from parts, and taking them exclusively into consideration; 'object' is here to be understood in the broadest sense. Accordingly, mathematical knowledge can be defined as knowledge based on the structural consideration of objects.[54]

The crucial questions are undoubtedly, what is the extension of *object*, and what kind of objects can be considered or viewed *structurally*. The second question implicitly concerns the boundary between mathematical knowledge secured by intuition (Anschauung), respectively obtained by thinking (Denken) and systematic extrapolation; it is to this question that Bernays turned.

Bernays's analysis of intuitive mathematical knowledge attempts to balance, uneasily, the philosophical demand for intuitive concreteness and the mathematical need for formal abstractness. The tension comes to the fore in first taking formal abstraction as the characteristic feature of intuitive mathematical knowledge, and in then claiming that it is naturally bound to finiteness and finds a principled delimitation only when facing the infinite.[55] Precisely this coextensiveness of finite and intuitive—and thus the sharp differentiation of the intuitive from the nonintuitive—was questioned by Bernays himself in the *Nachtrag* (1930), when arguing that the epistemological considerations underlying his paper should be revised in light of Gödel's results:

> Of course, the positive remarks, in particular those bringing out the mathematical element in logic and those highlighting elementary arithmetical evidence, are hardly in need of revision. It seems, however, that the sharp differentiation between the intuitive and non-intuitive, as used in treating the problem of the infinite, cannot be carried through this strictly. In this respect then, the view on the formation of mathematical ideas has to be worked out in further detail.[56]

Bernays refers to his later essays in *Abhandlungen zur Philosophie der Mathematik* as containing considerations to address this fundamental issue. His arguments of 1930 provide, it seems to me, an excellent starting point, as their detailed examination uncovers revealing difficulties. Thus, I will focus on them without relating them at this occasion to the broader philosophical framework in which they have their systematic place. That framework is deeply influenced by Kant, Fries, and Nelson, with Bernays keeping however a distinctive critical distance. The curious reader may consult Bernays's papers (1928, 1928a, 1930a, and 1937); to recognize that some of the issues discussed below are parallel to (still unresolved) problems in Kant's philosophy of arithmetic, see in particular the writings of Parsons, e.g., 1980, 1982, 1984, and 1994.

The arguments that support the uneasy balancing act between philosophically motivated concreteness and mathematically necessitated abstractness and generality are at crucial places strained. That is most evident, when formal abstraction is supposed to help us in going beyond the limits of our real power of representation (our "faktische Vorstellungskraft" or "tatsächliches Vorstellungsvermögen"); here, *intuitiveness of objects is to be secured by intuitiveness of processes generating them.* A similar step from objects to processes is taken when Bernays argues next that formal abstraction is essentially bound to the "moment of finiteness." Indeed, Bernays claims, finiteness is not at all a restrictive feature of objects from the standpoint of intuitive evidence: the finiteness of objects is obvious for formal abstraction ("die Endlichkeit der Objekte versteht sich für die formale Abstraktion ganz von selbst"). Why should that be? The answer to this question is given by Bernays paradigmatically for the case of numbers and appeals to their introduction as the "simplest formal objects" by iteration of a successor operation. This intuitive-structural introduction of numbers is appropriate, Bernays claims, only for finite numbers, as repetition is from the standpoint of "intuitive-formal considerations" *eo ipso* finite repetition. In short, and in parallel to the above italicized claim, *finiteness of these objects is to be secured by finiteness of the underlying generative process.*[57]

Finally, according to Bernays, it is the intuitive representation of the finite (die anschauliche Vorstellung des Endlichen) that provides the justification (Erkenntnisgrund) for the principle of complete induction and for the admissibility of recursive definitions, both in their elementary forms. Such a representation of the finite is thus explicitly used when reflecting on general characteristics of intuitive objects, and it is a crucial presupposition for the proof theoretic approach.

> The *intuitive representation of the finite* [my emphasis, WS] is forced on us, as soon as a formalism is turned into an object of investigation, thus especially in

the systematic theory of logical inferences. This brings out that finiteness is an essential moment of any formalism whatsoever.[58]

Bernays's analysis is consequently also basic for other domains whose elements are generated in elementary ways, especially for the domains of syntactic objects needed in proof theoretic investigations. Indeed, he continues by claiming that the limits of formalisms coincide with those of the general representability of intuitive combinations. (Die Grenzen des Formalismus sind aber keine anderen als die der Vorstellbarkeit überhaupt von anschaulichen Zusammensetzungen.)

How do these considerations compare with Hilbert and Bernays's earlier ones? Should the objects obtained through such elementary generation satisfy the demand articulated in the notes from 1921–22, namely, that "the figures we take as objects must be completely surveyable and only discrete determinations are to be considered for them"? Surveyability was then thought to insure that "our claims and considerations have the same reliability and clarity (Handgreiflichkeit) as in intuitive number theory." Against the backdrop of the generation of numerals we have here the same tension as in the considerations by Bernays, just replacing "surveyability" with "intuitive representability". In order to ground mathematical principles for finitist objects, the elementary and uniform generation of figures has to be appealed to—leading to *purely formal objects* of appropriate abstractness; in order to ground philosophical reflections on the primitive intuitive character of finitist mathematical knowledge, focus is shifted to the surveyability or intuitive representability of individual mathematical objects. Indeed, Bernays claims that we are free to represent (repräsentieren) the purely formal objects by concrete objects (e.g., numbers by numerals) in such a way that these representing concrete objects are intuitable and contain in their structure the essential properties of the represented objects, so that "the relations—to be investigated and holding— between the represented objects are found also between the representatives and can be ascertained by considering the representatives alone."[59]

The deep conflict that is apparent in this intricate discussion is not resolved by argument, but by *fiat*: numbers and other purely formal objects *just are* intuitively given—via representing concrete objects. It is most interesting to observe that Bernays (1934) contemplates narrower and admittedly vague boundaries for what is intuitive and distinguishes between numbers that are *reachable* (zugänglich) and those that are not. He does so in a critical discussion of intuitionism and intuitive evidence, viewing as reachable those numbers that do not outstrip our actual power of representation (Vorstellungskraft).[60] He suggests also a way of extending mathematical knowledge from reachable numbers to unreachable ones by the *general method of analogy* (die allgemeine Methode der Analogie), i.e., by extending the relations that can be verified for the former numbers to the latter. However one may want to interpret this, it seems clear that finitist mathematics is not secured by intuitive evidence alone. For an adequate conceptual analysis of finitist mathematics one has to go beyond (the representation of) finiteness, admit rather abstract means for capturing the arbitrary finite iteration of elementary steps, and grant in the end for potentially infinite domains what Bernays asserts for actually infinite ones, namely, that they can be characterized only by way of a lawful relation (gesetzliche Beziehung). However, the distinctive feature of domains with generated elements is that their lawful relation is not just assumed or claimed, but rooted in our understanding of the underlying generative process; that understanding allows us also to recognize induction principles for proofs and recursion principles for functions.

Bernays's broad *informal considerations* leading up to the natural numbers as *unique* (eindeutige) and *purely formal objects* (distinct from formal objects, i.e. types, that allow different concrete instantiations by tokens) are very appealing and, in a deep sense, similar to those of Dedekind, Helmholtz, and Kronecker.[61] However, only Dedekind, who reported in the letter to Keferstein on the informal reflections underlying his work of 1888 took the further step to a sharp and completely novel mathematical formulation. The latter makes crucial use of infinite sets and in particular of *simply infinite systems*. Following Dedekind, but avoiding infinite sets, Zermelo (1909) presented an analysis based on finite sets and "*simply finite systems.*" A central question is, *can Zermelo's considerations provide the mathematical basis for a detailed conceptual analysis of natural numbers?* (Bernays's natural numbers as purely formal objects might be obtained then by Tait's "Dedekind abstraction" applied to simply finite systems. Zermelo's work and its connections to that of others is described by Parsons 1987.)

4.2. I.d. Classes and Abstract Notions

Hilbert and Bernays's structuralism, when joined with their finitist methodological reflections, is really a structuralism of Parsons's *noneliminative* variety: it accepts basic, potentially infinite domains of constructed objects, in particular, of natural numbers and syntactic figures constituting formalisms.[62] I propose to call their structuralism and extensions thereof *reductive*, because of the special justificatory role the basic structures play. This is in analogy to "reductive proof theory." The systematic connection will become clear, I trust, from the following considerations that concern the challenging question, how to extend the preliminary and not unproblematic reflections of section 4.1 to appropriate infinitary configurations. Let me make this challenge concrete by describing one paradigmatic result of reductive proof theory. The domains of constructed objects are here the higher constructive number classes.

Brouwer (1927) considered infinite proofs[63] and treated them as well-founded trees, i.e., as constructive ordinals of the second number class $\mathbf{O}$. The latter are inductively generated according to the following clauses:

> 0 is in $\mathbf{O}$;
>
> if a is in $\mathbf{O}$, then the successor of a is in $\mathbf{O}$;
>
> if f is a function from $\mathbf{N}$ to $\mathbf{O}$ and, for all n in $\mathbf{N}$, f(n) is in $\mathbf{O}$, then the supremum of the f(n) is also in $\mathbf{O}$.

Even higher number classes were inductively defined by Brouwer (and by Hilbert in *Über das Unendliche*); the trees branch over $\mathbf{N}$ and over previously obtained number classes. These are quite complex i.d. classes, but acceptable to at least some constructivists, among them Bishop, Lorenzen, Myhill, but also Martin-Löf. Iterated i.d. classes were at the center of the foundational investigations in the Stanford Report and much subsequent work; see Feferman 1981. Church and Kleene had formulated already in the midthirties recursive analogues of the higher constructive number classes by requiring that the function f in the third defining clause be (partial) recursive. The elements of $\mathbf{O}$ can be pictured as infinitary trees that are uniformly and effectively generated; indeed, arbitrary finite subtrees can be effectively determined.

With that specification of constructive function it is quite direct to formulate proof and definition principles for the finite constructive number classes in the language of elementary number theory expanded by predicate symbols for the number classes. The resulting theory, based on intuitionistic logic, is denoted by $\text{ID}^i(\mathbf{O})_{<\omega}$. The paradigmatic result I want to discuss briefly reduces the impredicative subsystem of classical analysis $(\Pi_1^1\text{—CA})_0$ to the theory $\text{ID}^i(\mathbf{O})_{<\omega}$.[64] This reduction is pleasing for two reasons, especially, if one is affected by the implicit irony: (i) $(\Pi_1^1\text{—CA})_0$ suffices as a comprehensive formal framework for the development of mathematical analysis presented in the fourth supplement of Hilbert and Bernays's *Grundlagen der Mathematik II*, and (ii) Brouwer's constructive number classes provide the objective underpinnings for proving the consistency of a blatantly impredicative classical theory. There is a great deal of contemporary work in proof theory that extends this kind of result mainly by providing ordinal analyses for stronger theories. However, attention has been shifted from subsystems of analysis to a more uniform setting of subsystems of set theory, and the systems of ordinal notations needed for the proof theoretic analysis are connected rather directly with large cardinals in set theory.[65]

Aczel (1977) presented a very general notion of i.d. classes, that is, of classes given by inductive definitions in the broadest sense. All the examples I mentioned (the elementary i.d. classes of terms, formulas, and proofs constituting a formal theory, the Brouwerian constructive ordinals) fall under Aczel's notion. Indeed, Aczel's notion is so general that it encompasses also segments of the cumulative hierarchy of sets. These i.d. classes are in Aczel's terminology *deterministic* and, thus, guarantee the unique generation of the objects falling under them. Given an understanding of the uniform generation steps, the resulting processes allow us to understand the build-up of objects and to recognize proof and definition principles for the domains constituted by them. I call such i.d. classes *accessible domains*[66] and would like to see an abstract mathematical description that highlights their distinctive features. Joyal and Moerdijk's book *Algebraic Set Theory* and the subsequent paper (2000) by Moerdijk and Palmgren take, it seems, interesting steps that characterize some classes of accessible domains from the perspective of category theory. Here then is the general question, namely, *can one give a category theoretic characterization of accessible domains?*

The crucial task for Hilbert's proof theory is to insure the consistency of the "idea of the infinite totality of numbers and of number sets." That is formulated in Bernays 1930 in accord with the historical development sketched in sections 1 through 3, and it required then the use of finitist methods. The task is taken up again in Bernays 1934 with a broadened methodological perspective. The assumptions of totalities underlying mathematical theories are called *Platonist*; the condition that restricts their use, as well as the application of the principle of analogy, is described as follows:

> The assumptions we are dealing with amount to representations of totalities and to the principle of analogy or of the permanence of laws. And the condition restricting the application of these leading ideas is none other than the consistency of the consequences that can be drawn from those basic assumptions.[67]

Accessible domains have a foundational role in providing the means for consistency proofs, whether syntactic, proof-theoretic, or semantic, model-theoretic ones. Detailed mathematical and philosophical analyses of accessible domains will allow us to make informative distinctions that concern (constructive) generating operations and their (transfinite) iteration, but also fundamental deductive principles. This leads to a rather natural question, namely, *can theories for accessible domains be given in such a form that their classical versions are uniformly reducible to their intuitionistic variants?* As I consider suitable segments of the cumulative hierarchy as accessible domains, the investigations of axioms of infinity (i.e., large cardinal assumptions) are of deep conceptual interest and increasingly connected to proof theoretic work, as I mentioned above. This is a part of set theory, where wide-ranging Platonist assumptions in Bernays's sense are being made, and where their consequences on more concrete mathematical problems are being explored. This latter point was emphasized already by Gödel and has been pursued most vigorously by H. Friedman.

Accessible domains reflect the constructive or, if you wish, *quasi-constructive aspect* of mathematical experience; abstract notions like groups, fields, topological spaces reflect its *conceptional aspect*. These two aspects should be contrasted rather sharply. Accessible domains allow us to formulate correct fundamental principles, whereas abstract notions are distilled from mathematical practice to make precise analogies between different areas; that is done for the purpose of comprehending complex connections and obtaining a more profound understanding.[68] I have stressed (1990, 284–85) the broad significance of this distinction; in addition, I argued specifically that the notion of a complete ordered field (characterizing its model, the reals, up to isomorphism) is an abstract one. How different this case is from that of a simply infinite system is revealed by an analysis of the categoricity proofs: in the one case the desired isomorphism follows the build-up of the objects in the domain, whereas in the other the topological completeness has to be appealed to. Thus, there are two complementary important tasks: to analyze the principles for accessible domains and to establish the consistency of abstract notions relative to (theories for) appropriate domains.[69]

Let me return to Hilbert. As philosophers, mathematicians, and scientists we should explore Hilbert's broad insights into the complex workings of mathematics instead of keeping him shackled to a narrow foundational position that was taken for programmatic reasons in the twenties. Hilbert's particular proposal for mediating between constructivist and classical positions did not work out. However, the reductive program that emerged from it provides, in my view, an important perspective on mathematical experience. *Reductive structuralism* allows us to connect, in a prima facie coherent way, developments in (the foundations of) mathematics and more directly philosophical studies; it helps us to gain a better understanding of the distinctive character of modern mathematics and its role in our broader intellectual enterprises, in particular its role in the sciences.[70] Hilbert's modernized self can be taken as arguing for creative freedom along two dimensions: *constructions* and *abstract concepts*; the former call for abstract analysis, the latter for constructed models.

5. Concluding Remark

I want to end with a most appropriate comment of Stein's on Hilbert; it mirrors the remark I quoted from Hilbert's Paris Lecture in section 0. After complaining gently about Hilbert's insistence (in his later foundational investigations) that the statements of ordinary mathematics are meaningless and only finitist statements have meaning, Stein points out:

> Hilbert certainly never abandoned the view that mathematics is an organon for the sciences . . . ; and he surely did not think that physics is meaningless, or its discourse a play with "blind" symbols. His point is, I think, this rather: that the mathematical *logos* has no responsibility to any imposed *standard* of meaning: not to Kantian or Brouwerian "intuition", not to finite or effective decidability, not to anyone's metaphysical standards for "ontology"; its *sole* "formal" or "legal" responsibility is to be consistent (of course, it has also what one might call a "moral" or "aesthetic" responsibility: to be useful, or interesting, or beautiful; but to this it cannot be constrained—poetry is not produced through censorship).[71]

The mathematical (in particular, proof theoretic) and philosophical challenge is, of course, to analyze on what basis we can live up to the responsibility of being consistent.

APPENDIX

This appendix consists of a note by Bernays that is found in the Hilbert Nachlaß, cod. 685:9, 2 and was written, presumably, between 1925 and 1928. It is entitled *Existenz und Widerspruchsfreiheit.* The note is preceded by a brief note on the finitist standpoint (containing only well-known observations) and followed by a note analyzing the criticism of axiomatic set theory by Skolem and von Neumann. For our purposes the latter note is of interest only by the way in which Bernays explicates "consistency of the countable infinite." The consistency proof of arithmetic (including the transfinite axioms for the epsilon operator) establishes the consistency of the countable infinite in the following sense: "An axiom system has been recognized as consistent that cannot be satisfied by a finite system of objects." Clearly, this topic is taken up in a very illuminating way in Bernays 1950.

Existence and Consistency

The claim: "Existence = consistency" can only refer to a system *as a whole*. *Within* an axiomatic system the axioms decide about the existence of objects.

If, for a system as a whole, consistency is to be synonymous with existence, then the proof of consistency must consist in an exhibition [of a model].

(All consistency proofs up to now have been either direct exhibitions or indirect ones by reduction; in the latter case a certain other system is already taken as existent. *Frege* has defended with particular emphasis the view that any proof of consistency has to be given by the actual presentation of a system of objects.)

In proof theory, laying a new foundation of arithmetic, consistency proofs are *not* given by exhibition. From this foundational standpoint it does not hold any longer that existence equals consistency. Indeed, it is not the opinion that the possibility of an infinite system is to be proved, rather it is only to be shown that *operating with such a system* does not lead to contradictions in mathematical reasoning [beim Schliessen].

Existenz und Widerspruchsfreiheit

Die Behauptung: "Existenz = Widerspruchsfreiheit" kann sich immer nur auf ein System *als Ganzes* beziehen. *Innerhalb* eines axiomatischen Systems wird über die Existenz von Dingen durch die Axiome entschieden.

Soll für ein System als Ganzes die Widerspruchsfreiheit mit der Existenz gleichbedeutend sein, so muss der Beweis der Widerspruchsfreiheit in einer Aufweisung bestehen.

(Alle bisherigen Wf.-Beweise sind auch entweder direkte Aufweisungen oder indirekte durch Zurückführung, wobei dann ein gewisses anderes System schon als existent angenommen wird. *Frege* hat bes. nachdrücklich den Standpunkt vertreten, dass der Nachweis der Widerspruchsfreiheit durch wirkliche Aufzeigung eines Systems von Dingen geschehen müsse.)

In der neuen Grundlegung der Arithmetik durch die Beweistheorie geschieht der Wf.-Beweis *nicht* durch eine Aufweisung. Vom Standpunkt dieser Begründung gilt auch nicht mehr, dass Existenz = Widerspruchsfreiheit ist. In der Tat ist ja auch gar nicht die Meinung, dass die Möglichkeit eines unendlichen Systems erwiesen werden soll, vielmehr soll nur gezeigt werden, dass das *Operieren mit einem solchen System* beim Schliessen nicht zu Widersprüchen führt.

NOTES

1. In Paetzold 1995, p. 112. "Immer wieder schärft er seinen Zuhörern ein, daß alle historische Betrachtung letztlich—symbolisch—im Dienste der Gestaltung der Gegenwart und der Zukunft steht."

2. See p. 540 ff. in Weyl 1925, also pp. 482–84 in Weyl 1928.

3. Brouwer lists four basic insights. The first two are constitutive of Hilbert's proof theoretic program, and Brouwer emphasizes that they have been "understood and accepted in the formalistic literature," not without claiming that they have been taken over—without proper acknowledgement—from intuitionism. The first insight concerns the distinction between the construction of formalistic mathematics and the intuitive (contentual) metamathematics concerning this construction; the second insight points to the problematic character of the law of excluded middle (lem). The remaining two insights are formulated straightforwardly and with great clarity, namely, that the lem is identical with the claim that every mathematical problem is solvable, and that consistency does not guarantee correctness. The fourth insight can be reformulated as stating that consistency does not provide a contentual justification of formalistic mathematics. The third point would have been disputed, and the fourth was in this general formulation undoubtedly clear to Hilbert through his early work on non-Euclidean geometries. Brouwer made also a very specific claim concerning the lem as part of the fourth insight, namely, that its correctness can be justified only by the lem itself. This claim was taken back in Brouwer 1953, 14, fn. 1; see the editor's addition to footnote 8 in Brouwer 1927 on p. 460 of van Heijenoort. The substance of the claim had been refuted already earlier by the Gödel-Gentzen reduction of classical to intuitionistic arithmetic.

4. Brouwer 1927a, 490.

5. Hilbert 1920, 34. "Was aber diese beiden Forscher [Weyl and Brouwer] in ihren Untersuchungen über die Grundlagen der Mathematik an Positivem und Fruchtbarem leisten, das fügt sich der axiomatischen Methode durchaus ein und ist gerade im Sinne dieser Methode. Denn es wird hier untersucht wie sich ein Teil der Analysis durch ein gewisses engeres System von Voraussetzungen abgrenzen lässt."

6. It should be noted, however, that some literally identified finitism and intuitionism—before Gödel's and Gentzen's result. I am thinking of Bernays, von Neumann, and Herbrand "within" the Hilbert school, but also of Carnap and Fraenkel. As to the former, I am alluding to Carnap 1930, in particular, pp. 309–310; as to the latter, let me quote the ironic (but historically inaccurate) remark in his 1930, 294, "Wie mir scheint, hat *Brouwer* den größten Erfolg für seine Anschauungen dadurch erzielt, daß er als Anhänger seiner Ausgangsposition—*Hilbert* gewonnen hat!"

7. Stein 1988, 238.

8. Hilbert 1901, 1098.

9. Hilbert 1905, 136. The German text refers to principle I of mathematical thought: "In I. kommt das schöpferische Prinzip zum Ausdruck, das uns im freiesten Gebrauch zu immer neuen Begriffsbildungen berechtigt mit der einzigen Beschränkung der Vermeidung eines Widerspruchs."

10. Bernays 1930, 17. There Bernays locates first, in a most perspicuous way, the philosophical questions concerning mathematics. "Diese Fragen philosophischen Charakters haben eine besondere Dringlichkeit erhalten seit der Wandlung, welche die methodische Einstellung der Mathematik gegen Ende des 19. Jahrhunderts erfuhr." Then he describes the characteristic features of this transformation, namely, the advance of set theory, the emergence of existential axiomatics, and the forging of close connections between logic and mathematics.

11. The considerations in this paper have profited from critical reactions to a number of earlier presentations, namely, at the Boolos Conference at Notre Dame (April 16, 1998), the Steinfest at the University of Chicago (May 23, 1999), the Hilbert Workshop at the Sorbonne (May 26, 2000), and the Annual Meeting of the Association for Symbolic Logic in Urbana (June 3, 2000). Special thanks for helpful criticism go to Bernd Buldt, Michael Detlefsen, Jacques Dubucs, Sol Feferman, Carl Posy, Howard Stein, and Bill Tait.

12. The full German text is: "Trotz des hohen pädagogischen Wertes der genetischen Methode verdient doch zur endgültigen Darstellung und völligen logischen Sicherung des Inhaltes unserer Erkenntnis die axiomatische Methode den Vorzug."

13. Dedekind 1890, 101. The essay Dedekind refers to is obviously Dedekind 1888.

14. See Dedekind 1877, 268–69, in particular the long footnote on p. 269.

15. In Peckhaus 1990 the reader will find an informative, complementary discussion on pp. 29–33.

16. He had done so also in section 9 of *Grundlagen der Geometrie*, but established consistency there semantically—by an "inductive argument"—on pp. 19–20; see note 12 of my 1999. Notice that Hilbert did not specify in either of these works the character of the "steps."

17. See Bernays 1976, 46; footnote 11 makes an explicit terminological rec-

ommendation with regard to "Konsistenz": "Es mag hier angeregt sein, diesen von Cantor speziell in bezug auf Mengenbildungen gebrauchten Ausdruck allgemein mit Bezug auf irgendwelche theoretischen Ansätze zu verwenden."

18. Hilbert 1901, 1104. This is reemphasized in Bernays 1935, 198–99: "Zur Durchführung des Nachweises gedachte Hilbert mit einer geeigneten Modifikation der in der Theorie der reellen Zahlen angewandten Methoden auszukommen."

19. Cantor attended the Munich meeting and met with Hilbert. Cantor's views are carefully presented in his letter to Hilbert that was written on January 27, 1900. (The letter is contained in the Hilbert Nachlaß in Göttingen; cod. 54: 18.)

20. Hilbert 1927, 169. "Insbesondere war es ein von Zermelo und Russell gefundener Widerspruch, dessen Bekanntwerden in der mathematischen Welt geradezu von katastrophaler Wirkung war." Some indication of related activities in Göttingen from 1902 through 1904 is found in (Peckhaus 1990, 57.

21. Hilbert 1904, 166. "Er [Dedekind] drang zu der Ansicht durch, dass der Standpunkt mit der Selbstverständlichkeit der ganzen Zahlen nicht aufrecht zu erhalten ist; er erkannte, dass die Schwierigkeiten, die Kronecker bei der Definition der irrationalen Zahlen sah, schon bei den ganzen Zahlen auftreten und dass, wenn sie hier beseitigt sind, sie auch dort wegfallen. Diese Arbeit [*Was sind und was sollen die Zahlen*] war epochemachend, aber sie lieferte doch noch nichts definitives, es bleiben gewisse Schwierigkeiten übrig. Diese bestehen hier, wie bei der Definition der irrationalen Zahlen, vor allem im Begriff des Unendlichen."

22. Hilbert 1905, 130–31.

23. Hilbert 1905, 137–38.

24. Hilbert 1905, 131.

25. Hilbert 1905, 135.

26. On pp. 1042–43 in Ewald. This is part of Poincaré's review of contemporaneous investigations of the foundations of mathematics. The critical, but also sympathetic discussion of Hilbert is mainly found in Poincaré 1906, 1038–46. Brouwer pointed to Poincaré, when elaborating on the first insight in his *Intuitionistic Reflections on Formalism*, and suggested that this insight had been "strongly prepared" by him.

27. Poincaré 1905, 1026. Having discussed Mill's view of (mathematical) existence and calling it "inadmissible," Poincaré writes in the immediately preceding paragraph: "Mathematics is independent of the existence of material objects; in mathematics the word 'exist' can have only one meaning; it means free from contradiction. . . . in defining a thing, we affirm that the definition implies no contradiction."

28. Bernays 1935, 200. "Außerdem ist auch der methodische Standpunkt der Hilbertschen Beweistheorie in dem Heidelberger Vortrag noch nicht zur vollen Deutlichkeit entwickelt. Einige Stellen deuten darauf hin, daß Hilbert die anschauliche Zahlvorstellung vermeiden und durch die axiomatische Einführung des Zahlbegriffs ersetzen will. Ein solches Verfahren würde in den beweistheoretischen Überlegungen einen Zirkel ergeben."

29. Bernays 1935, 200. "Auch wird der Gesichtspunkt der Beschränkung in der inhaltlichen Anwendung der Formen des existentialen und des allgemeinen Urteils noch nicht ausdrücklich und restlos zur Geltung gebracht."

30. These lectures are discussed in detail by Peckhaus in 1990, 1994, 1994a, and 1995; a broad philosophical perspective is also provided by Hallett in 1994 and 1995.

31. That is supported, in a general way, by Bernays 1935. However, in Bernays's description there is a peculiar "smoothing" of the developments between 1904–1905 and 1917–22: Bernays does not mention that Hilbert gave lectures on the foundations of mathematics during that period; the 1917–18 lectures are not hinted at. Thus, he effectively creates the impression that the period is one of inactivity; e.g., on p. 200 one finds: "In diesem vorläufigen Stadium hat Hilbert seine Untersuchungen über die Grundlagen der Arithmetik für lange Zeit unterbrochen. Ihre Wiederaufnahme finden wir angekündigt in dem 1917 gehaltenen Vortrage 'Axiomatisches Denken'." This impression is reinforced by the footnote attached to the first sentence in this quote, where Bernays points to the work of others who pursued the research direction stimulated by Hilbert 1905.

32. The list of talks is found on pp. 304–5 of Mancosu 1999. Behmann is viewed by Mancosu as a central player and as the (indirect) source for some of Hilbert's views, i.e., as a conduit for Russellian views.

33. That is not at all reflected in Bernays's presentation in 1935; the logicist tendency is suppressed and the "ungelöste Problematik" of the consistency of *Principia Mathematica* is emphasized immediately; see p. 201.

34. How much direct continuity is there between *Principia Mathematica* and *Prinzipien der Mathematik*? That remains an important question for detailed investigation.

35. Hilbert 1917–18, 156. "Man darf dieses Ergebnis in seiner Bedeutung nicht überschätzen. Wir haben ja damit noch keine Gewähr, dass bei der symbolischen Einführung von inhaltlich einwandfreien Voraussetzungen das System der beweisbaren Formeln widerspruchsfrei bleibt."

36. See Sieg 1999, 23–27, for more details on the attempted radical constructive development.

37. The language of this fragment of arithmetic consists of variables a, b, . . . , nonlogical constants 1, +, and all numerals; = and $\neq$ are the only relation symbols, and -> is the sole logical symbol. The axioms are:

(1) $1 = 1$
(2) $a = b \rightarrow a + 1 = b + 1$
(3) $a + 1 = b + 1 \rightarrow a = b$
(4) $a = b \rightarrow (a = c \rightarrow b = c)$
(5) $a + 1 \neq 1$

As to inference rules we have modus ponens and a substitution rule for numerals.

38. Hilbert 1920, 38. "Betrachtet man einen Beweis in Hinsicht auf eine bestimmte, konkret aufweisbare Eigenschaft, welche er besitzt, so kann es sein, dass, nach Wegstreichung einiger Formeln in diesem Beweise noch immer ein Beweis (. . .) übrig bleibt, welcher auch noch jene Eigenschaft besitzt. In diesem Fall wollen wir sagen, dass der Beweis sich in bezug auf die betreffende Eigenschaft kürzen lässt."

39. On pages 7a–8a of Hilbert 1921–22 one finds the remark: "Diese Aufgabe [to show that it is impossible to derive in a given calculus certain formulas like $1 \neq$

1] liegt grundsätzlich ebenso im Bereich der anschaulichen Betrachtung wie etwa die Aufgabe des Beweises, dass es unmöglich ist, zwei Zahlzeichen a, b zu finden, welche in der Beziehung a2 = 2b2 stehen. Hier soll gezeigt werden, dass sich nicht zwei Zahlzeichen von einer gewissen Beschaffenheit angeben lassen. Entsprechend kommt es für uns darauf an zu zeigen, dass sich nicht ein *Beweis* von einer bestimmten Beschaffenheit angeben lässt. Ein formalisierter Beweis ist aber, ebenso wie ein Zahlzeichen, ein konkreter und überblickbarer Gegenstand. Er is (wenigstens grundsätzlich) von Anfang bis Ende mitteilbar. Auch die verlangte Eigenschaft der Endformel, z.B. dass sie "1 ≠ 1" lautet, ist eine konkret feststellbare Eigenschaft des Beweises." In the lecture notes from the following year this view is expressed again as follows (p. 33): "Hier kommt es zur Geltung, dass die Beweise, wenn sie auch inhaltlich sich im Transfiniten bewegen, doch, als Gegenstände genommen und formalisiert, von finiter Struktur sind. Aus diesem Grunde ist die Behauptung, dass aus bestimmten Aussagen nicht zwei Formeln A, ¬A bewiesen werden können, methodisch gleichzustellen mit inhaltlichen Behauptungen der anschaulichen Zahlentheorie, wie z.B. der, dass man nicht zwei Zahlzeichen a, b finden kann, für welche $a^2 = 2b^2$ gilt." That is also asserted in publications, for example, in *On the Infinite*, p. 383 of van Heijenoort.

In Bernays 1935, 76, one finds this remark about the character of consistency proofs: "Diese Unmöglichkeitsbehauptung, um deren Beweis es sich hier handelt, hat die gleiche Struktur wie z.B. die Behauptung, daß es unmöglich ist, die Gleichung $a^2 = 2b^2$ durch zwei ganze Zahlen a und b zu erfüllen."

40. Hilbert 1922, 161. "Sein [Poincarés] Einwand, dieses Prinzip [der vollständigen Induktion] könnte nicht anders als selbst durch vollständige Induktion bewiesen werden, ist durch meine Theorie widerlegt."

41. Bernays, in his contemporaneous paper (1922), acknowledges explicitly that a form of induction has to be used; indeed, the writing of Bernays 1922 preceded the 1921–22 lectures, where induction is discussed (in particular on p. 57) and used for the development of finitist arithmetic. Reflecting on the proof of the commutativity of addition, Hilbert and Bernays write (on pp. 56–57): "Bei diesem Beweis wenden wir eine Art von vollständiger Induktion an, die aber auch, in der Form, wie sie hier gebraucht wird, ganz dem Standpunkt unserer anschaulichen Betrachtungsweise entspricht. Das Beweisverfahren kommt auf einen *Abbau der Zahlzeichen* hinaus, d.h. wir benutzen die Tatsache, dass die Zahlzeichen, ebenso wie sie durch Zusammensetzung von 1 und + aufgebaut sind, sich auch umgekehrt durch Wegnahme von 1 und + abbauen lassen müssen." Bernays discusses matters in a very similar manner in his 1935, 203: "Was ferner die methodische Einstellung betrifft, welche Hilbert seiner Beweistheorie zugrunde legt und welche er an Hand der anschaulichen Zahlentheorie erläutert, so liegt darin—ungeachtet der Stellungnahme Hilberts gegen Kronecker—eine weitgehende Annäherung an den Standpunkt Kroneckers vor. Eine solche besteht insbesondere in der Anwendung des anschaulichen Begriffes der Ziffer und ferner darin, daß die anschauliche Form der vollständigen Induktion, d.h. die Schlußweise, welche sich auf die anschauliche Vorstellung von dem 'Aufbau' der Ziffern gründet, als einsichtig und keiner weiteren Zurückführung bedürftig anerkannt wird. Indem Hilbert sich zur Annahme dieser methodischen Voraussetzung entschloß, wurde auch der Grund der Einwendungen behoben, welche seinerzeit Poincaré gegen Hilberts Unternehmen in dem Heidelberger Vortrag gerichtet hatte."

42. For a more detailed discussion of this "transitional stage", see Sieg 1999, 26–27 and appendix A. The "proper" response to Poincaré is formulated in Hilbert's second Hamburg Talk, (Hilbert 1928, 473) see also the remarks by Bernays quoted at the end of the preceding note.

43. Hilbert 1921–22, 4a. "Wir müssen den Bereich der betrachteten Gegenstände erweitern, d.h. wir müssen unsere anschaulichen Überlegungen auch auf andere Figuren als Zahlzeichen anwenden. Wir sehen uns somit veranlasst, von dem früher herrschenden Grundsatz abzugehen, wonach jeder Satz der reinen Mathematik letzten Endes in einer Aussage über ganze Zahlen bestehen sollte. Dieses Prinzip, in welchem man eine grundlegende methodische Erkenntnis erblickt hat, müssen wir jetzt als Vorurteil preisgeben."

44. The text of the lecture was submitted to Mathematische Annalen on September 29, 1922, and published in 1923 as *Die logischen Grundlagen der Mathematik*.

45. Already in the lecture notes of 1921–22 we find, on p. 4a, this remark, after a discussion of the "incorrect application of the law of the excluded middle": "Wir sehen also, dass für den Zweck einer strengen Begründung der Mathematik die üblichen Schlussweisen der Analysis in der Tat nicht als logisch selbstverständlich übernommen werden dürfen. Vielmehr ist es gerade erst die Aufgabe für die Begründung, zu erkennen, warum und in wieweit die Anwendung der transfiniten Schlussweisen, so wie sie in der Analysis und in der (axiomatisch begründeten) Mengenlehre geschieht, stets richtige Resultate liefert."

46. Bernays 1935, 204. "Mit der Gestaltung der Beweistheorie, die uns in dem Leipziger Vortrag entgegentritt, war die grundsätzliche Form ihrer Anlage erreicht." Bernays gives on that page also a summary of the crucial features of the proof.

47. The 1921–22 lectures, p. 7a, emphasize the methodological point of formalization (and its relation to *existential axiomatics*) as follows:

"Diesen Formalismus können wir nun zum Gegenstand einer anschaulichen Betrachtung machen, und damit eröffnet sich uns die Möglichkeit einer strengen Begründung der Mathematik.

Denn das Problem der Widerspruchsfreiheit, welches ja die grundsätzlichen Schwierigkeiten bot, erhält durch den neuen Standpunkt eine ganz konkrete Fassung. Es handelt sich nicht mehr darum, ein System von unendlich vielen Dingen mit bestimmten Verknüpfungseigenschaften (eine stetige Mannigfaltigkeit von gewisser Art) als logisch möglich zu erweisen, sondern es kommt nur darauf an einzusehen, dass es unmöglich ist, aus den (in Formeln aufgeschriebenen) Axiomen nach den Regeln des logischen Kalküls gewisse Formeln wie z.B. $1 \neq 1$ abzuleiten."

48. Bernays 1970, 186. The same image is used in the almost contemporaneous correspondence with Gödel; Bernays, after describing how intuitionism considers proofs as proper objects of mathematics, remarks in his letter of March 16, 1972: "Gewiss macht auch die Hilbert'sche Metamathematik die mathematischen Beweise zum Gegenstand, aber doch nur, nachdem sie diese durch die Formalisierung gleichsam in die mathematische Gegenständlichkeit projiziert hat."

49. Parsons refers to Bernays 1950 and states with regard to Quine: "Quine is generally most explicit when speaking of natural numbers. For a very explicit gen-

eral statement, however, see *Ontological Relativity and Other Essays*, (Columbia University Press, New York, 1969), pp. 43–45.”

50. The first moment is beautifully formulated on pp. 3–4 of the 1921–22 lecture notes (and takes up the theme of the Paris Lecture quoted above): “Auf diese Weise bildete sich die Einsicht heraus, dass das Wesentliche an der axiomatischen Methode nicht in der Gewinnung einer absoluten Sicherheit besteht, die auf logischem Wege von den Axiomen auf die Lehrsätze übertragen wird, sondern darin, dass die Untersuchung der logischen Zusammenhänge von der Frage der sachlichen Wahrheit abgesondert wird.

Unter diesem Gesichtspunkt stellt sich die Methode des axiomatischen Aufbaues einer Theorie dar als ein Verfahren der Abbildung eines Wissensgebietes auf ein Fachwerk von Begriffen, welche so geschieht, dass den Gegenständen des Wissensgebietes die Begriffe und den Aussagen über die Gegenstände die logischen Beziehungen zwischen den Begriffen entsprechen.

Durch diese Abbildung wird die Untersuchung von der konkreten Wirklichkeit ganz losgelöst. Die Theorie hat mit den realen Objekten und mit dem anschaulichen Inhalt der Erkenntnis gar nichts mehr zu tun; sie ist ein reines Gedankengebilde, von dem man nicht sagen kann, dass es wahr oder falsch ist. Dennoch hat dieses Fachwerk von Begriffen eine Bedeutung für die Erkenntnis der Wirklichkeit, weil es eine mögliche Form von wirklichen Zusammenhängen darstellt. Die Aufgabe der Mathematik ist es, solche Begriffsfachwerke logisch zu entwickeln, sei es, dass man von der Erfahrung her oder durch systematische Spekulation auf sie geführt wird.

Hier erhebt sich nun die Frage, ob denn jedes beliebige Fachwerk ein Abbild wirklicher Zusammenhänge sein kann. Eine Bedingung ist dafür jedenfalls notwendig: Die Sätze der Theorie dürfen einander nicht widersprechen, das heisst, die Theorie muss in sich möglich sein, somit entsteht das *Problem der Widerspruchsfreiheit*.”

51. Bernays 1930, 21. “Ist eine solche Erfüllung undenkbar, d.h. logisch unmöglich, so führt das Axiomensystem zu gar keiner Theorie, und die einzige logisch belangvolle Aussage über das System [von Axiomen, WS] ist dann die Feststellung des aus den Axiomen sich ergebenden Widerspruchs. Aus diesem Grunde besteht für jede axiomatische Theorie die Erforderlichkeit eines Nachweises der *Erfüllbarkeit*, d.h. der *Widerspruchsfreiheit* ihrer Axiome.” It should be evident from my earlier discussion that Dedekind saw both of these moments very clearly and, consequently, formulated the consistency problem most appropriately and sharply. He tried to resolve it by model theoretic considerations within logic.

52. Bernays 1954, 11–12. “Die Mindest-Anforderung an eine verschärfte Axiomatik ist die, dass die Gegenstände nicht einem als vorgängig gedachten Bereich entnommen werden, sondern durch Erzeugungsprozesse konstituiert werden.” Bernays continues with a methodologically important remark: “Es kann aber dabei die Meinung sein, dass durch diese Erzeugungsprozesse der Umkreis der Gegenstände determiniert ist; bei dieser Auffassung erhält das *tertium non datur* seine Motivierung. In der Tat kann Offenheit eines Bereiches in zweierlei Sinn verstanden werden, einmal nur so, dass die Konstruktionsprozesse über jeden einzelnen Gegenstand hinausführen, und andererseits in dem Sinne, dass

der resultierende Bereich überhaupt nicht eine mathematisch bestimmte Mannigfaltigkeit darstellt. Je nachdem die Zahlenreihe in dem erstgenannten oder in dem zweiten Sinne aufgefasst wird, hat man die Anerkennung des *tertium non datur* in bezug auf die Zahlen oder den intuitionistischen Standpunkt. Bei dem finiten Standpunkt kommt noch die Anforderung hinzu, dass die Überlegungen an Hand der Betrachtung von endlichen Konfigurationen verlaufen, somit insbesondere Annahmen in der Form allgemeiner Sätze ausgeschlossen werden."

53. Gödel 1933, 51. The reflections in this paper are continued most directly in Gödel's *Lecture at Zilsel's* (1938); the latter notes contain in particular a detailed analysis of Gentzen's first consistency proof for elementary number theory.

54. Bernays 1930, 23. The German text: "Diese Abstraktion, welche als die *formale* oder *mathematische Abstraktion* bezeichnet werden möge, besteht darin, daß von einem Gegenstand—'Gegenstand' hier im weitesten Sinne genommen— die strukturellen Momente, d.h. die Art der Zusammensetzung aus Bestandteilen hervorgekehrt und ausschließlich in Betracht gezogen wird. Man kann demnach als mathematische Erkenntnis eine solche definieren, welche auf der strukturellen Betrachtung von Gegenständen beruht."

55. The first feature is expressed most clearly on p. 30 of Bernays 1930: "Als das Charakteristische an der mathematischen Erkenntnisweise haben wir die formale Abstraktion, d.h. die Einstellung auf die strukturelle Seite der Gegenstände festgestellt und damit das Feld des Mathematischen in grundsätzlicher Weise abgegrenzt." That is supplemented most forcefully on p. 40: "Die wesentliche Gebundenheit der formalen Abstraktion an das Moment der Endlichkeit macht sich insbesondere dadurch geltend, daß bei den Betrachtungen von Gesamtheiten und von Figuren die Eigenschaft der Endlichkeit für den Standpunkt der anschaulichen Evidenz gar kein besonderes beschränkendes Merkmal bildet. Die Beschränkung auf das Endliche wird von diesem Standpunkt aus ganz ohne weiteres, sozusagen *stillschweigend* vollzogen. Wir brauchen hier keine besondere Definition der Endlichkeit, denn die Endlichkeit der Objekte versteht sich für die formale Abstraktion ganz von selbst." The second aspect is emphasized on pages 38 and 39, where Bernays argues that formal abstraction helps us to transcend the limits of our "faktischen" or "wirklichen Vorstellungskraft": "An solche Grenzen für die Möglichkeit der Verwirklichung kehrt sich aber die anschauliche [sic!] Abstraktion nicht. Denn diese Grenzen sind vom Standpunkt der formalen Betrachtung zufällig. Die formale Abstraktion findet sozusagen keine frühere Stelle für eine prinzipielle Abgrenzung als bei dem Unterschied des Endlichen und Unendlichen." Bernays continues in the next paragraph: "Dieser Unterschied ist in der Tat ein grundsätzlicher. Wenn wir uns genauer besinnen, wie denn überhaupt eine unendliche Mannigfaltigkeit als solche charakterisiert sein kann, so finden wir, daß dieses gar nicht nach der Art einer anschaulichen Aufweisung möglich ist, sondern nur auf dem Wege der Behauptung (bzw. der Annahme oder der Feststellung) einer gesetzlichen Beziehung. Unendliche Mannigfaltigkeiten sind uns demnach nur durch das *Denken* zugänglich. Dieses Denken ist zwar auch eine Art des Vorstellens, aber es wird dadurch nicht die Mannigfaltigkeit als Gegenstand vorgestellt, sondern es werden Bedingungen vorgestellt, denen eine Mannigfaltigkeit genügt (bzw. zu genügen hat)."

56. Bernays 1976, 61. The German text: "Freilich, die positiven Ausführungen, insbesondere die Aufweisung des mathematischen Elementes in der Logik und die Herausstellung der elementaren arithmetischen Evidenz, bedürfen wohl kaum der Revision. Jedoch, die scharfe Unterscheidung des Anschaulichen und des Nicht-Anschaulichen, wie sie bei der Behandlung des Problems des Unendlichen angewandt wird, ist anscheinend nicht so strikt durchführbar, und die Betrachtung der mathematischen Ideenbildung bedarf wohl in dieser Hinsicht noch der näheren Ausarbeitung."

57. Two aspects are finite: the number of repetitions and what Bernays calls the "iteration figure" that formally represents the generating steps of the elementary operation. Bernays discusses on pp.30–32 the introduction of natural numbers as the simplest formal objects; the argument for the first italicized claim is presented on pp. 38–39, that for the second claim on p. 40. See also note 54. The importance of the "iterativistic tendency" was emphasized by Hand (1989 and 1990); see also section 3 of Zach 1998. The suggestion, however, to base a nonstandard semantics for numerical statements on this tendency runs into difficulty when trying to account for the meaningfulness of statements concerning syntactic objects, and that is crucial for proof theoretic investigations. It is an interesting suggestion, if one takes as the "explicit (and only) goal" of the finitist viewpoint "to give an account of truth for (a fragment of) arithmetic which is *secure*," as claimed in Zach 1998, 44.

58. Bernays 1930, 40. The German text: "Zwangsmäßig aber stellt sich die *anschauliche Endlichkeitsvorstellung* ein [my emphasis, WS], sobald man einen Formalismus selbst zum Gegenstand der Betrachtung macht, insbesondere also in der systematischen Theorie der logischen Schlüsse. Es kommt hiermit zum Ausdruck, daß die Endlichkeit ein wesentliches Moment an den Gebilden eines jeden Formalismus ist."

59. This is formulated in Bernays 1930, fn. 4 on p. 32. The reader should note that "represent" is here actually translating "repräsentieren"; earlier on and later on it is the translation for "vorstellen." The full German text of the note is: "Der Philosoph ist geneigt, dieses Verhältnis der Repräsentation als einen Bedeutungszusammenhang anzusprechen. Man hat aber zu beachten, daß gegenüber dem gewöhnlichen Verhältnis von Wort und Bedeutung hier der wesentliche Unterschied besteht, daß der repräsentierende Gegenstand in seiner Beschaffenheit die wesentlichen Eigenschaften des repräsentierten Objektes enthält, so daß die zu untersuchenden Beziehungen der repräsentierten Objekte sich auch an den Repräsentanten vorfinden und durch die Betrachtung der Repräsentanten selbst festgestellt werden können."

60. The discussion is found on p. 70. Parsons (1982, 496) makes a related distinction for Kant's philosophy of mathematics. He distinguishes between *weak* and *strong* intuitability as follows: "An object is strongly intuitable if it can be intuited, i.e., if it can be an object of intuition. An object is weakly intuitable if it can be represented in intuition without itself being intuitable. This notion is vague because we have not said what is meant by 'representing' an object in intuition. However, representation of abstract objects by concrete objects, or by objects relatively closer to the concrete, is a pervasive phenomenon and of great importance for understanding abstract objects."

61. I refer to Kronecker's *Über den Zahlbegriff* and Helmholtz's *Zählen und Messen, erkenntnistheoretisch betrachtet*; both papers were published in the Zeller-Festschrift, Leipzig 1887. Dedekind refers to these two papers when remarking in the first note to his 1888 *Vorwort*: "Das Erscheinen dieser Abhandlungen ist die Veranlassung, welche mich bewogen hat, nun auch mit meiner, in mancher Beziehung ähnlichen, aber durch ihre Begründung doch wesentlich verschiedenen Auffassung hervorzutreten, die ich mir seit vielen Jahren und ohne jede Beeinflussung von irgendwelcher Seite gebildet habe."

62. Parsons 1990, section 8.

63. Brouwer added in a famous footnote: "These mental mathematical proofs that in general contain infinitely many terms must not be confused with their linguistic accompaniments, which are finite and necessarily inadequate, hence do not belong to mathematics." Brouwer claimed that this remark contains his "main argument against the claims of Hilbert's metamathematics."

64. The result is obtained from work by Tait or by Feferman (in the Buffalo volume, Kino et al. and my 1977 Stanford dissertation; Tait's work goes back to early 1967 and is reported in 1968. For the detailed exposition of this and related results see Buchholz et al. and in particular my chapter "Inductive definitions, constructive ordinals, and normal derivations," pp. 143–87 of that volume.

65. For discussions of this part of advanced proof theory, see Jäger 1986, Pohlers 1989, Rathjen 1995, and Buchholz 2000.

66. A detailed presentation of the central features of Aczel's i.d. classes is found in Feferman and Sieg 1981, 18–25.

67. Bernays 1934, 75. The German text: "Die Annahmen, um die es sich dabei handelt, laufen hinaus auf Vorstellungen von Gesamtheiten und auf das Prinzip der Analogie oder der Permanenz der Gesetze. Und die Bedingung, welche die Anwendung dieser Leitgedanken einschränkt, ist keine andere als die der Widerspruchsfreiheit der Folgerungen, die sich aus jenen zugrundegelegten Annahmen ergeben."

68. Bernays (1930, 44) makes a related, but certainly not "identical" distinction, referring in a footnote to Fries; he distinguishes between "dem elementarmathematischen Standpunkt und einem darüber hinausgehenden systematischen Standpunkt." This more systematic standpoint covers not only analysis, but also set theory, whereas for me set theory with its iterative conception falls under the quasi-constructive aspect of mathematical experience.

69. I also pointed out (1990) that this second task (as well as the first one) cuts across the traditional foundational divides: if there is an abstract notion in intuitionistic mathematics, it is that of a choice sequence introduced by Brouwer to capture the essence of the continuum; the consistency proof of the theory of choice sequences relative to the theory of the second constructive number class ID(O) can be viewed as fulfilling exactly this task. The proof was given by Kreisel and Troelstra in their paper "Formal systems for some branches of intuitionistic analysis," Annals of Mathematical Logic 1, 1970, 229–387.

70. See Weyl 1925, 540ff and Weyl 1928, 482–84, in particular p. 484 where the earlier ideas are reexpressed.

71. Stein 1988, 255.

REFERENCES

The translations from the German are mine, unless the references are to English translations; I am grateful that I had access to the Hilbert Nachlaß at the University of Göttingen through the work on the Hilbert Edition.

Ackermann, W.
1924   Begründung des "tertium non datur" mittels der Hilbertschen Theorie
            der Widerspruchsfreiheit. *Mathematische Annalen* 93, 1–36.

Aczel, P.
1977   An introduction to inductive definitions. In *Handbook of Mathematical
            Logic*, edited by J. Barwise. Amsterdam, 739–82.

Aspray, W. and P. Kitcher (eds.).
1988   *History and Philosophy of Modern Mathematics*. Vol. 11. Minneapolis:
            Minnesota Studies in the Philosophy of Science.

Bernays, P.
1922   Über Hilberts Gedanken zur Grundlegung der Mathematik.
            Jahresberichte DMV 31: 10–19.
1928   Über Nelsons Stellungnahme in der Philosophie der Mathematik. *Die
            Naturwissenschaften,* 16 (9): 142–45.
1928a   Die Grundbegriffe der reinen Geometrie in ihrem Verhältnis zur
            Anschauung. *Die Naturwissenschaften,* 16 (12): 197–203.
1930   Die Philosophie der Mathematik und die Hilbertsche Beweistheorie. In
            Bernays 1976, 17–61.
1930a   Die Grundgedanken der Fries'schen Philosophie in ihrem Verhältnis zum
            heutigen Stand der Wissenschaft. *Abhandlungen der Fries'schen Schule,*
            Neue Folge, vol. 5 (2): 97–113. (Based on a talk presented on August
            10, 1928.)
1934   Über den Platonismus in der Mathematik. In Bernays 1976, 62–78.
1935   Hilberts Untersuchungen über die Grundlagen der Arithmetik. In
            Hilbert 1935, 196–216.
1937   Grundsätzliche Betrachtungen zur Erkenntnistheorie. *Abhandlungen der
            Fries'schen Schule,* Neue Folge, vol. 6 (3–4): 278–90.
1950   Mathematische Existenz und Widerspruchsfreiheit. In Bernays 1976,
            92–106.
1954   Zur Beurteilung der Situation in der beweistheoretischen Forschung.
            Revue internationale de philosophie 8, 9–13; Discussion, 15–21.
1970   Die schematische Korrespondenz und die idealisierten Stukturen. In
            Bernays 1976, 176–88.

1976   *Abhandlungen   zur   Philosophie   der   Mathematik*. Darmstadt: Wissenschaftliche Buchgesellschaft.

Brouwer, L. E. J.
1927   Über Definitionsbereiche von Funktionen. *Mathematische Annalen* 97: 60–75; translated in van Heijenoort, 446–63.
1927a   Intuitionistische Betrachtungen über den Formalismus. Koninklijke Akademie van wetenschappen te Amsterdam, Proceedings of the section of sciences, 31: 374–79; translated in van Heijenoort, 490–92.
1953   Points and spaces. *Canadian Journal of Mathematics* 6: 1–17.

Buchholz, W.
2000   Relating ordinals to proofs in a more perspicuous way. Forthcoming.

Buchholz, W., S. Feferman, W. Pohlers, and W. Sieg.
1981   *Iterated inductive definitions and subsystems of analysis: Recent prof-theoretical studies. Lecture Notes in Mathematics*, Springer Verlag.

Carnap, R.
1930   Die Mathematik als Zweig der Logik. *Blätter für Deutsche Philosophie* 4: 298–310.

Church, A. and S. Kleene
1936   Formal definitions in the theory of ordinal numbers. *Fund. Math.* 28: 11–21.

Dedekind, R.
1872   *Stetigkeit und irrationale Zahlen*. In Dedekind 1932, 315–24.
1877   Sur la théorie des nombres entiers algébriques. *Bulletin des Sciences mathématiques et astronomiques*, 1–121; partially reprinted in Dedekind 1932, 262–96.
1888   *Was sind und was sollen die Zahlen*. In Dedekind 1932, 335–91.
1890   Letter to Keferstein. In van Heijenoort, 98–103.
1932   *Gesammelte mathematische Werke*. Dritter Band. R. Fricke, E. Noether, and Ö. Ore, (eds.); Braunschweig: Vieweg.

Ewald, W. (ed.)
1996   *From Kant to Hilbert—A source book in the foundations of mathematics*. 2 vols. Oxford University Press.

Feferman, S.
1981   How we got from there to here. In Buchholz et al., 1–15.
1982   Inductively presented systems and the formalization of meta-mathematics. In *Logic Colloquium '80*, North Holland Publishing Company, 95–128.

Feferman, S. and W. Sieg.
1981   Iterated inductive definitions and subsystems of analysis. In Buchholz et. al., 16–77.

Fraenkel, A.
1930   Die heutigen Gegensätze in der Grundlegung der Mathematik. *Erkenntnis* 1, 286–302.

Gödel, K.
1933   The present situation in the foundations of mathematics. In *Collected Works III*, 36–53.
1938   Vortrag by Zilsel. In *Collected Works III*, 86–113.
1986   *Collected Works I*. Oxford University Press, Oxford, New York.
1990   *Collected Works II*. Oxford University Press, Oxford, New York.
1995   *Collected Works III*. Oxford University Press, Oxford, New York.

Hallett, M.
1994   Hilbert's axiomatic method and the laws of thought. In *Mathematics and Mind*, edited by A. George. Oxford University Press, 158–200.
1995   Hilbert and logic. In *Québec Studies in the Philosophy of Science I*, edited by M. Marion and R. S. Cohen. Dordrecht: Kluwer: 135–87.

Hand, M.
1989   A number in the exponent of an operation. *Synthese* 81: 243–65.
1990   Hilbert's iterativistic tendencies. *History and Philosophy of Logic* 11: 185–92.

Hilbert, D.
1894*   Quadratur des Kreises.
1897*   Zahlbegriff und Quadratur des Kreises.
1899*   Zahlbegriff und Quadratur des Kreises.
1900   Über den Zahlbegriff. *Jahresbericht der DMV* 8: 180–94; reprinted in: *Grundlagen der Geometrie*, 3. Leipzig: Auflage, 1909, 256–62.
1900a   *Grundlagen der Geometrie*, 5. edition. Leipzig and Berlin, 1922.
1901   Mathematische Probleme. Vortrag, gehalten auf dem internationalen Mathematiker-Kongress zu Paris 1900; Archiv der Mathematik und Physik, 3rd series, 1, 44–63, 213–37.
1904*   Zahlbegriff und Quadratur des Kreises.
1905   Über die Grundlagen der Logik und Arithmetik. In Hilbert 1900a, 243–58; translated in van Heijenoort, 129–38.
1917–8*   Prinzipien der Mathematik.
1920*   Probleme der mathematischen Logik.
1921–2*   Grundlagen der Mathematik.

[**N.B.** The starred items are unpublished lecture notes contained in the Hilbert Nachlaß at the University of Göttingen.]

1922   Neubegründung der Mathematik. *Abhandlungen aus dem mathematischen Seminar der Hamburgischen Universität* 1: 157–77.

1923   Die logischen Grundlagen der Mathematik. *Mathematische Annalen* 88: 151–65.

1925   Über das Unendliche. *Mathematische Annalen* 95, 1926: 161–90; translated in van Heijenoort, 367–92.

1928   Die Grundlagen der Mathematik. *Abhandlungen aus dem mathematischen Seminar der Hamburgischen Universität* 6 (1/2): 65–85.

1935   *Gesammelte Abhandlungen.* Volume 3. Berlin: Springer.

Hilbert, D. and P. Bernays.
1934   *Grundlagen der Mathematik.* Vol. 1. Springer Verlag.
1939   *Grundlagen der Mathematik.* Vol. 2. Springer Verlag.

Jäger, G.
1986   *Theories for admissible sets—a unifying approach to proof theory.* Naples: Bibliopolis.

Joyal, A. and I. Moerdijk.
1995   *Algebraic Set Theory. London Mathematical Society Lecture Notes Series* 220. Cambridge University Press.

Kino, A., J. Myhill, and R. E. Vesley. (eds.)
1970   *Intuitionism and Proof Theory.* Proceedings of the summer conference at Buffalo, N.Y., 1968.

Mancosu, P.
1999   Between Russell and Hilbert: Behmann on the foundations of mathematics. *Bulletin of Symbolic Logic* 5 (3): 303–330.

Moerdijk, I. and E. Palmgren.
2000   Type theories, toposes, and constructive set theory: Predicative aspects of AST. Manuscript, January 2000.

2000a   Wellfounded trees in categories. *Annals of Pure and Applied Logic* 104: 189–218.

Paetzold, H.
1995   *Ernst Cassirer—Von Marburg nach New York.* Darmstadt: Wissenschaftliche Buchgesellschaft.

Parsons, C. D.
1980   Mathematical intuition. *Proc. Aristotelian Society N.S.* 80 (1979–80): 145–68.
1982   Objects and logic. *The Monist* 65 (4): 491–516.
1984   Arithmetic and the categories. *Topoi* 3 (2): 109–121.

1987   Developing arithmetic in set theory without infinity: Some historical remarks. *History and Philosophy of Logic* 8: 201–213.
1990   The structuralist view of mathematical objects. *Synthese* 84, 303–346.
1994   Intuition and number. In *Mathematics and Mind*, edited by A. George. Oxford University Press, 141–57.

Peckhaus, V.
1990   *Hilbertprogramm und Kritische Philosophie*. Göttingen: Vandenhoek & Ruprecht.
1994   Hilbert's axiomatic programme and philosophy. In: *The History of Modern Mathematics*, vol. 3, edited by E. Knobloch and D. E. Rowe. Academic Press, 91–112.
1994a   Logic in transition: The logical calculi of Hilbert (1905) and Zermelo (1908). In *Logic and Philosophy of Science in Uppsala*, edited by D. Prawitz and D. Westrestahl. Kluwer, 311–23.
1995   Hilberts Logik. Von der Axiomatik zur Beweistheorie. *Intern. Zs. f. Gesch. u. Ethik der Naturwiss.*, Techn. U. Med. 3: 65–86.

Pohlers, W.
1989   *Proof Theory—an introduction. Lecture Notes in Mathematics* 1407, Springer Verlag.

Poincaré, H.
1905   Les mathématiques et la logique. *Revue de métaphysique et de morale* 13: 815–35; translated in Ewald, vol. 2, 1021–38.
1906   Les mathématiques et la logique. *Revue de métaphysique et de morale* 14: 17–34; translated in Ewald, vol. 2, 1038–52.
1906a   Les mathématiques et la logique. *Revue de métaphysique et de morale* 14: 294–317; translated in Ewald, vol. 2, 1052–71.

Rathjen, M.
1995   Recent advances in ordinal analysis. *Bulletin of Symbolic Logic* 1 (4): 468–85.

Shapiro, S.
1996   *Philosophy of mathematics—Structure and ontology*. Oxford University Press.

Sieg, W.
1977   *Trees in Metamathematics*. Ph.D. Thesis. Stanford.
1981   Inductive definitions, constructive ordinals, and normal derivations. In Buchholz et al., 143–87.
1984   Foundations for analysis and proof theory. *Synthese* 60 (2): 159–200.
1990   Relative consistency and accessible domains. *Synthese* 84, 259–97.
1997   Aspects of mathematical experience. In *Philosophy of mathematics today*, edited by E. Agazzi and G. Darvas. Kluwer Academic Publishers, 195–217.

1999   Hilbert's programs: 1917–1922. *Bulletin of Symbolic Logic* 5 (1): 1–44.

Stein, H.
1988   Logos, Logic, Logistiké: Some philosophical remarks on the 19th century transformation of mathematics. In Aspray and Kitcher, 238–59.

Tait, W. W.
1968   Constructive Reasoning. In *Proc. 3rd Int. Congress of Logic, Methodology, and Philosophy of Science.* Amsterdam, 185–99.
1981   Finitism. *Journal of Philosophy* 78, 524–46.
2000   Remarks on Finitism. Forthcoming in *Reflections on the Foundations of Mathematics: Essays in Honor of Solomon Feferman,* edited by Wilfried Sieg, Richard Sommer, and Carolyn Talcott. Association for Symbolikc Logic.

van Heijenoort, J. (ed.)
1967   *From Frege to Gödel, a source book in mathematical logic, 1879–1931.* Cambridge: Harvard University Press.

Weyl, H.
1925   Die heutige Erkenntnislage in der Mathematik. *Symposion* 1: 1–32. (Reprinted in volume 2 of Weyl's "Gesammelte Abhandlungen," Springer Verlag, 1968, 511–42.)
1928   Diskussionsbemerkungen zu dem zweiten Hilbertschen Vortrag über die Grundlagen der Mathematik. *Abhandlungen aus dem mathematischen Seminar der Hamburgischen Universität* 6: 86–88. (Translated in van Heijenoort, 482–84.

Whitehead, A. N. and B. Russell.
1910   *Principia Mathematica.* Volume 1. Cambridge: Cambridge University Press.
1912   ———. Volume 2.
1913   ———. Volume 3.

Zach, R.
1997   Numbers and Functions in Hilbert's Finitism. *Taiwanese Journal for Philosophy and History of Science* 10: 33–60.
Zermelo, E.
1909   Sur les ensembles finis et le principe de l'induction complète. *Acta Mathematica* 32, 185–93.

Bibliography of Howard Stein

ARTICLES

"An Examination of Some Aspects of Natural Science." Thesis, Department of Philosophy, The University of Chicago, 1958.

"Newtonian Space-Time." *The Texas Quarterly* 10 (Autumn 1967): 174–200; also in *The* Annus Mirabilis *of Sir Isaac Newton*, ed. Robert Palter. Cambridge, Mass.: MIT Press, 1970, 258–84.

"Limitations on Measurement" and "Alternative Schemes of Measurement," jointly with Abner Shimony. Two papers presented to the American Physical Society, meeting of November 16–18, 1967. Abstracts in *Bulletin of the American Physical Society*, ser. 2, 12 (1967): 1056.

"On Einstein-Minkowski Space-Time." *Journal of Philosophy* 65 (1968): 5–23.

"Comments on 'The Thesis of Parmenides'." *Review of Metaphysics* 22 (1969): 725–34.

"Is There a Problem of Interpreting Quantum Mechanics?" *Noûs* 4 (1970): 93–103.

"A Note on Time and Relativity Theory." *Journal of Philosophy* 67 (1970): 93–103.

"On the Paradoxical Time-Structures of Gödel." *Philosophy of Science* 37 (1970): 289–94.

"On the Notion of Field in Newton, Maxwell, and Beyond." In *Historical and Philosophical Perspectives of Science,* ed. Roger B. Stuewer. Minnesota Studies in the Philosophy of Science, vol. 5. University of Minnesota Press, 1970, 264–87. (Followed by comments, 287–99; and replies, 299–310.)

"Limitations on Measurement," jointly with Abner Shimony. In *Foundations of Quantum Mechanics,* ed. B. d'Espagnat. Proceedings of the International School of Physics "Enrico Fermi," course 49. New York: Academic Press, 1971, 56–76.

"On the Conceptual Structure of Quantum Mechanics." In *Paradigms and Paradoxes: The Philosophical Challenge of the Quantum Domain,* ed. Robert Colodny. University of Pittsburgh Series in the Philosophy of Science, vol. 5; Pittsburgh, Pa.: University of Pittsburgh Press, 1972, 367–438.

"Graves on the Philosophy of Physics." *Journal of Philosophy* 69 (1972): 621–34.

"Maurice Clavelin on Galileo's Natural Philosophy." *British Journal for the Philosophy of Science* 25 (1974): 375-97.

"Some Philosophical Prehistory of General Relativity." In *Foundations of Space-Time Theories,* ed. John Earman, Clark Glymour, and John Stachel. Minnesota Studies in the Philosophy of Science, vol. 8. University of Minnesota Press, 1977, 3–49.

"On Space-Time and Ontology: Extract from a Letter to Adolf Grünbaum." Minnesota Studies in the Philosophy of Science, vol. 8. University of Minnesota Press, 1977, 374–402.

"On Newton and Einstein." *The Library Chronicle of the University of Texas at Austin,* New Series, 12 (1979): 63–78.

"A Problem in Hilbert Space Theory Arising from the Quantum Theory of Measurement," jointly with Abner Shimony. *American Mathematical Monthly* 86 (1979): 292–93.

"'Subtler Forms of Matter' in the Period Following Maxwell." In *Conceptions of Ether: Studies in the History of Ether Theories, 1740-1900,* ed. G. N. Cantor and M. J. S. Hodge. Cambridge University Press, 1981, 309–40.

"On the Present State of the Philosophy of Quantum Mechanics." *PSA* 2 (1982): 563–81.

"The Everett Interpretation of Quantum Mechanics: Many Worlds or None?" *Noûs* 18 (1984): 635–52.

Introductory note to the paper of Kurt Gödel, "A Remark about the Relationship between Relativity Theory and Idealistic Philosophy." In Kurt Gödel, *Collected Works,* vol. 2, ed. Solomon Feferman et al. New York: Oxford University Press, 1990, 199–201.

"After the Baltimore Lectures: Some Philosophical Remarks on the Subsequent Development of Physics." In *Kelvin's Baltimore Lectures and Modern Theoretical*

Physics, ed. Robert Kargon and Peter Achinstein. Cambridge, Mass.: MIT Press, 1987, 375–98.

"*Logos,* Logic, and *Logistiké*: Some Philosophical Remarks on the Nineteenth-Century Transformation of Mathematics." In *History and Philosophy of Modern Mathematics,* ed. William Aspray and Philip Kitcher. Minnesota Studies in the Philosophy of Science, vol. 11. University of Minnesota Press, 1988, 238–59.

"Yes, but . . . : Some Skeptical Reflections on Realism and Anti-realism." *Dialectica* 43 (1989): 47–65.

"On Locke, 'The Great Huygenius, and the incomparable Mr. Newton'." In *Philosophical Perspectives on Newtonian Science,* ed. Phillip Bricker and R. I. G. Hughes. Cambridge, Mass.: MIT Press, 1990, 17–47.

"Eudoxos and Dedekind: On the Ancient Greek Theory of Ratios and Its Relation to Modern Mathematics." *Synthese* 84 (1990): 163–211. (Reprinted in part in William Demopoulis, ed., *Frege's Philosophy of Mathematics.* Cambridge, Mass.: Harvard University Press, 1995, 334–57.)

"On Relativity Theory and Openness of the Future." *Philosophy of Science* 58 (1991): 147–67.

"'From the Phenomena of Motions to the Forces of Nature': Hypothesis or Deduction?" *PSA* 2 (1990): 209–222.

"Was Carnap Entirely Wrong, After All?" *Synthese* 93 (1992): 275–95.

"On Philosophy and Natural Philosophy in the Seventeenth Century." *Midwest Studies in Philosophy* 18 (1993): 177–201.

"Newton." In *A Companion to Metaphysics,* ed. J. Kim and E. Sosa. Oxford: Blackwell, 1995, 353–55.

"Some Reflections on the Structure of our Knowledge in Physics." In *Logic, Methodology, and Philosophy of Science* 9. Proceedings of the Ninth International Congress of Logic, Methodology, and Philosophy of Science, ed. D. Prawitz, B. Skyrms, and D. Westerstahl. New York: Elsevier Science B.V., 1994, 633–55.

"Logicism." In *Routledge Encyclopedia of Philosophy,* ed. Edward Craig. London: Routledge, 1998.

"Dedekind, Julius Wilhelm Richard." In *Routledge Encyclopedia of Philosophy,* ed. Edward Craig. London: Routledge, 1998.

Introductory note to Kurt Gödel, "Some Observations about the Relationship between Theory of Relativity and Kantian Philosophy." In Kurt Gödel, *Collected*

Works, vol. 3. Oxford University Press, 1995, 202–229.

"Maximal Extension of an Impossibility Theorem Concerning Quantum Measurement." In *Potentiality, Entanglement, and Passion-at-a-Distance,* ed. R. S. Cohen et al. Dordrecht: Kluwer Academic Publishers, 1997, 231–43.

"Comment on 'Nonlocal Character of Quantum', by Henry P. Stapp," jointly with Abner Shimony, *American Journal of Physics* 69 (2001): 848–53.

"Newton's Metaphysics." To appear in the *Cambridge Companion to Newton,* ed. I. Bernard Cohen and George E. Smith. Cambridge: Cambridge University Press.

"On Quantum Non-locality, Special Relativity, and Counterfactual Reasoning," jointly with Abner Shimony. To appear in *Space-Time, Quantum Entanglement, and Critical Epistemology,* ed. Jürgen Renn et al. Dordrecht: Kluwer Academic Publishers.

PENDING (NOT YET SUBMITTED FOR PUBLICATION)

"Newtonus ab quibusdam nævibus vindicatus."

"On Metaphysics and Method in Newton."

"Further Considerations on Newton's Methods."

"Physics and Philosophy Meet: The Strange Case of Poincaré."

"How Does Physics Bear upon Metaphysics; and Why Did Plato Hold that Philosophy Cannot Be Written Down?"

"The Enterprise of Understanding and the Enterprise of Knowledge—for Isaac Levi's Seventieth Birthday."

BOOK REVIEWS

of M. Whiteman, *Philosophy of Space and Time. Journal of Philosophy* 66 (1969): 58–62.

of R. M. Gale, *The Language of Time.* Ibid., 350–55.

of J. L. Heilbron, *Electricity in the Seventeenth and Eighteenth Centuries. Philosophy of Science* 61 (1984).

Contributors

ROBERT DISALLE is Associate Professor of Philosophy at the University of Western Ontario. He completed his Ph.D. at the University of Chicago under Howard Stein's supervision. He works on the history and philosophical foundations of theories of space and time.

MICHAEL FRIEDMAN is currently Ruth N. Halls Professor of Arts and Humanities at Indiana University and Frederick P. Rehmus Family Professor of Humanities at Stanford University. His publications include: *Foundations of Space-Time Theories* (Princeton University Press, 1983), *Kant and the Exact Sciences* (Harvard University Press, 1992), *Reconsidering Logical Positivism* (Cambridge University Press, 1999), *A Parting of the Ways: Carnap, Cassirer, and Heidegger* (Open Court, 2000), and *Dynamics of Reason* (Center for the Study of Language and Information, 2001).

WILLIAM L. HARPER is Professor of Philosophy at the University of Western Ontario. He has written a number of papers on Newton's argument for universal gravitation, and the continuing use of Newton's methodology in relativistic gravitation theory. He is presently at work on a book to be titled *Newton's Argument for Universal Gravitation: A Case Study of Turning Data into Evidence*.

GEOFFREY HELLMAN is Professor of Philosophy at the University of Minnesota (Twin Cities). He is the author of *Mathematics without Numbers: Towards a Modal-Structural Interpretation* (Oxford University Press, 1989), the co-editor (with Richard Healey) of *Quantum Measurement: Beyond Paradox* (University of Minnesota Press, 1998), and the author of numerous articles on philosophy of mathematics, philosophy of logic, philosophy of quantum mechanics, and philosophy of science.

ISAAC LEVI is John Dewey Professor of Philosophy at Columbia University. He has worked on a decision theoretic approach to belief change, on an account of indeterminate probabilities and utilities together with an accompanying account of rational choice and statistical inference, and done exegetical work on the ideas of Charles Peirce and John Dewey. His books are: *Gambling with Truth* (Random House Press, 1967; reissued in paperback by MIT University Press 1973), *The Enterprise of Knowledge* (MIT University Press 1980), *Decisions and Revisions* (Cambridge University Press, 1984), *Hard Choices* (Cambridge University Press, 1986), *The Fixation of Belief and Its Undoing* (Cambridge University Press, 1991), *For the Sake of the Argument,* (Cambridge University Press, 1996), *The Covenant of Reason* (Cambridge University Press, 1997).

DAVID B. MALAMENT, the editor of this volume, is UCI Distinguished Professor in the Department of Logic and Philosophy of Science at the University of California, Irvine. Before moving to UCI in 1999, he taught at the University of Chicago, in the Department of Philosophy and the Committee on the Conceptual Foundations of Science, for 24 years. He works on the mathematical and philosophical foundations of Newtonian gravitation theory, general relativity, and quantum mechanics.

NANCY NERSESSIAN is Professor in the Program in Cognitive Science, in the College of Computing and the School of Public Policy, at the Georgia Institute of Technology. She is the author of *Faraday to Einstein: Constructing Meaning in Scientific Theories* and numerous articles dealing with cognitive studies in science and, in particular, the processes of conceptual innovation and change in science. Her historical research centers on late-nineteenth and early-twentieth century electrodynamics and relativity theory, especially on the work of Maxwell and Lorentz.

ROBERT PALTER is Dana Professor Emeritus of the History of Science at Trinity College in Hartford, Connecticut. His book, *The Duchess of Malfi's Apricots, and Other Literary Fruits,* will be published in 2002 by the University of South Carolina Press.

ABNER SHIMONY is Professor Emeritus of Philosophy and Physics at Boston University. His main research interests are the foundations of quantum mechanics, the interplay of philosophy and the natural sciences, and inductive logic. Many of his essays on these subjects are contained in *Search for a Naturalistic World View* (2 vols., Cambridge University Press, 1993). He is also the author of a children's book, *Tibaldo and the Hole in the Calendar* (Copernicus, 1998).

WILFRIED SIEG is Professor of Mathematical Logic and Philosophy at Carnegie Mellon University. He works in proof theory, philosophy and history of (modern) mathematics, and the foundations of cognitive science.

GEORGE E. SMITH is Professor of Philosophy at Tufts University, and Acting Director of the Dibner Institute for the History of Science and Technology. He has published several articles on Newton's *Principia* and is the co-editor, with I. Bernard Cohen, of the *Cambridge Companion to Newton*. A practicing engineer specializing in turbo-machinery failure analysis, he has also published articles on

J. J. Thomson's experimental research into the microphysics of electricity as well as on engineering epistemology.

JOHN STACHEL is Professor Emeritus of Physics, and Director of the Center for Einstein Studies, at Boston University. His book *Einstein from 'B' to 'Z'* has just appeared, and a collection of his essays on other topics, entitled *Going Critical,* is in preparation. He works in theoretical physics, history and philosophy of science.

W. W. TAIT is Professor Emeritus of Philosophy at the University of Chicago. His research interests are in logic, philosophy of mathematics and its history. He is presently completing preparation of a collection of essays, with the provisional title *The Provenance of Pure Reason: Essays in the Philosophy of Mathematics and its History,* to be published by Oxford University Press.

Index

abduction, 319–20
acceleration fields
combining, 89–91
for Sun and Jupiter, 88–89
accessible domains, 386–87
Ackermann, Wilhelm, 373
Aczel, Peter, 386
Adams, Ernest, 326
agreeing measurements, yielding
resiliency, 96–97
Alchourrón, Carlos E., 330
algebra, 248, 251
ampliative reasoning, 319–20
analogical modeling, 131–32,
137–38
abstraction in , 139
generating conceptual change,
138
antiinductivists, 319–20, 329,
330–31
Antony, Marc, 12
apple
Granny Smith, 121
as inspiring Newton, 118
as pedagogical motif, 113
Aristotle, 4, 17, 18, 300, 305, 307
on Forms, 16
on geometry, 16
Metaphysics, 19
Nicomachean Ethics, 308–9
Austen, Ralph, 123

automorphism group, 251

Bacon, Francis, 3
Baillet, Adrien, 114
Barbour, J., 184
Bar-Hillel, Yehoshua, 324
Batey, Mavis, 123
Bayesians, 322–23
Beeckman, Isaac, 115, 117
Behmann, Heinrich, 372
Benacerraf, Paul, 343
Bernays, Paul, 363, 365, 367, 369,
371, 373, 374, 376–80, 385
on abstraction, 381, 382
eliminative structuralism of, 379
on finiteness, 382–83
on formalized theory, 377–78
on intuitive mathematics, 381, 382
on modern mathematics, 378–79
as structuralist, 384
on totalities, 386
Bernoulli, Johann, 33
Bertotti, Giorgio, 184
Bernstein, Felix, 372
Bhaskar, Roy, 239
Bishop, Errett, 385
Bohr, Niels, 151
Boolos, George, 344, 350, 352
Boulliau, Ismael, 33
Bower, T., 303
Bradley, James, 49, 50